Graduate Texts in Mathematics 65

Graduate Texts in Mathematics

1 TAKEUTI/ZARING. Introduction to Axiomatic Set Theory. 2nd ed.
2 OXTOBY. Measure and Category. 2nd ed.
3 SCHAEFER. Topological Vector Spaces. 2nd ed.
4 HILTON/STAMMBACH. A Course in Homological Algebra. 2nd ed.
5 MAC LANE. Categories for the Working Mathematician. 2nd ed.
6 HUGHES/PIPER. Projective Planes.
7 J.-P. SERRE. A Course in Arithmetic.
8 TAKEUTI/ZARING. Axiomatic Set Theory.
9 HUMPHREYS. Introduction to Lie Algebras and Representation Theory.
10 COHEN. A Course in Simple Homotopy Theory.
11 CONWAY. Functions of One Complex Variable I. 2nd ed.
12 BEALS. Advanced Mathematical Analysis.
13 ANDERSON/FULLER. Rings and Categories of Modules. 2nd ed.
14 GOLUBITSKY/GUILLEMIN. Stable Mappings and Their Singularities.
15 BERBERIAN. Lectures in Functional Analysis and Operator Theory.
16 WINTER. The Structure of Fields.
17 ROSENBLATT. Random Processes. 2nd ed.
18 HALMOS. Measure Theory.
19 HALMOS. A Hilbert Space Problem Book. 2nd ed.
20 HUSEMOLLER. Fibre Bundles. 3rd ed.
21 HUMPHREYS. Linear Algebraic Groups.
22 BARNES/MACK. An Algebraic Introduction to Mathematical Logic.
23 GREUB. Linear Algebra. 4th ed.
24 HOLMES. Geometric Functional Analysis and Its Applications.
25 HEWITT/STROMBERG. Real and Abstract Analysis.
26 MANES. Algebraic Theories.
27 KELLEY. General Topology.
28 ZARISKI/SAMUEL. Commutative Algebra. Vol. I.
29 ZARISKI/SAMUEL. Commutative Algebra. Vol. II.
30 JACOBSON. Lectures in Abstract Algebra I. Basic Concepts.
31 JACOBSON. Lectures in Abstract Algebra II. Linear Algebra.
32 JACOBSON. Lectures in Abstract Algebra III. Theory of Fields and Galois Theory.
33 HIRSCH. Differential Topology.
34 SPITZER. Principles of Random Walk. 2nd ed.
35 ALEXANDER/WERMER. Several Complex Variables and Banach Algebras. 3rd ed.
36 KELLEY/NAMIOKA et al. Linear Topological Spaces.
37 MONK. Mathematical Logic.
38 GRAUERT/FRITZSCHE. Several Complex Variables.
39 ARVESON. An Invitation to C^*-Algebras.
40 KEMENY/SNELL/KNAPP. Denumerable Markov Chains. 2nd ed.
41 APOSTOL. Modular Functions and Dirichlet Series in Number Theory. 2nd ed.
42 J.-P. SERRE. Linear Representations of Finite Groups.
43 GILLMAN/JERISON. Rings of Continuous Functions.
44 KENDIG. Elementary Algebraic Geometry.
45 LOÈVE. Probability Theory I. 4th ed.
46 LOÈVE. Probability Theory II. 4th ed.
47 MOISE. Geometric Topology in Dimensions 2 and 3.
48 SACHS/WU. General Relativity for Mathematicians.
49 GRUENBERG/WEIR. Linear Geometry. 2nd ed.
50 EDWARDS. Fermat's Last Theorem.
51 KLINGENBERG. A Course in Differential Geometry.
52 HARTSHORNE. Algebraic Geometry.
53 MANIN. A Course in Mathematical Logic.
54 GRAVER/WATKINS. Combinatorics with Emphasis on the Theory of Graphs.
55 BROWN/PEARCY. Introduction to Operator Theory I: Elements of Functional Analysis.
56 MASSEY. Algebraic Topology: An Introduction.
57 CROWELL/FOX. Introduction to Knot Theory.
58 KOBLITZ. p-adic Numbers, p-adic Analysis, and Zeta-Functions. 2nd ed.
59 LANG. Cyclotomic Fields.
60 ARNOLD. Mathematical Methods in Classical Mechanics. 2nd ed.
61 WHITEHEAD. Elements of Homotopy Theory.
62 KARGAPOLOV/MERIZJAKOV. Fundamentals of the Theory of Groups.
63 BOLLOBAS. Graph Theory.

(continued after the subject index)

Raymond O. Wells, Jr.

Differential Analysis on Complex Manifolds

Third Edition

New Appendix
By Oscar Garcia-Prada

Springer

Raymond O. Wells, Jr.
Jacobs University Bremen
Campus Ring 1
28759 Bremen
Germany

Mathematics Subject Classification 2000: 58-01, 32-01

Library of Congress Control Number: 2007935275

ISBN: 978-0-387-73891-8

Printed on acid-free paper.

9 8 7 6 5 4 3 2 1

springer.com

PREFACE TO THE FIRST EDITION

This book is an outgrowth and a considerable expansion of lectures given at Brandeis University in 1967–1968 and at Rice University in 1968–1969. The first four chapters are an attempt to survey in detail some recent developments in four somewhat different areas of mathematics: geometry (manifolds and vector bundles), algebraic topology, differential geometry, and partial differential equations. In these chapters, I have developed various tools that are useful in the study of compact complex manifolds. My motivation for the choice of topics developed was governed mainly by the applications anticipated in the last two chapters. Two principal topics developed include Hodge's theory of harmonic integrals and Kodaira's characterization of projective algebraic manifolds.

This book should be suitable for a graduate level course on the general topic of complex manifolds. I have avoided developing any of the theory of several complex variables relating to recent developments in Stein manifold theory because there are several recent texts on the subject (*Gunning and Rossi, Hörmander*). The text is relatively self-contained and assumes familiarity with the usual first year graduate courses (including some functional analysis), but since geometry is one of the major themes of the book, it is developed from first principles.

Each chapter is prefaced by a general survey of its content. Needless to say, there are numerous topics whose inclusion in this book would have been appropriate and useful. However, this book is not a treatise, but an attempt to follow certain threads that interconnect various fields and to culminate with certain key results in the theory of compact complex manifolds. In almost every chapter I give formal statements of theorems which are understandable in context, but whose proof oftentimes involves additional machinery not developed here (e.g., the Hirzebruch Riemann-Roch Theorem); hopefully, the interested reader will be sufficiently prepared (and perhaps motivated) to do further reading in the directions indicated.

Text references of the type (4.6) refer to the 6th equation (or theorem, lemma, etc.) in Sec. 4 of the chapter in which the reference appears. If the reference occurs in a different chapter, then it will be prefixed by the Roman numeral of that chapter, e.g., (II.4.6.).

I would like to express appreciation and gratitude to many of my colleagues and friends with whom I have discussed various aspects of the book during its development. In particular I would like to mention M. F. Atiyah, R. Bott, S. S. Chern, P. A. Griffiths, R. Harvey, L. Hörmander, R. Palais, J. Polking, O. Riemenschneider, H. Rossi, and W. Schmid whose comments were all very useful. The help and enthusiasm of my students at Brandeis and Rice during the course of my first lectures, had a lot to do with my continuing the project. M. Cowen and A. Dubson were very helpful with their careful reading of the first draft. In addition, I would like to thank two of my students for their considerable help. M. Windham wrote the first three chapters from my lectures in 1968–69 and read the first draft. Without his notes, the book almost surely would not have been started. J. Drouilhet read the final manuscript and galley proofs with great care and helped eliminate numerous errors from the text.

I would like to thank the Institute for Advanced Study for the opportunity to spend the year 1970–71 at Princeton, during which time I worked on the book and where a good deal of the typing was done by the excellent Institute staff. Finally, the staff of the Mathematics Department at Rice University was extremely helpful during the preparation and editing of the manuscript for publication.

Houston
December 1972

Raymond O. Wells, Jr.

PREFACE TO THE SECOND EDITION

In this second edition I have added a new section on the classical finite-dimensional representation theory for $\mathfrak{sl}(2, \mathbb{C})$. This is then used to give a natural proof of the Lefschetz decomposition theorem, an observation first made by S. S. Chern. H. Hecht observed that the Hodge $*$-operator is essentially a representation of the Weyl reflection operator acting on $\mathfrak{sl}(2, \mathbb{C})$ and this fact leads to new proofs (due to Hecht) of some of the basic Kähler identities which we incorporate into a completely revised Chapter V. The remainder of the book is generally the same as the first edition, except that numerous errors in the first edition have been corrected, and various examples have been added throughout.

I would like to thank my many colleagues who have commented on the first edition, which helped a great deal in getting rid of errors. Also, I would like to thank the graduate students at Rice who went carefully through the book with me in a seminar. Finally, I am very grateful to David Yingst and David Johnson who both collated errors, made many suggestions, and helped greatly with the editing of this second edition.

Houston
July 1979

Raymond O. Wells, Jr.

PREFACE TO THE THIRD EDITION

In the almost four decades since the first edition of this book appeared, many of the topics treated there have evolved in a variety of interesting manners. In both the 1973 and 1980 editions of this book, one finds the first four chapters (vector bundles, sheaf theory, differential geometry and elliptic partial differential equations) being used as fundamental tools for solving difficult problems in complex differential geometry in the final two chapters (namely the development of Hodge theory, Kodaira's embedding theorem, and Griffiths' theory of period matrix domains). In this new edition of the book, I have not changed the contents of these six chapters at all, as they have proved to be good building blocks for many other mathematical developments during these past decades.

I have asked my younger colleague Oscar García-Prada to add an Appendix to this edition which highlights some aspects of mathematical developments over the past thirty years which depend substantively on the tools developed in the first six chapters. The title of the Appendix, "Moduli spaces and geometric structures" and its introduction gives the reader a good overview to what is covered in this appendix.

The object of this appendix is to report on some topics in complex geometry that have been developed since the book's second edition appeared about 25 years ago. During this period there have been many important developments in complex geometry, which have arisen from the extremely rich interaction between this subject and different areas of mathematics and theoretical physics: differential geometry, algebraic geometry, global analysis, topology, gauge theory, string theory, etc. The number of topics that could be treated here is thus immense, including Calabi-Yau manifolds and mirror symmetry, almost-complex geometry and symplectic manifolds, Gromov-Witten theory, Donaldson and Seiberg-Witten theory, to mention just a few, providing material for several books (some already written).

However, since already the original scope of the book was not to be a treatise, "but an attempt to follow certain threads that interconnect various fields and to culminate with certain key results in the theory of compact complex manifolds...", as I said in the Preface to the first edition, in the Appendix we have chosen to focus on a particular set of topics in the theory of moduli spaces and geometric structures on Riemann surfaces. This is a subject which has played a central role in complex geometry in the last 25 years, and which, very much in the spirit of the book, reflects another instance of the powerful interaction between differential analysis (differential geometry and partial differential equations), algebraic topology and complex geometry. In choosing the topic, we have also taken into account that the book provides much of the background material needed (Chern classes, theory of connections on Hermitian vector bundles, Sobolev spaces, index theory, sheaf theory, etc.), making the appendix (in combination with the book) essentially self-contained.

It is my hope that this book will continue to be useful for mathematicians for some time to come, and I want to express my gratitude to Springer-Verlag for undertaking this new edition and for their patience in waiting for our revision and the new Appendix. One note to the reader: the *Subject Index* and the *Author Index* of the book refer to the original six chapters of this book and not to the new Appendix (which has its own bibliographical references).

Finally, I want to thank Oscar García-Prada so very much for the painstaking care and elegance in which he has summarized some of the most exciting results in the past years concerning the moduli spaces of vector bundles and Higgs' fields, their relation to representations of the fundamental group of a compact Riemann surface (or more generally of a compact Kähler manifold) in Lie groups, and to the solutions of differential equations which have their roots in the classical Laplace and Einstein equations, yielding a type of non-Abelian Hodge theory.

Bremen Raymond O. Wells, Jr.
June 2007

CONTENTS

CHAPTER I

MANIFOLDS AND VECTOR BUNDLES

There are many classes of manifolds which are under rather intense investigation in various fields of mathematics and from various points of view. In this book we are primarily interested in *differentiable manifolds* and *complex manifolds*. We want to study (a) the "geometry" of manifolds, (b) the analysis of functions (or more general objects) which are defined on manifolds, and (c) the interaction of (a) and (b). Our basic interest will be the application of techniques of real analysis (such as differential geometry and differential equations) to problems arising in the study of complex manifolds. In this chapter we shall summarize some of the basic definitions and results (including various examples) of the elementary theory of manifolds and vector bundles. We shall mention some nontrivial embedding theorems for differentiable and real-analytic manifolds as motivation for Kodaira's characterization of projective algebraic manifolds, one of the principal results which will be proved in this book (see Chap. VI). The "geometry" of a manifold is, from our point of view, represented by the behavior of the tangent bundle of a given manifold. In Sec. 2 we shall develop the concept of the tangent bundle (and derived bundles) from, more or less, first principles. We shall also discuss the continuous and C^∞ classification of vector bundles, which we shall not use in any real sense but which we shall meet a version of in Chap. III, when we study Chern classes. In Sec. 3 we shall introduce almost complex structures and the calculus of differential forms of type (p, q), including a discussion of integrability and the Newlander-Nirenberg theorem.

General background references for the material in this chapter are Bishop and Crittenden [1], Lang [1], Narasimhan [1], and Spivak [1], to name a few relatively recent texts. More specific references are given in the individual sections. The classical reference for calculus on manifolds is de Rham [1]. Such concepts as differential forms on differentiable manifolds, integration on chains, orientation, Stokes' theorem, and partition of unity are all covered adequately in the above references, as well as elsewhere, and in this book we shall assume familiarity with these concepts, although we may review some specific concept in a given context.

1. Manifolds

We shall begin this section with some basic definitions in which we shall use the following standard notations. Let **R** and **C** denote the fields of real and complex numbers, respectively, with their usual topologies, and let K denote either of these fields. If D is an open subset of K^n, we shall be concerned with the following function spaces on D:

(a) $K = \mathbf{R}$:

(1) $\mathcal{E}(D)$ will denote the real-valued *indefinitely differentiable* functions on D, which we shall simply call C^∞ functions on D; i.e., $f \in \mathcal{E}(D)$ if and only if f is a real-valued function such that partial derivatives of all orders exist and are continuous at all points of D [$\mathcal{E}(D)$ is often denoted by $C^\infty(D)$].

(2) $\mathcal{A}(D)$ will denote the real-valued *real-analytic functions* on D; i.e., $\mathcal{A}(D) \subset \mathcal{E}(D)$, and $f \in \mathcal{A}(D)$ if and only if the Taylor expansion of f converges to f in a neighborhood of any point of D.

(b) $K = C$:

(1) $\mathcal{O}(D)$ will denote the complex-valued *holomorphic functions* on D, i.e., if $(z_1, \ldots, z_n)$ are coordinates in C^n, then $f \in \mathcal{O}(D)$ if and only if near each point $z^0 \in D$, f can be represented by a convergent power series of the form

$$f(z) = f(z_1, \ldots, z_n) = \sum_{\alpha_1, \ldots, \alpha_n = 0}^{\infty} a_{\alpha_1, \ldots, \alpha_n} \left(z_1 - z_1^0\right)^{\alpha_1} \cdots \left(z_n - z_n^0\right)^{\alpha_n}.$$

(See, e.g., Gunning and Rossi [1], Chap. I, or Hörmander [2], Chap. II, for the elementary properties of holomorphic functions on an open set in $\mathbf{C}^n$). These particular classes of functions will be used to define the particular classes of manifolds that we shall be interested in.

A *topological n-manifold* is a Hausdorff topological space with a countable basis† which is locally homeomorphic to an open subset of $\mathbf{R}^n$. The integer n is called the *topological dimension* of the manifold. Suppose that $\mathcal{S}$ is one of the three K-valued families of functions defined on the open subsets of K^n described above, where we let $\mathcal{S}(D)$ denote the functions of $\mathcal{S}$ defined on D, an open set in K^n. [That is, $\mathcal{S}(D)$ is either $\mathcal{E}(D)$, $\mathcal{A}(D)$, or $\mathcal{O}(D)$. We shall only consider these three examples in this chapter. The concept of a family of functions is formalized by the notion of a *presheaf* in Chap. II.]

Definition 1.1: An $\mathcal{S}$-*structure*, $\mathcal{S}_M$, on a K-manifold M is a family of K-valued continuous functions defined on the open sets of M such that

†The additional assumption of a countable basis ("countable at infinity") is important for doing analysis on manifolds, and we incorporate it into the definition, as we are less interested in this book in the larger class of manifolds.

(a) For every $p \in M$, there exists an open neighborhood U of p and a homeomorphism $h : U \to U'$, where U' is open in K^n, such that for any open set $V \subset U$

$$f\colon V \longrightarrow K \in \mathcal{S}_M \text{ if and only if } f \circ h^{-1} \in \mathcal{S}(h(V)).$$

(b) If $f\colon U \to K$, where $U = \cup_i U_i$ and U_i is open in M, then $f \in \mathcal{S}_M$ if and only if $f|_{U_i} \in \mathcal{S}_M$ for each i.

It follows clearly from (a) that if $K = \mathbf{R}$, the dimension, k, of the topological manifold is equal to n, and if $K = \mathbf{C}$, then $k = 2n$. In either case n will be called the *K-dimension* of M, denoted by $\dim_K M = n$ (which we shall call *real-dimension* and *complex-dimension*, respectively). A manifold with an $\mathcal{S}$-structure is called an *$\mathcal{S}$-manifold*, denoted by $(M, \mathcal{S}_M)$, and the elements of $\mathcal{S}_M$ are called *$\mathcal{S}$-functions* on M. An open subset $U \subset M$ and a homeomorphism $h : U \to U' \subset K^n$ as in (a) above is called an *$\mathcal{S}$-coordinate system*.

For our three classes of functions we have defined

(a) $\mathcal{S} = \mathcal{E}$: *differentiable* (or C^∞) *manifold*, and the functions in $\mathcal{E}_M$ are called C^∞ *functions* on open subsets of M.

(b) $\mathcal{S} = \mathcal{A}$: *real-analytic manifold*, and the functions in $\mathcal{A}_M$ are called *real-analytic functions* on open subsets of M.

(c) $\mathcal{S} = \mathcal{O}$: *complex-analytic* (or simply *complex*) *manifold*, and the functions in $\mathcal{O}_M$ are called *holomorphic* (or *complex-analytic functions*) on open subsets of M.

We shall refer to $\mathcal{E}_M, \mathcal{A}_M$, and $\mathcal{O}_M$ as *differentiable, real-analytic*, and *complex structures* respectively.

Definition 1.2:

(a) An *$\mathcal{S}$-morphism* $F : (M, \mathcal{S}_M) \to (N, \mathcal{S}_N)$ is a continuous map, $F : M \to N$, such that

$$f \in \mathcal{S}_N \text{ implies } f \circ F \in \mathcal{S}_M.$$

(b) An *$\mathcal{S}$-isomorphism* is an $\mathcal{S}$-morphism $F : (M, \mathcal{S}_M) \to (N, \mathcal{S}_N)$ such that $F : M \to N$ is a homeomorphism, and

$$F^{-1} : (N, \mathcal{S}_N) \to (M, \mathcal{S}_M) \text{ is an } \mathcal{S}\text{-morphism.}$$

It follows from the above definitions that if on an $\mathcal{S}$-manifold $(M, \mathcal{S}_M)$ we have two coordinate systems $h_1\colon U_1 \to K^n$ and $h_2\colon U_2 \to K^n$ such that $U_1 \cap U_2 \neq \varnothing$, then

(1.1) $h_2 \circ h_1^{-1} : h_1(U_1 \cap U_2) \to h_2(U_1 \cap U_2)$ is an $\mathcal{S}$-isomorphism on open subsets of $(K^n, \mathcal{S}_{K^n})$.

Conversely, if we have an open covering $\{U_\alpha\}_{\alpha\in A}$ of M, a topological manifold, and a family of homeomorphisms $\{h_\alpha\colon U_\alpha \to U'_\alpha \subset K^n\}_{\alpha\in A}$ satisfying (1.1), then this defines an $\mathcal{S}$-structure on M by setting $\mathcal{S}_M = \{f\colon U \to K\}$ such that U is open in M and $f\circ h_\alpha^{-1} \in \mathcal{S}(h_\alpha(U\cap U_\alpha))$ for all $\alpha \in A$; i.e., the functions in $\mathcal{S}_M$ are pullbacks of functions in $\mathcal{S}$ by the homeomorphisms $\{h_\alpha\}_{\alpha\in A}$. The collection $\{(U_\alpha, h_\alpha)\}_{\alpha\in A}$ is called an *atlas* for $(M, \mathcal{S}_M)$.

In our three classes of functions, the concept of an $\mathcal{S}$-morphism and $\mathcal{S}$-isomorphism have special names:

(a) $\mathcal{S} = \mathcal{E}$: *differentiable mapping* and *diffeomorphism* of M to N.

(b) $\mathcal{S} = \mathcal{A}$: *real-analytic mapping* and *real-analytic isomorphism* (or *bianalytic* mapping) of M to N.

(c) $\mathcal{S} = \mathcal{O}$: *holomorphic mapping* and *biholomorphism* (*biholomorphic mapping*) of M to N.

It follows immediately from the definition above that a differentiable mapping

$$f\colon M \longrightarrow N,$$

where M and N are differentiable manifolds, is a continuous mapping of the underlying topological space which has the property that in local coordinate systems on M and N, f can be represented as a matrix of C^∞ functions. This could also be taken as the definition of a differentiable mapping. A similar remark holds for the other two categories.

Let N be an arbitrary subset of an $\mathcal{S}$-manifold M; then an *$\mathcal{S}$-function on N* is defined to be the restriction to N of an $\mathcal{S}$-function defined in some open set containing N, and $\mathcal{S}_M|_N$ consists of all the functions defined on relatively open subsets of N which are restrictions of $\mathcal{S}$-functions on the open subsets of M.

Definition 1.3: Let N be a closed subset of an $\mathcal{S}$-manifold M; then N is called an *$\mathcal{S}$-submanifold* of M if for each point $x_0 \in N$, there is a coordinate system $h\colon U \to U' \subset K^n$, where $x_0 \in U$, with the property that $U \cap N$ is mapped onto $U' \cap K^k$, where $0 \le k \le n$. Here $K^k \subset K^n$ is the standard embedding of the linear subspace K^k into K^n, and k is called the *K-dimension* of N, and $n-k$ is called the *K-codimension* of N.

It is easy to see that an $\mathcal{S}$-submanifold of an $\mathcal{S}$-manifold M is itself an $\mathcal{S}$-manifold with the $\mathcal{S}$-structure given by $\mathcal{S}_M|_N$. Since the implicit function theorem is valid in each of our three categories, it is easy to verify that the above definition of submanifold coincides with the more common one that an $\mathcal{S}$-submanifold (of k dimensions) is a closed subset of an $\mathcal{S}$-manifold M which is locally the common set of zeros of $n-k$ $\mathcal{S}$-functions whose Jacobian matrix has maximal rank.

It is clear that an n-dimensional complex structure on a manifold induces a $2n$-dimensional real-analytic structure, which, likewise, induces a $2n$-dimensional differentiable structure on the manifold. One of the questions we shall be concerned with is how many different (i.e., nonisomorphic) complex-analytic structures induce the same differentiable structure on a given manifold? The analogous question of how many different differentiable structures exist on a given topological manifold is an important problem in differential topology.

What we have actually defined is a category wherein the objects are $\mathcal{S}$-manifolds and the morphisms are $\mathcal{S}$-morphisms. We leave to the reader the proof that this actually is a category, since it follows directly from the definitions. In the course of what follows, then, we shall use three categories—the differentiable ($\mathcal{S} = \mathcal{E}$), the real-analytic ($\mathcal{S} = \mathcal{A}$), and the holomorphic ($\mathcal{S} = \mathcal{O}$) categories—and the above remark states that each is a subcategory of the former.

We now want to give some examples of various types of manifolds.

Example 1.4 (Euclidean space): K^n, $(\mathbf{R}^n, \mathbf{C}^n)$. For every $p \in K^n$, $U = K^n$ and $h =$ identity. Then $\mathbf{R}^n$ becomes a real-analytic (hence differentiable) manifold and $\mathbf{C}^n$ is a complex-analytic manifold.

Example 1.5: If $(M, \mathcal{S}_M)$ is an $\mathcal{S}$-manifold, then any open subset U of M has an $\mathcal{S}$-structure, $\mathcal{S}_U = \{f|_U : f \in \mathcal{S}_M\}$.

Example 1.6 (Projective space): If V is a finite dimensional vector space over K, then† $\mathbf{P}(V) :=$ {the set of one-dimensional subspaces of V} is called the *projective space* of V. We shall study certain special projective spaces, namely

$$\mathbf{P}_n(\mathbf{R}) := \mathbf{P}(\mathbf{R}^{n+1})$$

$$\mathbf{P}_n(\mathbf{C}) := \mathbf{P}(\mathbf{C}^{n+1}).$$

We shall show how $\mathbf{P}_n(\mathbf{R})$ can be made into a differentiable manifold.

There is a natural map $\pi: \mathbf{R}^{n+1} - \{0\} \to \mathbf{P}_n(\mathbf{R})$ given by

$$\pi(x) = \pi(x_0, \ldots, x_n) := \{\text{subspace spanned by } x = (x_0, \ldots, x_n) \in \mathbf{R}^{n+1}\}.$$

The mapping π is onto; in fact, $\pi|_{S^n = \{x \in \mathbf{R}^{n+1} : |x| = 1\}}$ is onto. Let $\mathbf{P}_n(\mathbf{R})$ have the quotient topology induced by the map π; i.e., $U \subset \mathbf{P}_n(\mathbf{R})$ is open if and only if $\pi^{-1}(U)$ is open in $\mathbf{R}^{n+1} - \{0\}$. Hence π is continuous and $\mathbf{P}_n(\mathbf{R})$ is a Hausdorff space with a countable basis. Also, since

$$\pi|_{S^n} : S^n \longrightarrow \mathbf{P}_n(\mathbf{R})$$

is continuous and surjective, $\mathbf{P}_n(\mathbf{R})$ is compact.

If $x = (x_0, \ldots, x_n) \in \mathbf{R}^{n+1} - \{0\}$, then set

$$\pi(x) = [x_0, \ldots, x_n].$$

We say that $(x_0, \ldots, x_n)$ are *homogeneous coordinates* of $[x_0, \ldots, x_n]$. If $(x_0', \ldots, x_n')$ is another set of homogeneous coordinates of $[x_0, \ldots, x_n]$,

†$:=$ means that the object on the left is defined to be equal to the object on the right.

then $x_i = tx'_i$ for some $t \in \mathbf{R} - \{0\}$, since $[x_0, \ldots, x_n]$ is the one-dimensional subspace spanned by $(x_0, \ldots, x_n)$ or $(x'_0, \ldots, x'_n)$. Hence also $\pi(x) = \pi(tx)$ for $t \in \mathbf{R} - \{0\}$. Using homogeneous coordinates, we can define a differentiable structure (in fact, real-analytic) on $\mathbf{P}_n(\mathbf{R})$ as follows. Let

$$U_\alpha = \{S \in \mathbf{P}_n(\mathbf{R}) \colon S = [x_0, \ldots, x_n] \text{ and } x_\alpha \neq 0\}, \quad \text{for } \alpha = 0, \ldots, n.$$

Each U_α is open and $\mathbf{P}_n(\mathbf{R}) = \bigcup_{\alpha=0}^n U_\alpha$ since $(x_0, \ldots, x_n) \in \mathbf{R}^{n+1} - \{0\}$. Also, define the map $h_\alpha \colon U_\alpha \to \mathbf{R}^n$ by setting

$$h_\alpha([x_0, \ldots, x_n]) = \left(\frac{x_0}{x_\alpha}, \ldots, \frac{x_{\alpha-1}}{x_\alpha}, \frac{x_{\alpha+1}}{x_\alpha}, \ldots, \frac{x_n}{x_\alpha}\right) \in \mathbf{R}^n.$$

Note that both U_α and h_α are well defined by the relation between different choices of homogeneous coordinates. One shows easily that h_α is a homeomorphism and that $h_\alpha \circ h_\beta^{-1}$ is a diffeomorphism; therefore, this defines a differentiable structure on $\mathbf{P}_n(\mathbf{R})$. In exactly this same fashion we can define a differentiable structure on $\mathbf{P}(V)$ for any finite dimensional $\mathbf{R}$-vector space V and a complex-analytic structure on $\mathbf{P}(V)$ for any finite dimensional $\mathbf{C}$-vector space V.

Example 1.7 (Matrices of fixed rank): Let $\mathfrak{M}_{k,n}(\mathbf{R})$ be the $k \times n$ matrices with real coefficients. Let $M_{k,n}(\mathbf{R})$ be the $k \times n$ matrices of rank $k (k \leq n)$. Let $M_{k,n}^m(\mathbf{R})$ be the elements of $\mathfrak{M}_{k,n}(\mathbf{R})$ of rank $m (m \leq k)$. First, $\mathfrak{M}_{k,n}(\mathbf{R})$ can be identified with $\mathbf{R}^{kn}$, and hence it is a differentiable manifold. We know that $M_{k,n}(\mathbf{R})$ consists of those $k \times n$ matrices for which at least one $k \times k$ minor is nonsingular; i.e.,

$$M_{k,n}(\mathbf{R}) = \bigcup_{i=1}^{l} \{A \in \mathfrak{M}_{k,n}(\mathbf{R}) : \det\ A_i \neq 0\},$$

where for each $A \in \mathfrak{M}_{k,n}(\mathbf{R})$ we let $\{A_1, \ldots, A_l\}$ be a fixed ordering of the $k \times k$ minors of A. Since the determinant function is continuous, we see that $M_{k,n}(\mathbf{R})$ is an open subset of $\mathfrak{M}_{k,n}(\mathbf{R})$ and hence has a differentiable structure induced on it by the differentiable structure on $\mathfrak{M}_{k,n}(\mathbf{R})$ (see Example 1.5). We can also define a differentiable structure on $M_{k,n}^m(\mathbf{R})$. For convenience we delete the $\mathbf{R}$ and refer to $M_{k,n}^m$. For $X_0 \in M_{k,n}^m$, we define a coordinate neighborhood at X_0 as follows. Since the rank of X is m, there exist permutation matrices P, Q such that

$$PX_0Q = \begin{bmatrix} A_0 & B_0 \\ C_0 & D_0 \end{bmatrix},$$

where A_0 is a nonsingular $m \times m$ matrix. Hence there exists an $\epsilon > 0$ such that $\|A - A_0\| < \epsilon$ implies A is nonsingular, where $\|A\| = \max_{ij} |a_{ij}|$, for $A = [a_{ij}]$. Therefore let

$$W = \{X \in \mathfrak{M}_{k,n} : PXQ = \begin{bmatrix} A & B \\ C & D \end{bmatrix} \quad \text{and} \quad \|A - A_0\| < \epsilon\}.$$

Then W is an open subset of $\mathfrak{M}_{k,n}$. Since this is true, $U := W \cap M_{k,n}^m$ is an open neighborhood of X_0 in $M^m_{k,n}$ and will be the necessary coordinate neighborhood of X_0. Note that

$$X \in U \text{ if and only if } D = CA^{-1}B, \quad \text{where } PXQ = \begin{bmatrix} A & B \\ C & D \end{bmatrix}.$$

This follows from the fact that

$$\begin{bmatrix} I_m & 0 \\ -CA^{-1} & I_{k-m} \end{bmatrix} \begin{bmatrix} A & B \\ C & D \end{bmatrix} = \begin{bmatrix} A & B \\ 0 & D - CA^{-1}B \end{bmatrix}$$

and

$$\begin{bmatrix} I_m & 0 \\ -CA^{-1} & I_{k-m} \end{bmatrix}$$

is nonsingular (where I_j is the $j \times j$ identity matrix). Therefore

$$\begin{bmatrix} A & B \\ C & D \end{bmatrix} \quad \text{and} \quad \begin{bmatrix} A & B \\ 0 & D - CA^{-1}B \end{bmatrix}$$

have the same rank, but

$$\begin{bmatrix} A & B \\ 0 & D - CA^{-1}B \end{bmatrix}$$

has rank m if and only if $D - CA^{-1}B = 0$.

We see that $M^m_{k,n}$ actually becomes a manifold of dimension $m(n+k-m)$ by defining

$$h\colon U \longrightarrow \mathbf{R}^{m^2+(n-m)m+(k-m)m} = \mathbf{R}^{m(n+k-m)},$$

where

$$h(X) = \begin{bmatrix} A & B \\ C & 0 \end{bmatrix} \in \mathbf{R}^{m(n+k-m)} \quad \text{for } PXQ = \begin{bmatrix} A & B \\ C & D \end{bmatrix},$$

as above. Note that we can define an inverse for h by

$$h^{-1}\left(\begin{bmatrix} A & B \\ C & 0 \end{bmatrix}\right) = P^{-1}\begin{bmatrix} A & B \\ C & CA^{-1}B \end{bmatrix} Q^{-1}.$$

Therefore h is, in fact, bijective and is easily shown to be a homeomorphism. Moreover, if h_1 and h_2 are given as above,

$$h_2 \circ h_1^{-1}\left(\begin{bmatrix} A_1 & B_1 \\ C_1 & 0 \end{bmatrix}\right) = \begin{bmatrix} A_2 & B_2 \\ C_2 & 0 \end{bmatrix},$$

where

$$P_2 P_1^{-1}\begin{bmatrix} A_1 & B_1 \\ C_1 & C_1 A_1^{-1} B_1 \end{bmatrix} Q_1^{-1} Q_2 = \begin{bmatrix} A_2 & B_2 \\ C_2 & D_2 \end{bmatrix},$$

and these maps are clearly diffeomorphisms (in fact, real-analytic), and so $M^m_{k,n}(\mathbf{R})$ is a differentiable submanifold of $\mathfrak{M}_{k,n}(\mathbf{R})$. The same procedure can be used to define complex-analytic structures on $\mathfrak{M}_{k,n}(\mathbf{C})$, $M_{k,n}(\mathbf{C})$, and $M^m_{k,n}(\mathbf{C})$, the corresponding sets of matrices over $\mathbf{C}$.

Example 1.8 (Grassmannian manifolds): Let V be a finite dimensional K-vector space and let $G_k(V) := \{$the set of k-dimensional subspaces of $V\}$, for $k < \dim_K V$. Such a $G_k(V)$ is called a *Grassmannian manifold*. We shall use two particular Grassmannian manifolds, namely

$$G_{k,n}(\mathbf{R}) := G_k(\mathbf{R}^n) \quad \text{and} \quad G_{k,n}(\mathbf{C}) := G_k(\mathbf{C}^n).$$

The Grassmannian manifolds are clearly generalizations of the projective spaces [in fact, $\mathbf{P}(V) = G_1(V)$; see Example 1.6] and can be given a manifold structure in a fashion analogous to that used for projective spaces.

Consider, for example, $G_{k,n}(\mathbf{R})$. We can define the map

$$\pi\colon M_{k,n}(\mathbf{R}) \longrightarrow G_{k,n}(\mathbf{R}),$$

where

$$\pi(A) = \pi\begin{pmatrix} a_1 \\ \cdot \\ \cdot \\ \cdot \\ a_k \end{pmatrix} := \{k\text{-dimensional subspace of } \mathbf{R}^n \text{ spanned by the row vectors } \{a_j\} \text{ of } A\}.$$

We notice that for $g \in GL(k, \mathbf{R})$ (the $k \times k$ nonsingular matrices) we have $\pi(gA) = \pi(A)$ (where gA is matrix multiplication), since the action of g merely changes the basis of $\pi(A)$. This is completely analogous to the projection $\pi\colon \mathbf{R}^{n+1} - \{0\} \to \mathbf{P}_n(\mathbf{R})$, and, using the same reasoning, we see that $G_{k,n}(\mathbf{R})$ is a compact Hausdorff space with the quotient topology and that π is a surjective, continuous open map.†

We can also make $G_{k,n}(\mathbf{R})$ into a differentiable manifold in a way similar to that used for $\mathbf{P}_n(\mathbf{R})$. Consider $A \in M_{k,n}$ and let $\{A_1, \ldots, A_l\}$ be the collection of $k \times k$ minors of A (see Example 1.7). Since A has rank k, A_α is nonsingular for some $1 \le \alpha \le l$ and there is a permutation matrix P_α such that

$$AP_\alpha = [A_\alpha \tilde{A}_\alpha],$$

where $\tilde{A}_\alpha$ is a $k \times (n-k)$ matrix. Note that if $g \in GL(k, \mathbf{R})$, then gA_α is a nonsingular minor of gA and $gA_\alpha = (gA)_\alpha$. Let $U_\alpha = \{S \in G_{k,n}(\mathbf{R})\colon S = \pi(A)$, where A_α is nonsingular$\}$. This is well defined by the remark above concerning the action of $GL(k, \mathbf{R})$ on $M_{k,n}(\mathbf{R})$. The set U_α is defined by the condition $\det A_\alpha \neq 0$; hence it is an open set in $G_{k,n}(\mathbf{R})$, and $\{U_\alpha\}_{\alpha=1}^{l}$ covers $G_{k,n}(\mathbf{R})$. We define a map

$$h_\alpha : U_\alpha \longrightarrow \mathbf{R}^{k(n-k)}$$

by setting

$$h_\alpha(\pi(A)) = A_\alpha^{-1}\tilde{A}_\alpha \in \mathbf{R}^{k(n-k)},$$

where $AP_\alpha = [A_\alpha \tilde{A}_\alpha]$. Again this is well defined and we leave it to the reader to show that this does, indeed, define a differentiable structure on $G_{k,n}(\mathbf{R})$.

†Note that the compact set $\{A \in M_{k,n}(\mathbf{R}) : A^t A = I\}$ is analogous to the unit sphere in the case $k = 1$ and is mapped surjectively onto $G_{k,n}(R)$.

Example 1.9 (Algebraic submanifolds): Consider $\mathbf{P}_n = \mathbf{P}_n(\mathbf{C})$, and let

$$H = \{[z_0, \ldots, z_n] \in \mathbf{P}_n \colon a_0 z_0 + \cdots + a_n z_n = 0\},$$

where $(a_0, \ldots, a_n) \in \mathbf{C}^{n+1} - \{0\}$. Then H is called a *projective hyperplane*. We shall see that H is a submanifold of $\mathbf{P}_n$ of dimension $n-1$. Let U_α be the coordinate systems for $\mathbf{P}_n$ as defined in Example 1.6. Let us consider $U_0 \cap H$, and let $(\zeta_1, \ldots, \zeta_n)$ be coordinates in $\mathbf{C}^n$. Suppose that $[z_0, \ldots, z_n] \in H \cap U_0$; then, since $z_0 \neq 0$, we have

$$a_1 \frac{z_1}{z_0} + \cdots + a_n \frac{z_n}{z_0} = -a_0,$$

which implies that if $\zeta = (\zeta_1, \ldots, \zeta_n) = h_0([z_0, \ldots, z_n])$, then ζ satisfies

$$a_1\zeta_1 + \cdots + a_n\zeta_n = -a_0, \tag{1.2}$$

which is an affine linear subspace of $\mathbf{C}^n$, provided that at least one of $a_1, \ldots, a_n$ is not zero. If, however, $a_0 \neq 0$ and $a_1 = \cdots = a_n = 0$, then it is clear that there is no point $(\zeta_1, \ldots, \zeta_n) \in \mathbf{C}^n$ which satisfies (1.2), and hence in this case $U_0 \cap H = \varnothing$ (however, H will then necessarily intersect all the other coordinate systems $U_1, \ldots, U_n$). It now follows easily that H is a submanifold of dimension $n-1$ of $\mathbf{P}_n$ (using equations similar to (1.2) in the other coordinate systems as a representation for H). More generally, one can consider

$$V = \{[z_0, \ldots, z_n] \in \mathbf{P}_n(\mathbf{C}) \colon p_1(z_0, \ldots, z_n) = \cdots = p_r(z_0, \ldots, z_n) = 0\},$$

where $p_1, \ldots, p_r$ are homogeneous polynomials of varying degrees. In local coordinates, one can find equations of the form (for instances, in U_0)

$$\begin{aligned} p_1\left(1, \frac{z_1}{z_0}, \ldots, \frac{z_n}{z_0}\right) &= 0 \\ p_r\left(1, \frac{z_1}{z_0}, \ldots, \frac{z_n}{z_0}\right) &= 0, \end{aligned} \tag{1.3}$$

and V will be a *submanifold* of $\mathbf{P}_n$ if the Jacobian matrix of these equations in the various coordinate systems has maximal rank. More generally, V is called a *projective algebraic variety*, and points where the Jacobian has less than maximal rank are called *singular points* of the variety.

We say that an $\mathcal{S}$-morphism

$$f\colon (M, \mathcal{S}_M) \longrightarrow (N, \mathcal{S}_N)$$

of two $\mathcal{S}$-manifolds is an $\mathcal{S}$-*embedding* if f is an $\mathcal{S}$-isomorphism onto an $\mathcal{S}$-submanifold of $(N, \mathcal{S}_N)$. Thus, in particular, we have the concept of differentiable, real-analytic, and holomorphic embeddings. Embeddings are most often used (or conceived of as) embeddings of an "abstract" manifold as a submanifold of some more concrete (or more elementary) manifold. Most common is the concept of embedding in Euclidean space and in projective space, which are the simplest geometric models (noncompact and compact, respectively). We shall state some results along this line to give the reader some feeling for the differences among the three categories we have been dealing with. Until now they have behaved very similarly.

Theorem 1.10 (Whitney [1]): Let M be a differentiable n-manifold. Then there exists a differentiable embedding f of M into $\mathbf{R}^{2n+1}$. Moreover, the image of M, $f(M)$ can be realized as a real-analytic submanifold of $\mathbf{R}^{2n+1}$.

This theorem tells us that all differentiable manifolds (compact and non-compact) can be considered as submanifolds of Euclidean space, such submanifolds having been the motivation for the definition and concept of manifold in general. The second assertion, which is a more difficult result, tells us that on any differentiable manifold M one can find a subfamily of the family ε of differentiable functions on M so that this subfamily gives a real-analytic structure to the manifold M; i.e., every differentiable manifold admits a real-analytic structure. It is strictly false that differentiable manifolds admit complex structures in general, since, in particular, complex manifolds must have even topological dimension. We shall discuss this question somewhat more in Sec. 3. We shall not prove Whitney's theorem since we do not need it later (see, e.g., de Rham [1], Sternberg [1], or Whitney's original paper for a proof of Whitney's theorems).

A deeper result is the theorem of Grauert and Morrey (see Grauert [1] and Morrey [1]) that any real-analytic manifold can be embedded, by a real-analytic embedding, into $\mathbf{R}^N$, for some N (again either compact or non-compact). However, when we turn to complex manifolds, things are completely different. First, we have the relatively elementary result.

Theorem 1.11: Let X be a connnected compact complex manifold and let $f \in \mathcal{O}(X)$. Then f is constant; i.e., global holomorphic functions are necessarily constant.

Proof: Suppose that $f \in \mathcal{O}(X)$. Then, since f is a continuous function on a compact space, $|f|$ assumes its maximum at some point $x_0 \in X$ and $S = \{x: f(x) = f(x_0)\}$ is closed. Let $z = (z_1, \ldots, z_n)$ be local coordinates at $x \in S$, with $z = 0$ corresponding to the point x. Consider a small ball B about $z = 0$ and let $z \in B$. Then the function $g(\lambda) = f(\lambda z)$ is a function of one complex variable (λ) which assumes its maximum absolute value at $\lambda = 0$ and is hence constant by the maximum principle. Therefore, $g(1) = g(0)$ and hence $f(z) = f(0)$, for all $z \in B$. By connectedness, $S = X$, and f is constant.

Q.E.D.

Remark: The maximum principle for holomorphic functions in domains in $\mathbf{C}^n$ is also valid and could have been applied (see Gunning and Rossi [1]).

Corollary 1.12: There are no compact complex submanifolds of $\mathbf{C}^n$ of positive dimension.

Proof: Otherwise at least one of the coordinate functions $z_1, \ldots, z_n$ would be a nonconstant function when restricted to such a submanifold.

Q.E.D.

Therefore, we see that not all complex manifolds admit an embedding into Euclidean space in contrast to the differentiable and real-analytic situations, and of course, there are many examples of such complex manifolds [e.g., $\mathbf{P}_n(\mathbf{C})$]. One *can* characterize the (necessarily noncompact) complex manifolds which admit embeddings into $\mathbf{C}^n$, and these are called *Stein manifolds*, which have an abstract definition and have been the subject of much study during the past 20 years or so (see Gunning and Rossi [1] and Hörmander [2] for an exposition of the theory of Stein manifolds). In this book we want to develop the material necessary to provide a characterization of the compact complex manifolds which admit an embedding into projective space. This was first accomplished by Kodaira in 1954 (see Kodaira [2]) and the material in the next several chapters is developed partly with this characterization in mind. We give a formal definition.

Definition 1.13: A compact complex manifold X which admits an embedding into $\mathbf{P}_n(\mathbf{C})$ (for some n) is called a *projective algebraic manifold.*

Remark: By a theorem of Chow (see, e.g., Gunning and Rossi [1]), every complex submanifold V of $\mathbf{P}_n(\mathbf{C})$ is actually an *algebraic* submanifold (hence the name projective algebraic manifold), which means in this context that V can be expressed as the zeros of homogeneous polynomials in homogeneous coordinates. Thus, such manifolds can be studied from the point of view of algebra (and hence algebraic geometry). We will not need this result since the methods we shall be developing in this book will be analytical and not algebraic. As an example, we have the following proposition.

Proposition 1.14: The Grassmannian manifolds $G_{k,n}(\mathbf{C})$ are projective algebraic manifolds.

Proof: Consider the following map:

$$\tilde{F}\colon M_{k,n}(\mathbf{C}) \longrightarrow \wedge^k \mathbf{C}^n$$

defined by

$$\tilde{F}(A) = \tilde{F}\begin{pmatrix} a_1 \\ \cdot \\ \cdot \\ \cdot \\ a_k \end{pmatrix} = a_1 \wedge \cdots \wedge a_k.$$

The image of this map is actually contained in $\wedge^k \mathbf{C}^n - \{0\}$ since $\{a_j\}$ is an independent set. We can obtain the desired embedding by completing the following diagram by F:

$$\begin{array}{ccc} M_{k,n}(\mathbf{C}) & \xrightarrow{\tilde{F}} & \wedge^k \mathbf{C}^n - \{0\} \\ \downarrow{\scriptstyle \pi_G} & & \downarrow{\scriptstyle \pi_P} \\ G_{k,n}(\mathbf{C}) & \overset{F}{\dashrightarrow} & P(\wedge^k \mathbf{C}^n), \end{array}$$

where π_G, π_P are the previously defined projections. We must show that F is well defined; i.e.,

$$\pi_G(A) = \pi_G(B) \Longrightarrow \pi_P \circ \tilde{F}(A) = \pi_P \circ \tilde{F}(B).$$

But $\pi_G(A) = \pi_G(B)$ implies that $A = gB$ for $g \in GL(k, \mathbf{C})$, and so

$$a_1 \wedge \cdots \wedge a_k = \det\, g(b_1 \wedge \cdots \wedge b_k),$$

where

$$A = \begin{pmatrix} a_1 \\ \cdot \\ \cdot \\ \cdot \\ a_k \end{pmatrix} \quad \text{and} \quad B = \begin{pmatrix} b_1 \\ \cdot \\ \cdot \\ \cdot \\ b_k \end{pmatrix},$$

but

$$\pi_P(a_1 \wedge \cdots \wedge a_k) = \pi_P(\det\, g(b_1 \wedge \cdots \wedge b_k)) = \pi_P(b_1 \wedge \cdots \wedge b_k),$$

and so the map F is well defined. We leave it to the reader to show that F is also an embedding.

Q.E.D.

2. Vector Bundles

The study of vector bundles on manifolds has been motivated primarily by the desire to linearize nonlinear problems in geometry, and their use has had a profound effect on various modern fields of mathematics. In this section we want to introduce the concept of a vector bundle and give various examples. We shall also discuss some of the now classical results in differential topology (the classification of vector bundles, for instance) which form a motivation for some of our constructions later in the context of holomorphic vector bundles.

We shall use the same notation as in Sec. 1. In particular $\mathcal{S}$ will denote one of the three structures on manifolds ($\mathcal{E}, \mathcal{A}, \mathcal{O}$) studied there, and $K = \mathbf{R}$ or $\mathbf{C}$.

Definition 2.1: A continuous map $\pi: E \to X$ of one Hausdorff space, E, onto another, X, is called a *K-vector bundle of rank r* if the following conditions are satisfied:

(a) $E_p := \pi^{-1}(p)$, for $p \in X$, is a K-vector space of dimension r (E_p is called the *fibre* over p).

(b) For every $p \in X$ there is a neighborhood U of p and a homeomorphism

$$h: \pi^{-1}(U) \longrightarrow U \times K^r \text{ such that } h(E_p) \subset \{p\} \times K^r,$$

and h^p, defined by the composition

$$h^p: E_p \xrightarrow{h} \{p\} \times K^r \xrightarrow{\text{proj.}} K^r,$$

is a K-vector space isomorphism [the pair (U, h) is called a *local trivialization*].

For a K-vector bundle $\pi: E \to X$, E is called the *total space* and X is called the *base space*, and we often say that E is a vector bundle over X. Notice that for two local trivializations (U_α, h_α) and (U_β, h_β) the map

$$h_\alpha \circ h_\beta^{-1} : (U_\alpha \cap U_\beta) \times K^r \longrightarrow (U_\alpha \cap U_\beta) \times K^r$$

induces a map

(2.1) $$g_{\alpha\beta}: U_\alpha \cap U_\beta \longrightarrow GL(r, K),$$

where

$$g_{\alpha\beta}(p) = h_\alpha^p \circ (h_\beta^p)^{-1} : K^r \longrightarrow K^r.$$

The functions $g_{\alpha\beta}$ are called the *transition functions* of the K-vector bundle $\pi: E \to X$ (with respect to the two local trivializations above).†

The transition functions $g_{\alpha\beta}$ satisfy the following compatibility conditions:

(2.2a) $$g_{\alpha\beta} \cdot g_{\beta\gamma} \cdot g_{\gamma\alpha} = I_r \quad \text{on } U_\alpha \cap U_\beta \cap U_\gamma,$$

and

(2.2b) $$g_{\alpha\alpha} = I_r \quad \text{on } U_\alpha,$$

where the product is a matrix product and I_r is the identity matrix of rank r. This follows immediately from the definition of the transition functions.

Definition 2.2: A K-vector bundle of rank $r, \pi: E \to X$, is said to be an $\mathcal{S}$-*bundle* if E and X are $\mathcal{S}$-manifolds, π is an $\mathcal{S}$-morphism, and the local trivializations are $\mathcal{S}$-isomorphisms.

Note that the fact that the local trivializations are $\mathcal{S}$-isomorphisms is equivalent to the fact that the transition functions are $\mathcal{S}$-morphisms. In particular, then, we have *differentiable vector bundles, real-analytic vector bundles*, and *holomorphic vector bundles* (K must equal **C**).

Remark: Suppose that on an $\mathcal{S}$-manifold we are given an open covering $\mathfrak{A} = \{U_\alpha\}$ and that to each ordered nonempty intersection $U_\alpha \cap U_\beta$ we have assigned an $\mathcal{S}$-function

$$g_{\alpha\beta}: U_\alpha \cap U_\beta \longrightarrow GL(r, K)$$

satisfying the compatibility conditions (2.2). Then one can construct a vector bundle $E \xrightarrow{\pi} X$ having these transition functions. An outline of the construction is as follows: Let

$$\tilde{E} = \bigcup_\alpha U_\alpha \times K^r \quad \text{(disjoint union)}$$

equipped with the natural product topology and $\mathcal{S}$-structure. Define an equivalence relation in $\tilde{E}$ by setting

$$(x, v) \sim (y, w), \text{ for } (x, v) \in U_\beta \times K^r, (y, w) \in U_\alpha \times K^r$$

if and only if

$$y = x \quad \text{and} \quad w = g_{\alpha\beta}(x)v.$$

†Note that the transition function $g_{\alpha\beta}(p)$ is a linear mapping from the U_β trivialization to the U_α trivialization. The order is significant.

The fact that this is a well-defined equivalence relation is a consequence of the compatibility conditions (2.2). Let $E = \tilde{E}/\sim$ (the set of equivalence classes), equipped with the quotient topology, and let $\pi: E \to X$ be the mapping which sends a representative (x, v) of a point $p \in E$ into the first coordinate. One then shows that an E so constructed carries on $\mathcal{S}$-structure and is an $\mathcal{S}$-vector bundle. In the examples discussed below we shall see more details of such a construction.

Example 2.3 (Trivial bundle): Let M be an $\mathcal{S}$-manifold. Then

$$\pi: M \times K^n \longrightarrow M,$$

where π is the natural projection, is an $\mathcal{S}$-bundle called a *trivial bundle.*

Example 2.4 (Tangent bundle): Let M be a differentiable manifold. Then we want to construct a vector bundle over M whose fibre at each point is the linearization of the manifold M, to be called the *tangent bundle* to M. Let $p \in M$. Then we let

$$\mathcal{E}_{M,p} := \varinjlim_{\substack{p \in U \subset M \\ \text{open}}} \mathcal{E}_M(U)$$

be the *algebra* (over $\mathbf{R}$) *of germs of differentiable functions at the point* $p \in M$, where the inductive limit† is taken with respect to the partial ordering on open neighborhoods of p given by inclusion. Expressed differently, we can say that if f and g are defined and C^∞ near p and they coincide on some neighborhood of p, then they are equivalent. The set of equivalence classes is easily seen to form an algebra over $\mathbf{R}$ and is the same as the inductive limit algebra above; an equivalence class (element of $\mathcal{E}_{M,p}$) is called a *germ* of a C^∞ function at p. A *derivation of the algebra* $\mathcal{E}_{M,p}$ is a vector space homomorphism $D: \mathcal{E}_{M,p} \to \mathbf{R}$ with the property that $D(fg) = D(f) \cdot g(p) + f(p) \cdot D(g)$, where $g(p)$ and $f(p)$ denote evaluation of a germ at a point p (which clearly makes sense). The *tangent space* to M at p is the vector space of all derivations of the algebra $\mathcal{E}_{M,p}$, which we denote by $T_p(M)$. Since M is a differentiable manifold, we can find a diffeomorphism h defined in a neighborhood U of p where

$$h: U \longrightarrow U' \underset{\text{open}}{\subset} \mathbf{R}^n$$

and where, letting $h^* f(x) = f \circ h(x)$, h has the property that, for $V \subset U'$,

$$h^*: \mathcal{E}_{\mathbf{R}^n}(V) \longrightarrow \mathcal{E}_M(h^{-1}(V))$$

is an algebra isomorphism. It follows that h^* induces an algebra isomorphism on germs, i.e., (using the same notation),

$$h^*: \mathcal{E}_{\mathbf{R}^n, h(p)} \xrightarrow{\cong} \mathcal{E}_{M,p},$$

†We denote by $\varinjlim$ the inductive (or direct) limit and by $\varprojlim$ the projective (or inverse) limit of a partially ordered system.

and hence induces an isomorphism on derivations:

$$h_*\colon T_p(M) \xrightarrow{\cong} T_{h(p)}(\mathbf{R}^n).$$

It is easy to verify that

(a) $\partial/\partial x_j$ are derivations of $\mathcal{E}_{\mathbf{R}^n, h(p)}$, $j = 1, \ldots, n$, and that
(b) $\{\partial/\partial x_1, \ldots, \partial/\partial x_n\}$ is a basis for $T_{h(p)}(\mathbf{R}^n)$,

and thus that $T_p(M)$ is an n-dimensional vector space over $\mathbf{R}$, for each point $p \in M$ [the derivations are, of course, simply the classical directional derivatives evaluated at the point $h(p)$]. Suppose that $f\colon M \to N$ is a differentiable mapping of differentiable manifolds. Then there is a natural map

$$df_p\colon T_p(M) \longrightarrow T_{f(p)}(N)$$

defined by the following diagram:

$$\begin{array}{ccc} \mathcal{E}_{M,p} & \xleftarrow{\;f^*\;} & \mathcal{E}_{N,f(p)} \\ & {\scriptstyle D_p} \searrow \quad \swarrow & \\ & \mathbf{R} & \end{array} \qquad D_p \circ f^* = df_p(D_p),$$

for $D_p \in T_p(M)$. The mapping df_p is a linear mapping and can be expressed as a matrix of first derivatives with respect to local coordinates. The coefficients of such a matrix representation will be C^∞ functions of the local coordinates. Classically, the mapping df_p (the *derivative mapping, differential mapping*, or *tangent mapping*) is called the *Jacobian* of the differentiable map f. The tangent map represents a first-order linear approximation (at p) to the differentiable map f. We are now in a position to construct the tangent bundle to M. Let

$$T(M) = \bigcup_{p \in M} T_p(M) \quad \text{(disjoint union)}$$

and define

$$\pi\colon T(M) \longrightarrow M$$

by

$$\pi(v) = p \quad \text{if } v \in T_p(M).$$

We can now make $T(M)$ into a vector bundle. Let $\{(U_\alpha, h_\alpha)\}$ be an atlas for M, and let $T(U_\alpha) = \pi^{-1}(U_\alpha)$ and

$$\psi_\alpha\colon T(U_\alpha) \longrightarrow U_\alpha \times \mathbf{R}^n$$

be defined as follows: Suppose that $v \in T_p(M) \subset T(U_\alpha)$. Then $dh_{\alpha,p}(v) \in T_{h\alpha(p)}(\mathbf{R}^n)$. Thus

$$dh_{\alpha,p}(v) = \sum_{j=1}^{n} \xi_j(p) \frac{\partial}{\partial x_j}\bigg|_{h\alpha(p)},$$

where $\xi_j \in \mathcal{E}_M(U_\alpha)$ (the fact that the coefficients are C^∞ follows easily from the proof that $\{\partial/\partial x_1, \ldots, \partial/\partial x_n\}$ is a basis for the tangent vectors at a point in $\mathbf{R}^n$). Now let

$$\psi_\alpha(v) = (p, \xi_1(p), \ldots, \xi_n(p)) \in U_\alpha \times \mathbf{R}^n.$$

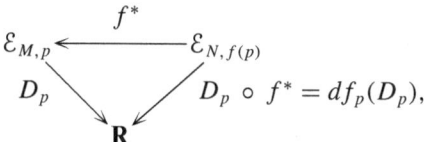

It is easy to verify that ψ_α is bijective and fibre-preserving and moreover that

$$\psi_\alpha^p\colon T_p(M)\xrightarrow{\psi_\alpha}\{p\}\times\mathbf{R}^n\xrightarrow{\text{proj.}}\mathbf{R}^n$$

is an $\mathbf{R}$-linear isomorphism. We can define transition functions

$$g_{\alpha\beta}\colon U_\beta\cap U_\alpha\longrightarrow GL(n,\mathbf{R})$$

by setting

$$g_{\alpha\beta}(p)=\psi_\alpha^p\circ(\psi_\beta^p)^{-1}:\mathbf{R}^n\longrightarrow\mathbf{R}^n.$$

Moreover, it is easy to check that the coefficients of the matrices $\{g_{\alpha\beta}\}$ are C^∞ functions in $U_\alpha\cap U_\beta$, since $g_{\alpha\beta}$ is a matrix representation for the composition $dh_\alpha\circ dh_\beta^{-1}$ with respect to the basis $\{\partial/\partial x_1,\ldots,\partial/\partial x_n\}$ at $T_{h_\beta(p)}(\mathbf{R}^n)$ and $T_{h_\alpha(p)}(\mathbf{R}^n)$, and that the tangent maps are differentiable functions of local coordinates. Thus the $\{(U_\alpha,\psi_\alpha)\}$ become the desired trivializations. We have only to put the right topology on $T(M)$ so that $T(M)$ becomes a differentiable manifold. We simply require that $U\subset T(M)$ be open if and only if $\psi_\alpha(U\cap T(U_\alpha))$ is open in $U_\alpha\times\mathbf{R}^n$ for every α. This is well defined since

$$\psi_\alpha\circ\psi_\beta^{-1}:(U_\alpha\cap U_\beta)\times\mathbf{R}^n\longrightarrow(U_\alpha\cap U_\beta)\times\mathbf{R}^n$$

is a diffeomorphism for any α and β such that $U_\alpha\cap U_\beta\neq\varnothing$ (since $\psi_\alpha\circ\psi_\beta^{-1}=id\times g_{\alpha\beta}$, where *id* is the identity mapping). Because the transition functions are diffeomorphisms, this defines a differentiable structure on $T(M)$ so that the projection π and the local trivializations ψ_α are differentiable maps.

Example 2.5 (Tangent bundle to a complex manifold): Let $X=(X,\mathcal{O}_x)$ be a complex manifold of complex dimension n, let

$$\mathcal{O}_{X,x}:=\varinjlim_{\substack{x\in U\subset X\\ \text{open}}}\mathcal{O}(U)$$

be the $\mathbf{C}$-algebra of germs of holomorphic functions at $x\in X$, and let $T_x(X)$ be the derivations of this $\mathbf{C}$-algebra (defined exactly as in Example 2.4). Then $T_x(X)$ is the *holomorphic* (or *complex*) *tangent space* to X at x. In local coordinates, we see that $T_x(X)\cong T_x(\mathbf{C}^n)$ (abusing notation) and that the complex partial derivatives $\{\partial/\partial z_1,\ldots,\partial/\partial z_n\}$ form a basis over $\mathbf{C}$ for the vector space $T_x(\mathbf{C}^n)$ (see also Sec. 3). In the same manner as in Example 2.4 we can make the union of these tangent spaces into a holomorphic vector bundle over X, i.e,. $T(X)\to X$, where the fibres are all isomorphic to $\mathbf{C}^n$.

Remark: The same technique used to construct the tangent bundles in the above examples can be used to construct other vector bundles. For instance, suppose that we have $\pi\colon E\to X$, where X is an $\mathcal{S}$-manifold and π is a surjective map, so that

(a) E_p is a K-vector space,
(b) For each $p\in X$ there is a neighborhood U of p and a bijective map:

$$h\colon\pi^{-1}(U)\longrightarrow U\times K^r\text{such that } h(E_p)\subset\{p\}\times K^r.$$

(c) $h_p\colon E_p \to \{p\} \times K^r \xrightarrow{\text{proj.}} K^r$ is a K-vector space isomorphism.

Then, if for every (U_α, h_α), (U_β, h_β) as in (b) $h_\alpha \circ h_\beta^{-1}$ is an $\mathcal{S}$-isomorphism, we can make E into an $\mathcal{S}$-bundle over X by giving it the topology that makes h_α a homeomorphism for every α.

Example 2.6 (Universal bundle): Let $U_{r,n}$ be the disjoint union of the r-planes (r-dimensional K-linear subspaces) in K^n. Then there is a natural projection

$$\pi\colon U_{r,n} \to G_{r,n},$$

where $G_{r,n} = G_{r,n}(K)$, given by $\pi(v) = S$, if v is a vector in the r-plane S, and S is considered as a point in the Grassmannian manifold $G_{r,n}$. Thus the inverse image under π of a point p in the Grassmannian is the subspace of K^n which is the point p, and we may regard $U_{r,n}$ as a subset of $G_{r,n} \times K^n$. We can make $U_{r,n}$ into an $\mathcal{S}$-bundle by using the coordinate systems of $G_{r,n}$ to define transition functions, as was done with the tangent bundle in Example 2.4, and by then applying the remark following Example 2.5. To simplify things somewhat consider $U_{1,n} \to G_{1,n} = \mathbf{P}_{n-1}(\mathbf{R})$. First we note that any point $v \in U_{1,n}$ can be represented (not in a unique manner) in the form

$$v = (tx_0, \dots, tx_{n-1}) = t(x_0, \dots, x_{n-1}) \in \mathbf{R}^n,$$

where $(x_0, \dots, x_{n-1}) \in \mathbf{R}^n - \{0\}$, and $t \in \mathbf{R}$. Moreover, the projection $\pi\colon U_{1,n} \to \mathbf{P}_{n-1}$ is given by

$$\pi(t(x_0, \dots, x_{n-1})) = \pi(x_0, \dots, x_{n-1}) = [x_0, \dots, x_{n-1}] \in \mathbf{P}_{n-1}.$$

Letting $U_\alpha = \{[x_0, \dots, x_{n-1}] \in \mathbf{P}_{n-1}\colon x_\alpha \neq 0\}$, (cf. Example 1.6), we see that

$$\pi^{-1}(U_\alpha) = \{v = t(x_0, \dots, x_{n-1}) \in \mathbf{R}^n\colon t \in \mathbf{R}, x_\alpha \neq 0\}.$$

Now if $v = t(x_0, \dots, x_{n-1}) \in \pi^{-1}(U_\alpha)$, then we can write v in the form

$$v = t_\alpha \left(\frac{x_0}{x_\alpha}, \dots, \underset{(\alpha)}{1}, \dots, \frac{x_{n-1}}{x_\alpha} \right),$$

and $t_\alpha = tx_\alpha \in \mathbf{R}$ is uniquely determined by v. Then we can define the mapping

$$h_\alpha\colon \pi^{-1}(U_\alpha) \to U_\alpha \times \mathbf{R}$$

by setting

$$h_\alpha(v) = h_\alpha(t(x_0, \dots, x_{n-1})) = ([x_0, \dots, x_{n-1}], t_\alpha).$$

The mapping h_α is bijective and is $\mathbf{R}$-linear from the fibres of $\pi^{-1}(U_\alpha)$ to the fibres of $U_\alpha \times \mathbf{R}$. Suppose now that $v = t(x_0, \dots, x_{n-1}) \in \pi^{-1}(U_\alpha \cap U_\beta)$, then we have two different representations for v and we want to compute the relationship. Namely,

$$h_\alpha(v) = ([x_0, \dots, x_{n-1}], t_\alpha)$$

$$h_\beta(v) = ([x_0, \dots, x_{n-1}], t_\beta)$$

and then $t_\alpha = tx_\alpha, t_\beta = tx_\beta$, Therefore

$$t = \frac{t_\alpha}{x_\alpha} = \frac{t_\beta}{x_\beta},$$

which implies that

$$t_\alpha = \frac{x_\alpha}{x_\beta} t_\beta.$$

Thus if we let $g_{\alpha\beta} = x_\alpha / x_\beta$, then it follows that $g_{\alpha\beta} \bullet g_{\beta\gamma} \bullet g_{\gamma\alpha} = 1$, and thus by the remark following Example 2.5, we see that $U_{1,n}$ can be given the structure of a vector bundle by means of the functions $\{h_\alpha\}$, (the trivializations), and the transition functions of $U_{1,n}$

$$g_{\alpha\beta}([x_0, \ldots, x_{n-1}]) = \frac{x_\alpha}{x_\beta},$$

are mappings of $U_\alpha \cap U_\beta \to GL(1, \mathbf{R}) = \mathbf{R} - \{0\}$. These are the standard transition functions for the universal bundle over $\mathbf{P}_{n-1}$. Exactly the same relation holds for $U_{1,n}(\mathbf{C}) \to \mathbf{P}_{n-1}(\mathbf{C})$, which we meet again in later chapters. Namely, for complex homogeneous coordinates $[z_0, \ldots, z_{n-1}]$ we have the transition functions for the universal bundle over $P_{n-1}(\mathbf{C})$:

$$g_{\alpha\beta}([z_0, \ldots, z_{n-1}]) = \frac{z_\alpha}{z_\beta}.$$

The more general case of $U_{r,n} \to G_{r,n}$ can be treated in a similar manner, using the coordinate systems developed in Sec. 1. We note that $U_{r,n}(\mathbf{R}) \to G_{r,n}(\mathbf{R})$ is a real-analytic (and hence also differentiable) **R**-vector bundle and that $U_{r,n}(\mathbf{C}) \to G_{r,n}(\mathbf{C})$ is a holomorphic vector bundle. The reason for the name "universal bundle" will be made more apparent later in this section.

Definition 2.7: Let $\pi\colon E \to X$ be an $\mathcal{S}$-bundle and U an open subset of X. Then the restriction of E to U, denoted by $E|_U$ is the $\mathcal{S}$-bundle

$$\pi|_{\pi^{-1}(U)}\colon \pi^{-1}(U) \longrightarrow U.$$

Definition 2.8: Let E and F be $\mathcal{S}$-bundles over X; i.e., $\pi_E\colon E \to X$ and $\pi_F\colon F \to X$. Then a *homomorphism of $\mathcal{S}$-bundles*,

$$f\colon E \longrightarrow F,$$

is an $\mathcal{S}$-morphism of the total spaces which preserves fibres and is K-linear on each fibre; i.e., f commutes with the projections and is a K-linear mapping when restricted to fibres. An *$\mathcal{S}$-bundle isomorphism* is an $\mathcal{S}$-bundle homomorphism which is an $\mathcal{S}$-isomorphism on the total spaces and a K-vector space isomorphism on the fibres. Two $\mathcal{S}$-bundles are *equivalent* if there is some $\mathcal{S}$-bundle isomorphism between them. This clearly defines an equivalence relation on the $\mathcal{S}$-bundles over an $\mathcal{S}$-manifold, X.

The statement that a bundle is locally trivial now becomes the following:

For every $p \in X$ there is an open neighborhood U of p and a bundle isomorphism

$$h\colon E|_U \xrightarrow{\sim} U \times K^r.$$

Suppose that we are given two K-vector spaces A and B. Then from them we can form new K-vector spaces, for example,

(a) $A \oplus B$, the direct sum.
(b) $A \otimes B$, the tensor product.
(c) $\mathrm{Hom}(A, B)$, the linear maps from A to B.
(d) A^*, the linear maps from A to K.
(e) $\wedge^k A$, the antisymmetric tensor products of degree k (exterior algebra of A).
(f) $S^k(A)$, the symmetric tensor products of degree k (symmetric algebra of A).

Using the remark following Example 2.5, we can extend all the above algebraic constructions to vector bundles. For example, suppose that we have two vector bundles

$$\pi_E\colon E \longrightarrow X \quad \text{and} \quad \pi_F\colon F \longrightarrow X.$$

Then define

$$E \oplus F = \bigcup_{p \in X} E_p \oplus F_p.$$

We then have the natural projection

$$\pi\colon E \oplus F \longrightarrow X$$

given by

$$\pi^{-1}(p) = E_p \oplus F_p.$$

Now for any $p \in X$ we can find a neighborhood U of p and local trivializations

$$h_E\colon E|_U \xrightarrow{\sim} U \times K^n$$

$$h_F\colon F|_U \xrightarrow{\sim} U \times K^m,$$

and we define

$$h_{E\oplus F}\colon E \oplus F|_U \longrightarrow U \times (K^n \oplus K^m)$$

by $h_{E\oplus F}(v+w) = (p, h_E^p(v)+h_F^p(w))$ for $v \in E_p$ and $w \in F_p$. Then this map is bijective and K-linear on fibres, and for intersections of local trivializations we obtain the transition functions

$$g_{\alpha\beta}^{E\oplus F}(p) = \begin{bmatrix} g_{\alpha\beta}^E(p) & 0 \\ 0 & g_{\alpha\beta}^F(p) \end{bmatrix}.$$

So by the remark and the fact that $g_{\alpha\beta}^E$ and $g_{\alpha\beta}^F$ are bundle transition functions, $\pi\colon E \oplus F \to X$ is a vector bundle. Note that if E and F were $\mathcal{S}$-bundles over an $\mathcal{S}$-manifold X, then $g_{\alpha\beta}^E$ and $g_{\alpha\beta}^F$ would be $\mathcal{S}$-isomorphisms, and so

$E \oplus F$ would then be an $\mathcal{S}$-bundle over X. The same is true for all the other possible constructions induced by the vector space constructions listed above. Transition functions for the algebraically derived bundles are easily determined by knowing the transition functions for the given bundle.

The above examples lead naturally to the following definition.

Definition 2.9: Let $E \xrightarrow{\pi} X$ be an $\mathcal{S}$-bundle. An $\mathcal{S}$-submanifold $F \subset E$ is said to be an $\mathcal{S}$-subbundle of E if

(a) $F \cap E_x$ is a vector subspace of E_x.

(b) $\pi|_F \colon F \longrightarrow X$ has the structure of an $\mathcal{S}$-bundle induced by the $\mathcal{S}$-bundle structure of E, i.e., there exist local trivializations for E and F which are compatible as in the following diagram:

$$\begin{array}{ccc} E|_U & \xrightarrow{\sim} & U \times K^r \\ \uparrow i & & \uparrow id \times j \\ F|_U & \xrightarrow{\sim} & U \times K^s, \end{array} \qquad s \le r,$$

where the map j is the natural inclusion mapping of K^s as a subspace of K^r and i is the inclusion of F in E.

We shall frequently use the language of linear algebra in discussing homomorphisms of vector bundles. As an example, suppose that $E \xrightarrow{f} F$ is a vector bundle homomorphism of K-vector bundles over a space X. We define

$$\text{Ker } f = \bigcup_{x \in X} \text{Ker} f_x$$

$$\text{Im } f = \bigcup_{x \in X} \text{Im} f_x,$$

where $f_x = f|_{E_x}$. Moreover, we say that f has *constant rank* on X if rank f_x (as a K-linear mapping) is constant for $x \in X$.

Proposition 2.10: Let $E \xrightarrow{f} F$ be an $\mathcal{S}$-homomorphism of $\mathcal{S}$-bundles over X. If f has constant rank on X, then Ker f and Im f are $\mathcal{S}$-subbundles of E and F, respectively. In particular, f has constant rank if f is injective or surjective.

We leave the proof of this simple proposition to the reader.

Suppose now that we have a sequence of vector bundle homomorphisms over a space X,

$$\cdots \longrightarrow E \xrightarrow{f} F \xrightarrow{g} G \longrightarrow \cdots,$$

then the sequence is said to be *exact* at F if Ker $g = \text{Im} f$. A *short exact sequence* of vector bundles is a sequence of vector bundles (and vector bundle homomorphisms) of the following form,

$$0 \longrightarrow E' \xrightarrow{f} E \xrightarrow{g} E'' \longrightarrow 0,$$

which is exact at E', E, and E''. In particular, f is injective and g is surjective, and Im $f = \text{Ker } g$ is a subbundle of E. We shall see examples of short exact sequences and their utility in the next two chapters.

As we have stated before, vector bundles represent the geometry of the underlying base space. However, to get some understanding via analysis of vector bundles, it is necessary to introduce a generalized notion of function (reflecting the geometry of the vector bundle) to which we can apply the tools of analysis.

Definition 2.11: An $\mathcal{S}$-*section* of an $\mathcal{S}$-bundle $E \xrightarrow{\pi} X$ is an $\mathcal{S}$-morphism $s: X \longrightarrow E$ such that

$$\pi \circ s = 1_X,$$

where 1_X is the identity on X; i.e., s maps a point in the base space into the fibre over that point. $\mathcal{S}(X, E)$ will denote the $\mathcal{S}$-sections of E over X. $\mathcal{S}(U, E)$ will denote the $\mathcal{S}$-sections of $E|_U$ over $U \subset X$; i.e., $\mathcal{S}(U, E) = \mathcal{S}(U, E|_U)$ [we shall also occasionally use the common notation $\mathbf{\Gamma}(X, E)$ for sections, provided that there is no confusion as to which category we are dealing with].

Example 2.12: Consider the trivial bundle $M \times \mathbf{R}$ over a differentiable manifold M. Then $\mathcal{E}(M, M \times \mathbf{R})$ can be identified in a natural way with $\mathcal{E}(M)$, the global real-valued functions on M. Similarly, $\mathcal{E}(M, M \times \mathbf{R}^n)$ can be identified with global differentiable mappings of M into $\mathbf{R}^n$ (i.e., vector-valued functions). Since vector bundles are locally of the form $U \times \mathbf{R}^n$, we see that sections of a vector bundle can be viewed as vector-valued functions (locally), where two different local representations are related by the transition functions for the bundle. Therefore sections can be thought of as "twisted" vector-valued functions.

Remarks: (a) A section s is often identified with its image $s(X) \subset E$; for example, the term *zero section* is used to refer to the section $0: X \longrightarrow E$ given by $0(x) = 0 \in E_x$ and is often identified with its image, which is, in fact, $\mathcal{S}$-isomorphic with the base space X.

(b) For $\mathcal{S}$-bundles $E \xrightarrow{\pi} X$ and $E' \xrightarrow{\pi'} X$ we can identify the set of $\mathcal{S}$-bundle homomorphisms of E into E', with $\mathcal{S}(X, \text{Hom}(E, E'))$. A section $s \in \mathcal{S}(X, \text{Hom}(E, E'))$ picks out for each point $x \in X$ a K-linear map $s(x)$: $E_x \longrightarrow E'_x$, and s is identified with $f_s: E \longrightarrow E'$ which is defined by

$$f_s|_{E\pi(e)} = s(\pi(e)) \qquad \text{for } e \in E.$$

(c) If $E \longrightarrow X$ is an $\mathcal{S}$-bundle of rank r with transition functions $\{g_{\alpha\beta}\}$ associated with a trivializing cover $\{U_\alpha\}$, then let $f_\alpha: U_\alpha \longrightarrow K^r$ be $\mathcal{S}$-morphisms satisfying the compatibility conditions

$$f_\alpha = g_{\alpha\beta} f_\beta \qquad \text{on } U_\alpha \cap U_\beta \neq \emptyset.$$

Here we are using matrix multiplication, considering f_α and f_β as column vectors. Then the collection $\{f_\alpha\}$ defines an $\mathcal{S}$-section f of E, since each f_α gives a section of $U_\alpha \times K^r$, and this pulls back by the trivialization to a section of $E|_{U_\alpha}$. These sections of $E|_{U_\alpha}$ agree on the overlap regions $U_\alpha \cap U_\beta$

by the compatibility conditions imposed on $\{f_\alpha\}$, and thus define a global section. Conversely, any $\mathcal{S}$-section of E has this type of representation. We call each f_α a *trivialization* of the section f.

Example 2.13: We use remark (c) above to compute the global sections of the holomorphic line bundles $E^k \longrightarrow \mathbf{P}_1(\mathbf{C})$, which we define as follows, using the transition function g_{01} for the universal bundle $U_{1,2}(\mathbf{C}) \longrightarrow \mathbf{P}_1(\mathbf{C})$ of Example 2.6. Let the $\mathbf{P}_1(\mathbf{C})$ coordinate maps (Example 1.6) $\tilde{\varphi} : U_\alpha \longrightarrow \mathbf{C}$ be denoted by $\varphi_0([z_0, z_1]) = z_1/z_0 = z$ and $\varphi_1([z_0, z_1]) = z_0/z_1 = w$ so that $z = \varphi_0 \circ \varphi_1^{-1}(w) = 1/w$ for $w \neq 0$. For a fixed integer k define the line bundle $E^k \longrightarrow \mathbf{P}_1(\mathbf{C})$ by the transition function $g_{01}^k : U_0 \cap U_1 \longrightarrow GL(1, \mathbf{C})$ where $g_{01}^k([z_0, z_1]) = (z_0/z_1)^k$. E^k is the kth tensor power of $U_{1,2}(\mathbf{C})$ for $k > 0$, the kth tensor power of the dual bundle $U_{1,2}(\mathbf{C})^*$ for $k < 0$, and trivial for $k = 0$. If $f \in \mathcal{O}(\mathbf{P}_1(\mathbf{C}), E^k)$, then each trivialization of f, f_α is in $\mathcal{O}(U_\alpha, U_\alpha \times \mathbf{C}) = \mathcal{O}(U_\alpha)$ and the $f_\alpha \circ \varphi_\alpha^{-1}$ are entire functions, say $f_0 \circ \varphi_0^{-1}(z) = \sum_{n=0}^{\infty} a_n z^n$ and $f_1 \circ \varphi_1^{-1}(w) = \sum_{n=0}^{\infty} b_n w^n$. If $z = \varphi_0(p)$ and $w = \varphi_1(p)$, then by remark (c), $f_0(p) = g_{01}^k(p) f_1(p)$, for $p \in U_0 \cap U_1$ in the w-coordinate plane, and this becomes

$$\sum_{n=0}^{\infty} a_n (1/w)^n = f_0 \circ \varphi_0^{-1}(\varphi_0 \circ \varphi_1^{-1}(w)) = g_{01}^k(\varphi_1^{-1}(w)) f_1 \circ \varphi_1^{-1}(w) = w^k \sum_{n=0}^{\infty} b_n w^n.$$

Hence,

$$\mathcal{O}(\mathbf{P}_1(\mathbf{C}), E^k) = \begin{cases} 0 & \text{for } k > 0, \\ \mathbf{C} & \text{for } k = 0 \text{ (Theorem 1.11)}, \\ \text{homogeneous polynomials} & \text{for } k < 0. \\ \text{in } \mathbf{C}^2 \text{ of degree } -k & \end{cases}$$

When dealing with certain categories of $\mathcal{S}$-manifolds, it is possible to define algebraic structures on $\mathcal{S}(X, E)$. First, $\mathcal{S}(X, E)$ can be made into a K-vector space under the following operations:

(a) For $s, t \in \mathcal{S}(X, E)$,

$$(s + t)(x) := s(x) + t(x) \quad \text{for all } x \in X.$$

(b) For $s \in \mathcal{S}(X, E)$ and $\alpha \in K$, $(\alpha s)(x) := \alpha(s(x))$ for all $s \in X$. Moreover, $\mathcal{S}(X, E)$ can be given the structure of an $\mathcal{S}_X(X)$ module [where the $\mathcal{S}_X(X)$ are the globally defined K-valued $\mathcal{S}$-functions on X] by defining

(c) For $s \in \mathcal{S}(X, E)$ and $f \in \mathcal{S}_X(X)$,

$$fs(x) := f(x)s(x) \quad \text{for all } x \in X.$$

To ensure that the above maps actually are $\mathcal{S}$-morphisms and thus $\mathcal{S}$-sections, it is necessary that the vector space operations on K^n be $\mathcal{S}$-morphisms in the $\mathcal{S}$-structure on K^n. But this is clearly the case for the three categories with which we are dealing.

Let M be a differentiable manifold and let $T(M) \longrightarrow M$ be its tangent bundle. Using the techniques outlined above, we would like to consider new differentiable vector bundles over M, derived from $T(M)$. We have

(a) The *cotangent bundle*, $T^*(M)$, whose fibre at $x \in M$, $T_x^*(M)$, is the **R**-linear dual to $T_x(M)$.

(b) The *exterior algebra bundles*, $\wedge^p T(M), \wedge^p T^*(M)$, whose fibre at $x \in M$ is the antisymmetric tensor product (of degree p) of the vector spaces $T_x(M)$ and $T_x^*(M)$, respectively, and

$$\wedge T(M) = \bigoplus_{p=0}^{n} \wedge^p T(M)$$

$$\wedge T^*(M) = \bigoplus_{p=0}^{n} \wedge^p T^*(M).$$

(c) The *symmetric algebra bundles*, $S^k(T(M)), S^k(T^*(M))$, whose fibres are the symmetric tensor products (of degree k) of $T_x(M)$ and $T_x^*(M)$, respectively.

We define

$$\mathcal{E}^p(U) = \mathcal{E}(U, \wedge^p T^*(M)),$$

the C^∞ *differential forms of degree* p on the open set $U \subset M$. As usual, we can define the exterior derivative

$$d\colon \mathcal{E}^p(U) \longrightarrow \mathcal{E}^{p+1}(U).$$

We recall how this is done. First, consider $U \subset \mathbf{R}^n$ and recall that the derivations $\{\partial/\partial x_1, \ldots, \partial/\partial x_n\}$ form a basis for $T_x(\mathbf{R}^n)$ at $x \in U$. Let $\{dx_1, \ldots, dx_n\}$ be a *dual basis* for $T_x^*(R^n)$. Then the maps

$$dx_j\colon U \longrightarrow T^*(R^n)|_U$$

given by

$$dx_j(x) = dx_j|_x$$

form a basis for the $\mathcal{E}(U)(= \mathcal{E}_{\mathbf{R}^n}(U))$-module $\mathcal{E}(U, T^*(\mathbf{R}^n)) = \mathcal{E}^1(U)$. Moreover, $\{dx_I = dx_{i_1} \wedge \cdots \wedge dx_{i_p}\}$, where $I = (i_1, \ldots, i_p)$ and $1 \le i_1 < i_2 < \cdots < i_p \le n$, form a basis for the $\mathcal{E}(U)$-module $\mathcal{E}^p(U)$. We defined $d\colon \mathcal{E}^p(U) \longrightarrow \mathcal{E}^{p+1}(U)$ as follows:

Case 1 $(p = 0)$: Suppose that $f \in \mathcal{E}^0(U) = \mathcal{E}(U)$. Then let

$$df = \sum_{j=1}^{n} \frac{\partial f}{\partial x_j} dx_j \in \mathcal{E}^1(U).$$

Case 2 $(p > 0)$: Suppose that $f \in \mathcal{E}^p(U)$. Then

$$f = \sum_{|I|=p}{}' f_I dx_I,$$

where $f_I \in \mathcal{E}(U), I = (i_1, \ldots, i_p), |I| =$ the number of indices, and $\sum'$ signifies that the sum is taken over strictly increasing indices. Then

$$df = \sum_{|I|=p}{}' df_I \wedge dx_I = \sum_{|I|=p}{}' \sum_{j=1}^{n} \frac{\partial f_I}{\partial x_j} dx_j \wedge dx_I.$$

Suppose now that (U, h) is a coordinate system on a differentiable manifold M. Then we have that $T(M)|_U \xrightarrow{\sim} T(\mathbf{R}^n)|_{h(U)}$; hence $\mathcal{E}^p(U) \xleftarrow{\sim} \mathcal{E}^p(h(U))$, and the mapping

$$d\colon \mathcal{E}^p(h(U)) \longrightarrow \mathcal{E}^{p+1}(h(U))$$

defined above induces a mapping (also denoted by d)

$$d\colon \mathcal{E}^p(U) \longrightarrow \mathcal{E}^{p+1}(U).$$

This defines the exterior derivative d locally on M, and it is not difficult to show, using the chain rule, that the definition is independent of the choice of local coordinates. It follows that the exterior derivative is well defined globally on the manifold M.

We have previously defined a bundle homomorphism of two bundles over the same base space (Definition 2.8). We now would like to define a mapping between bundles over different base spaces.

Definition 2.14: An $\mathcal{S}$-*bundle morphism* between two $\mathcal{S}$-bundles $\pi_E\colon E \longrightarrow X$ and $\pi_F\colon F \longrightarrow Y$ is an $\mathcal{S}$-morphism $f\colon E \longrightarrow F$ which takes fibres of E isomorphically (as vector spaces) onto fibres in F. An $\mathcal{S}$-bundle morphism $f\colon E \longrightarrow F$ induces an $\mathcal{S}$-morphism $\bar{f}(\pi_E(e)) = \pi_F(f(e))$; in other words, the following diagram commutes:

$$\begin{array}{ccc} E & \xrightarrow{f} & F \\ {\scriptstyle \pi_E}\downarrow & & \downarrow{\scriptstyle \pi_F} \\ X & \xrightarrow{\bar{f}} & Y. \end{array}$$

If X is identified with $0(X)$, the zero section, then $\bar{f}$ may be identified with

$$\bar{f} = f|_X\colon X \longrightarrow Y = 0(Y)$$

since f is a homomorphism on fibres and maps the zero section of X into the zero section of Y, which can likewise be identified with Y. If E and F are bundles over the same space X and $\bar{f}$ is the identity, then E and F are said to be *equivalent* (which implies that the two vector bundles are $\mathcal{S}$-isomorphic and hence equivalent in the sense of Definition 2.8).

Proposition 2.15: Given an $\mathcal{S}$-morphism $f\colon X \longrightarrow Y$ and an $\mathcal{S}$-bundle $\pi\colon E \longrightarrow Y$, then there exists an $\mathcal{S}$-bundle $\pi'\colon E' \longrightarrow X$ and an $\mathcal{S}$-bundle morphism g such that the following diagram commutes:

$$\begin{array}{ccc} E' & \xrightarrow{g} & E \\ {\scriptstyle \pi'}\downarrow & & \downarrow{\scriptstyle \pi} \\ X & \xrightarrow{f} & Y. \end{array}$$

Moreover, E' is unique up to equivalence. We call E' the *pullback of* E *by* f and denote it by f^*E.

Proof: Let

(2.3) $$E' = \{(x, e) \in X \times E\colon f(x) = \pi(e)\}.$$

We have the natural projections

$$\begin{array}{lll} g\colon E' \longrightarrow E & \text{and} & \pi'\colon E' \longrightarrow X \\ (x, e) \longrightarrow e & & (x, e) \longrightarrow x. \end{array}$$

Giving $E'_x = \{x\} \times E_{f(x)}$ the structure of a K-vector space induced by $E_{f(x)}$, E' becomes a fibered family of vector spaces over X.

If (U, h) is a local trivialization for E, i.e.,

$$E|_U \overset{h}{\underset{\sim}{\longrightarrow}} U \times K^n,$$

then it is easy to show that

$$E'|_{f^{-1}(U)} \overset{\sim}{\longrightarrow} f^{-1}(U) \times K^n$$

is a local trivialization of E'; hence E' is the necessary bundle.

Suppose that we have another bundle $\tilde{\pi}\colon \tilde{E} \longrightarrow X$ and a bundle morphism $\tilde{g}$ such that

$$\begin{array}{ccc} \tilde{E} & \overset{\tilde{g}}{\longrightarrow} & E \\ \tilde{\pi} \downarrow & & \downarrow \pi \\ X & \overset{f}{\longrightarrow} & Y \end{array}$$

commutes. Then define the bundle homomorphism $h\colon \tilde{E} \longrightarrow E'$ by

$$h(\tilde{e}) = (\tilde{\pi}(\tilde{e}), \tilde{g}(\tilde{e})) \in \{\pi(\tilde{e})\} \times E.$$

Note that $h(\tilde{e}) \in E'$ since the commutativity of the above diagram yields $f(\tilde{\pi}(\tilde{e})) = \pi(\tilde{g}(\tilde{e}))$; hence this is a bundle homomorphism. Moreover, it is a vector space isomorphism on fibres and hence an $\mathcal{S}$-bundle morphism inducing the identity $1_X\colon X \longrightarrow X$, i.e., an equivalence.

Q.E.D.

Remark: In the diagram in Proposition 2.15, the vector bundle E' and the maps π' and g depend on f and π, and we shall sometimes denote this relation by

$$\begin{array}{ccc} f^*E & \overset{f_*}{\longrightarrow} & E \\ \pi_f \downarrow & & \downarrow \pi \\ X & \overset{f}{\longrightarrow} & Y \end{array}$$

to indicate the dependence on the map f of the pullback. For convenience, we assume from now on that f^*E is given by (2.3) and that the maps π_f and f_* are the natural projections.

The concepts of $\mathcal{S}$-bundle homomorphism and $\mathcal{S}$-bundle morphism are related by the following proposition.

Proposition 2.16: Let $E \overset{\pi}{\longrightarrow} X$ and $E' \overset{\pi'}{\longrightarrow} Y$ be $\mathcal{S}$-bundles. If $f\colon E \longrightarrow E'$ is an $\mathcal{S}$-morphism of the total spaces which maps fibres to fibres and which is a vector space homomorphism on each fibre, then f can be expressed as the composition of an $\mathcal{S}$-bundle homomorphism and an $\mathcal{S}$-bundle morphism.

Proof: Let $\bar{f}$ be the map on base spaces $\bar{f}\colon X \longrightarrow Y$ induced by f. Let $\bar{f}^*E'$ be the pullback of E' by $\bar{f}$, and consider the following diagram,

$$\begin{array}{ccccc} E & \xrightarrow{\;h\;} & \bar{f}^*E' & \xrightarrow{\;\bar{f}_*\;} & E' \\ & \searrow^{\pi} & \downarrow{\scriptstyle \pi'_{\bar{f}}} & & \downarrow{\scriptstyle \pi'} \\ & & X & \xrightarrow{\;\bar{f}\;} & Y, \end{array}$$

where h is defined by $h(e) = (\pi(e), f(e))$ [see (2.3)]. It is clear that $f = \bar{f}_* \circ h$. Moreover, $\bar{f}_*$ is an $\mathcal{S}$-bundle morphism, and h is an $\mathcal{S}$-bundle homomorphism.

Q.E.D.

There are two basic problems concerning vector bundles on a given space: first, to determine, up to equivalence, how many different vector bundles there are on a given space, and second, to decide how "twisted" or how far from being trivial a given vector bundle is. The second question is the motivation for the theory of characteristic classes, which will be studied in Chap. III. The first question has different "answers," depending on the category. A special important case is the following theorem. Let $U = U_{r,n}$ denote the universal bundle over $G_{r,n}$ (see Example 2.6).

Theorem 2.17: Let X be a differentiable manifold and let $E \longrightarrow X$ be a differentiable vector bundle of rank r. Then there exists an $N > 0$ (depending only on X) and a differentiable mapping $f\colon X \longrightarrow G_{r,N}(\mathbf{R})$), so that $f^*U \cong E$. Moreover, any mapping $\tilde{f}$ which is homotopic to f has the property that $\tilde{f}^*U \cong E$.

We recall that f and $\tilde{f}$ are homotopic if there is a one-parameter family of mappings $F\colon [0, 1] \times X \longrightarrow G_{r,N}$ so that $F|_{\{0\}\times X} = f$ and $F|_{\{1\}\times X} = \tilde{f}$. The content of the theorem is that the different isomorphism classes of differentiable vector bundles over X are classified by homotopy classes of maps into the Grassmannian $G_{r,N}$. For certain spaces, these are computable (e.g., if X is a sphere, see Steenrod [1]). If one assumes that X is compact, one can actually require that the mapping f in Theorem 2.17 be an *embedding* of X into $G_{r,N}$ (by letting N be somewhat larger). One could have phrased the above result in another way: Theorem 2.17 is valid in the category of continuous vector bundles, and there is a one-to-one correspondence between isomorphism classes of continuous and differentiable (and also real-analytic) vector bundles. However, such a result is *not* true in the case of holomorphic vector bundles over a compact complex manifold unless additional assumptions (*positivity*) are made. This is studied in Chap. VI. In fact, the problem of finding a projective algebraic embedding of a given compact complex manifold (mentioned in Sec. 1) is reduced to finding a class of holomorphic bundles over X so that Theorem 2.17 holds for *these* bundles and the mapping f gives an embedding into $G_{r,n}(\mathbf{C})$, which by Proposition 1.14 is itself projective algebraic. We shall not need the classification given by Theorem 2.17 in our later chapters and we refer the reader to the classical reference Steenrod

[1] (also see Proposition III.4.2). A thorough and very accessible discussion of the topics in this section can be found in Milnor [2].

The set of all vector bundles on a space X (in a given category) can be made into a *ring* by considering the free abelian group generated by the set of all vector bundles and introducing the equivalence relation that $E - (E' + E'')$ is equivalent to zero if there is a short exact sequence of the form $0 \longrightarrow E' \longrightarrow E \longrightarrow E'' \longrightarrow 0$. The set of equivalence classes form a ring $K(X)$ (using tensor product as multiplication), which was first introduced by Grothendieck in the context of algebraic geometry (Borel and Serre [2]) and generalized by Atiyah and Hirzebruch [1]. For an introduction to this area, as well as a good introduction to vector bundles which is more extensive than our brief summary, see the text by Atiyah [1]. The subject of K-theory plays an important role in the Atiyah-Singer theorem (Atiyah and Singer [1]) and in modern differential topology. We shall not develop this in our book, as we shall concentrate more on the analytical side of the subject.

3. Almost Complex Manifolds and the $\bar{\partial}$-Operator

In this section we want to introduce certain first-order differential operators which act on differential forms on a complex manifold and which intrinsically reflect the complex structure. The most natural context in which to discuss these operators is from the viewpoint of almost complex manifolds, a generalization of a complex manifold which has the first-order structure of a complex manifold (i.e., at the tangent space level). We shall first discuss the concept of a **C**-linear structure on an **R**-linear vector space and will apply the (linear algebra) results obtained to the real tangent bundle of a differentiable manifold.

Let V be a real vector space and suppose that J is an **R**-linear isomorphism $J\colon V \xrightarrow{\sim} V$ such that $J^2 = -I$ (where I = identity). Then J is called a *complex structure* on V. Suppose that V and a complex structure J are given. Then we can equip V with the structure of a complex vector space in the following manner:

$$(\alpha + i\beta)v := \alpha v + \beta J v, \quad \alpha, \beta \in \mathbf{R}, \quad i = \sqrt{-1}.$$

Thus scalar multiplication on V by complex numbers is defined, and it is easy to check that V becomes a complex vector space. Conversely, if V is a complex vector space, then it can also be considered as a vector space over **R**, and the operation of multiplication by i is an **R**-linear endomorphism of V onto itself, which we can call J, and is a complex structure. Moreover, if $\{v_1, \ldots, v_n\}$ is a basis for V over **C**, then $\{v_1, \ldots, v_n, Jv_1, \ldots, Jv_n\}$ will be a basis for V over **R**.

Example 3.1: Let $\mathbf{C}^n$ be the usual Euclidean space of n-tuples of complex numbers, $\{z_1, \ldots, z_n\}$, and let $z_j = x_j + iy_j$, $j = 1, \ldots, n$, be the real and imaginary parts. Then $\mathbf{C}^n$ can be identified with $\mathbf{R}^{2n} = \{x_1, y_1, \ldots, x_n, y_n\}$,

$x_j, y_j \in \mathbf{R}$. Scalar multiplication by i in $\mathbf{C}^n$ induces a mapping $J: \mathbf{R}^{2n} \longrightarrow \mathbf{R}^{2n}$ given by

$$J(x_1, y_1, \ldots, x_n, y_n) = (-y_1, x_1, \ldots, -y_n, x_n),$$

and, moreover, $J^2 = -1$. This is the *standard complex structure* on $\mathbf{R}^{2n}$. The coset space $GL(2n, \mathbf{R})/GL(n, \mathbf{C})$ determines all complex structures on $\mathbf{R}^{2n}$ by the mapping $[A] \longrightarrow A^{-1}JA$, where $[A]$ is the equivalence class of $A \in GL(2n, \mathbf{R})$.

Example 3.2: Let X be a complex manifold and let $T_x(X)$ be the (complex) tangent space to X at x. Let X_0 be the underlying differentiable manifold of X (i.e., X induces a differentiable structure on the underlying topological manifold of X) and let $T_x(X_0)$ be the (real) tangent space to X_0 at x. Then we claim that $T_x(X_0)$ is canonically isomorphic with the underlying real vector space of $T_x(X)$ and that, in particular, $T_x(X)$ induces a complex structure J_x on the real tangent space $T_x(X_0)$. To see this, we let (h, U) be a holomorphic coordinate system near x. Then $h: U \longrightarrow U' \subset \mathbf{C}^n$, and hence, by taking real and imaginary parts of the vector-valued function h, we obtain

$$\tilde{h}: U \longrightarrow \mathbf{R}^{2n}$$

given by

$$\tilde{h}(x) = (\text{Re } h_1(x), \text{Im } h_1(x), \ldots, \text{Re } h_n(x), \text{Im } h_n(x)),$$

which is a real-analytic (and, in particular, differentiable) coordinate system for X_0 near x. Then it suffices to consider the claim above for the vector spaces $T_0(\mathbf{C}^n)$ and $T_0(\mathbf{R}^{2n})$ at $0 \in \mathbf{C}^n$, where $\mathbf{R}^{2n}$ has the standard complex structure. Let $\{\partial/\partial z_1, \ldots, \partial/\partial z_n\}$ be a basis for $T_0(\mathbf{C}^n)$ and let $\{\partial/\partial x_1, \partial/\partial y_1, \ldots, \partial/\partial x_n, \partial/\partial y_n\}$ be a basis for $T_0(\mathbf{R}^{2n})$. Then we have the diagram

$$\begin{array}{ccc} T_0(\mathbf{C}^n) & \cong_{\mathbf{C}} & \mathbf{C}^n \\ \alpha \| \wr_{\mathbf{R}} & & \| \wr_{\mathbf{R}} \\ T_0(\mathbf{R}^{2n}) & \cong_{\mathbf{R}} & \mathbf{R}^{2n}, \end{array}$$

where α is the $\mathbf{R}$-linear isomorphism between $T_0(\mathbf{R}^{2n})$ and $T_0(\mathbf{C}^n)$ induced by the other maps, and thus the complex structure of $T_0(\mathbf{C}^n)$ induces a complex structure on $T_0(\mathbf{R}^{2n})$, just as in Example 3.1. We claim that the complex structure J_x induced on $T_x(X_0)$ in this manner is independent of the choice of local holomorphic coordinates. To check that this is the case, consider a biholomorphism f defined on a neighborhood N of the origin in $\mathbf{C}^n$, $f: N \longrightarrow N$, where $f(0) = 0$. Then, letting $\zeta = f(z)$ and writing in terms of real and imaginary coordinates, we have the corresponding diffeomorphism expressed in real coordinates:

$$(3.1) \qquad \begin{aligned} \xi &= u(x, y) \\ \eta &= v(x, y), \end{aligned}$$

where $\xi, \eta, x, y \in \mathbf{R}^n$ and $\xi + i\eta = \zeta \in \mathbf{C}^n, x + iy = z \in \mathbf{C}^n$. The map $f(z)$ corresponds to a holomorphic change of coordinates on the complex manifold X; the pair of mappings u, v corresponds to the change of coordinates for the underlying differentiable manifold. The Jacobian matrix (differential) of these mappings corresponds to the transition functions for the corresponding trivializations for $T(X)$ and $T(X_0)$, respectively. Let J denote the standard complex structure in $\mathbf{C}^n$, and we shall show that J commutes with the Jacobian of the real mapping. The real Jacobian of (3.1) has the form of an $n \times n$ matrix of 2×2 blocks,

$$M = \begin{bmatrix} \dfrac{\partial u_\alpha}{\partial x_\beta} & \dfrac{\partial u_\alpha}{\partial y_\beta} \\ \dfrac{\partial v_\alpha}{\partial x_\beta} & \dfrac{\partial v_\alpha}{\partial y_\beta} \end{bmatrix}, \quad \alpha, \beta = 1, \ldots, n,$$

which, by the Cauchy-Riemann equations (since f is a holomorphic mapping), is the same as

$$\begin{bmatrix} \dfrac{\partial v_\alpha}{\partial y_\beta} & \dfrac{\partial u_\alpha}{\partial y_\beta} \\ -\dfrac{\partial u_\alpha}{\partial y_\beta} & \dfrac{\partial v_\alpha}{\partial y_\beta} \end{bmatrix}, \quad \alpha, \beta = 1, \ldots, n.$$

Thus the Jacobian is an $n \times n$ matrix consisting of 2×2 blocks of the form

$$\begin{bmatrix} a & b \\ -b & a \end{bmatrix}.$$

Moreover, J can be expressed in matrix form as an $n \times n$ matrix of 2×2 blocks with matrices of the form

$$\begin{bmatrix} 0 & 1 \\ -1 & 0 \end{bmatrix}$$

along the diagonal and zero elsewhere. It is now easy to check that $MJ = JM$. It follows then that J induces the same complex structure on $T_x(X_0)$ for each choice of local holomorphic coordinates at x.

Let V be a real vector space with a complex structure J, and consider $V \otimes_{\mathbf{R}} \mathbf{C}$, the complexification of V. The $\mathbf{R}$-linear mapping J extends to a $\mathbf{C}$-linear mapping on $V \otimes_{\mathbf{R}} \mathbf{C}$ by setting $J(v \otimes \alpha) = J(v) \otimes \alpha$ for $v \in V, \alpha \in \mathbf{C}$. Moreover, the extension still has the property that $J^2 = -I$, and it follows that J has two eigenvalues $\{i, -i\}$. Let $V^{1,0}$ be the eigenspace corresponding to the eigenvalue i and let $V^{0,1}$ be the eigenspace corresponding to $-i$. Then we have

$$V \otimes_{\mathbf{R}} \mathbf{C} = V^{1,0} \oplus V^{0,1}.$$

Moreover, conjugation on $V \otimes_{\mathbf{R}} \mathbf{C}$ is defined by $\overline{v \otimes \alpha} = v \otimes \bar{\alpha}$ for $v \in V$ and $\alpha \in \mathbf{C}$. Thus $V^{1,0} \cong_{\mathbf{R}} V^{0,1}$ (conjugation is a conjugate-linear mapping). It is easy to see that the complex vector space obtained from V by means of the complex structure J, denoted by V_J, is $\mathbf{C}$-linearly isomorphic to $V^{1,0}$, and we shall identify V_J with $V^{1,0}$ from now on.

We now want to consider the exterior algebras of these complex vector spaces. Namely, denote $V \otimes_{\mathbf{R}} \mathbf{C}$ by V_c and consider the exterior algebras

$$\wedge V_c, \quad \wedge V^{1,0}, \quad \text{and} \quad \wedge V^{0,1}.$$

Then we have natural injections

$$\begin{array}{c} \wedge V^{1,0} \searrow \\ \qquad\qquad \wedge V_c, \\ \wedge V^{0,1} \nearrow \end{array}$$

and we let $\wedge^{p,q} V$ be the subspace of $\wedge V_c$ generated by elements of the form $u \wedge w$, where $u \in \wedge^p V^{1,0}$ and $w \in \wedge^q V^{0,1}$. Thus we have the direct sum (letting $n = \dim_c V^{1,0}$)

$$\wedge V_c = \sum_{r=0}^{2n} \sum_{p+q=r} \wedge^{p,q} V.$$

We now want to carry out the above algebraic construction on the tangent bundle to a manifold. First, we have a definition.

Definition 3.3: Let X be a differentiable manifold of dimension $2n$. Suppose that J is a differentiable vector bundle isomorphism

$$J: T(X) \longrightarrow T(X)$$

such that $J_x: T_x(X) \longrightarrow T_x(X)$ is a complex structure for $T_x(X)$; i.e., $J^2 = -I$, where I is the identity vector bundle isomorphism acting on $T(X)$. Then J is called an *almost complex structure* for the differentiable manifold X. If X is equipped with an almost complex structure J, then (X, J) is called an *almost complex manifold.*

We see that a differentiable manifold having an almost complex structure is equivalent to prescribing a **C**-vector bundle structure on the **R**-linear tangent bundle.

Proposition 3.4: A complex manifold X induces an almost complex structure on its underlying differentiable manifold.

Proof: As we saw in Example 3.2, for each point $x \in X$ there is a complex structure induced on $T_x(X_0)$, where X_0 is the underlying differentiable manifold. What remains to check is that the mapping

$$J_x: T_x(X_0) \longrightarrow T_x(X_0), \quad x \in X_0,$$

is, in fact, a C^∞ mapping with respect to the parameter x. To see that J is a C^∞ vector bundle mapping, choose local holomorphic coordinates (h, U) and obtain a trivialization for $T(X_0)$ over U, i.e.,

$$T(X_0)|_U \cong h(U) \times \mathbf{R}^{2n},$$

where we let $z_j = x_j + iy_j$ be the coordinates in $h(U)$ and $(\xi_1, \eta_1, \ldots, \xi_n, \eta_n)$

be the coordinates in $\mathbf{R}^{2n}$. Then the mapping $J|_U$ is defined by (with respect to this trivialization)

$$id \times J\colon h(U) \times \mathbf{R}^{2n} \longrightarrow h(U) \times \mathbf{R}^{2n},$$

where

$$J(\xi_1, \eta_1, \ldots, \xi_n, \eta_n) = (-\eta_1, \xi_1, \ldots, -\eta_n, \xi_n),$$

as before, That is, in this trivialization J is a *constant* mapping, and hence C^∞. Since differentiability is a local property, it follows that J is a differentiable bundle mapping.

Q.E.D.

Remark: There are various examples of almost complex structures which do not arise from complex structures. The 2-sphere S^2 carries a complex structure [$\cong \mathbf{P}_1(\mathbf{C})$], and the 6-sphere S^6 carries an almost complex structure induced on it by the unit Cayley numbers in S^7 (see Steenrod [1]). However, this almost complex structure does not come from a complex structure (it is not integrable; see the discussion below). Moreover, it is unknown whether S^6 carries a complex structure. A theorem of Borel and Serre [1] asserts that only S^2 and S^6 admit almost complex structures among the even dimensional real spheres. For more information about almost complex structures on manifolds, consult, e.g., Kobayashi and Nomizu [1] or Helgason [1].

Let X be a differentiable m-manifold, let $T(X)_c = T(X) \otimes_{\mathbf{R}} \mathbf{C}$ be the complexification of the tangent bundle, and let $T^*(X)_c$ be the complexification of the cotangent bundle. We can form the exterior algebra bundle $\wedge T^*(X)_c$, and we let

$$\mathcal{E}^r(X)_c = \mathcal{E}(X, \wedge^r T^*(X)_c).$$

These are the *complex-valued differential forms of total degree* r on X. We shall usually drop the subscript c and denote them simply by $\mathcal{E}^r(X)$ when there is no chance of confusion with the real-valued forms discussed in Sec. 2. In local coordinates we have $\varphi \in \mathcal{E}^r(X)$ if and only if φ can be expressed in a coordinate neighborhood by

$$\varphi(x) = \sum_{|I|=r}{}' \varphi_I(x) dx_I,$$

where we use the multiindex notation of Sec. 2. and $\varphi_I(x)$ is a C^∞ complex-valued function on the neighborhood. The exterior derivative d is extended by complex linearity to act on complex-valued differential forms, and we have the sequence

$$\mathcal{E}^0(X) \xrightarrow{d} \mathcal{E}^1(X) \xrightarrow{d} \cdots \xrightarrow{d} \mathcal{E}^m(X) \longrightarrow 0,$$

where $d^2 = 0$.

Suppose now that (X, J) is an almost complex manifold. Then we can apply the linear algebra above to $T(X)_c$. Namely, J extends to a $\mathbf{C}$-linear bundle isomorphism on $T(X)_c$ and has (fibrewise) eigenvalues $\pm i$. Let $T(X)^{1,0}$ be the bundle of $(+i)$-eigenspaces for J and let $T(X)^{0,1}$ be the bundle of

$(-i)$-eigenspaces for J [note that these are differentiable subbundles of $T(X)_c$]. We can define a conjugation on $T(X)_c$,

$$Q\colon T(X)_c \longrightarrow T(X)_c,$$

by fibrewise conjugation, and, as before,

$$Q\colon T(X)^{1,0} \longrightarrow T(X)^{0,1}$$

is a conjugate-linear isomorphism. Moreover, there is a C-linear isomorphism

$$T(X)_J \cong T(X)^{1,0},$$

where $T(X)_J$ is the C-linear bundle constructed from $T(X)$ by means of J. Let $T^*(X)^{1,0}, T^*(X)^{0,1}$ denote the **C**-dual bundles of $T(X)^{1,0}$ and $T(X)^{0,1}$, respectively. Consider the exterior algebra bundles $\wedge T^*(X)_c$, $\wedge T^*(X)^{1,0}$, and $\wedge T^*(X)^{0,1}$, and, as in the case of vector spaces, we have

$$T^*(X)_c = T^*(X)^{1,0} \oplus T^*(X)^{0,1}$$

and natural bundle injections

$$\begin{matrix} \wedge T^*(X)^{1,0} \searrow & \\ & \wedge T^*(X)_c, \\ \wedge T^*(X)^{0,1} \nearrow & \end{matrix}$$

and we let $\wedge^{p,q} T^*(X)$ be the bundle whose fibre is $\wedge^{p,q} T^*_x(X)$. This bundle is the one we are interested in, since its sections are the *complex-valued differential forms of type* (p, q) *on* X, which we denote by

$$\mathcal{E}^{p,q}(X) = \mathcal{E}(X, \wedge^{p,q} T^*(X)).$$

Moreover, we have that

$$\mathcal{E}^r(X) = \sum_{p+q=r} \mathcal{E}^{p,q}(X).$$

Note that the differential forms of degree r do not reflect the almost complex structure J, whereas its decomposition into subspaces of type (p, q) does.

We want to obtain local representations for differential forms of type (p, q). To do this, we make the following general definition.

Definition 3.5: Let $E \longrightarrow X$ be an $\mathcal{S}$-bundle of rank r and let U be an open subset of X. A *frame for* E *over* U is a set of r $\mathcal{S}$-sections $\{s_1, \ldots, s_r\}$, $s_j \in \mathcal{S}(U, E)$, such that $\{s_1(x), \ldots, s_r(x)\}$ is a basis for E_x for any $x \in U$.

Any $\mathcal{S}$-bundle E admits a frame in some neighborhood of any given point in the base space. Namely, let U be a trivializing neighborhood for E so that

$$h\colon E|_U \xrightarrow{\sim} U \times K^r,$$

and thus we have an isomorphism

$$h_*\colon \mathcal{S}(U, E|_U) \xrightarrow{\sim} \mathcal{S}(U, U \times K^r).$$

Consider the vector-valued functions

$$e_1 = (1, 0, \ldots, 0), e_2 = (0, 1, \ldots, 0), \ldots, e_r = (0, \ldots, 0, 1),$$

which clearly form a (constant) frame for $U \times K^n$, and thus $\{(h_*)^{-1}(e_1), \ldots, (h_*)^{-1}(e_r)\}$ forms a frame for $E|_U$, since the bundle mapping h is an isomorphism on fibres, carrying a basis to a basis. Therefore we see that having a frame is equivalent to having a trivialization and that the existence of a global frame (defined over X) is equivalent to the bundle being trivial.

Let now (X, J) be an almost complex manifold as before and let $\{w_1, \ldots, w_n\}$ be a local frame (defined over some open set U) for $T^*(X)^{1,0}$. It follows that $\{\bar{w}_1, \ldots, \bar{w}_n\}$† is a local frame for $T^*(X)^{0,1}$. Then a local frame for $\wedge^{p,q}T^*(X)$ is given by (using the multiindex notation of Sec. 2)

$$\{w^I \wedge \bar{w}^J\}, \quad |I| = p, \quad |J| = q, \quad (I, J \text{ strictly increasing}).$$

Therefore any section $s \in \mathcal{E}^{p,q}(X)$ can be written (in U) as

$$s = \sum_{\substack{|I|=p\\|J|=q}}' a_{IJ} w^I \wedge \bar{w}^J, \quad a_{IJ} \in \mathcal{E}^0(U).$$

Note that

$$ds = \sum_{\substack{|I|=p\\|J|=q}}' da_{IJ} \wedge w^I \wedge \bar{w}^J + a_{IJ} d(w^I \wedge \bar{w}^J),$$

where the second term is not necessarily zero, since $w_i(x)$ is not necessarily a constant function of the local coordinates in the base space (which will, however, be the case for a complex manifold and certain canonical frames defined with respect to local holomorphic coordinates, as will be seen below).

We now have, based on the almost complex structure, a direct sum decomposition of $\mathcal{E}^r(X)$ into subspaces $\{\mathcal{E}^{p,q}(X)\}$. Let $\pi_{p,q}$ denote the natural projection operators

$$\pi_{p,q}\colon \mathcal{E}^r(X) \longrightarrow \mathcal{E}^{p,q}(X), \quad p + q = r.$$

We have in general

$$d\colon \mathcal{E}^{p,q}(X) \longrightarrow \mathcal{E}^{p+q+1}(X) = \sum_{r+s=p+q+1} \mathcal{E}^{r,s}(X)$$

by restricting d to $\mathcal{E}^{p,q}$. We define

$$\partial\colon \mathcal{E}^{p,q}(X) \longrightarrow \mathcal{E}^{p+1,q}(X)$$
$$\bar{\partial}\colon \mathcal{E}^{p,q}(X) \longrightarrow \mathcal{E}^{p,q+1}(X)$$

by setting

$$\partial = \pi_{p+1,q} \circ d$$
$$\bar{\partial} = \pi_{p,q+1} \circ d.$$

We then extend ∂ and $\bar{\partial}$ to all

$$\mathcal{E}^*(X) = \sum_{r=0}^{\dim X} \mathcal{E}^r(X)$$

by complex linearity.

Recalling that Q denotes complex conjugation, we have the following elementary results.

Proposition 3.6: $Q\bar{\partial}(Qf) = \partial f$, for $f \in \mathcal{E}^*(X)$.

†We shall use both Q and overbars to denote the conjugation, depending on the context.

Proof: One has to verify that if $f \in \mathcal{E}^r(X)$ and $p+q = r$, then $Q\pi_{p,q} f = \pi_{q,p} Qf$ and $Q(df) = dQf$, which are simple, and we shall omit the details. Q.E.D.

In general, we know that $d^2 = 0$, but it is not necessarily the case that $\bar{\partial}^2 = 0$. However, it follows from Proposition 3.6 that $\bar{\partial}^2 = 0$ if and only if $\partial^2 = 0$.

In general

$$d\colon \mathcal{E}^{p,q}(X) \longrightarrow \mathcal{E}^{p+q+1}(X)$$

can be decomposed as

$$d = \sum_{r+s=p+q+1} \pi_{r,s} \circ d = \partial + \bar{\partial} + \cdots .$$

If, however, $d = \partial + \bar{\partial}$, then

$$d^2 = \partial^2 + \partial\bar{\partial} + \bar{\partial}\partial + \bar{\partial}^2,$$

and since each operator projects to a different summand of $\mathcal{E}^{p+q+2}(X)$ (in which case the operators are said to be of different type), we obtain

$$\partial^2 = \partial\bar{\partial} + \bar{\partial}\partial = \bar{\partial}^2 = 0.$$

If $d = \partial + \bar{\partial}$ then we say that the almost complex structure is *integrable*.

Theorem 3.7: The induced almost complex structure on a complex manifold is integrable.

Proof: Let X be a complex manifold and let (X_0, J) be the underlying differentiable manifold with the induced almost complex structure J. Since $T(X)$ is **C**-linear isomorphic to $T(X_0)$ equipped with the **C**-bundle structure induced by J, it follows that, as **C**-bundles,

$$T(X) \cong T(X_0)^{1,0},$$

and similarly for the dual bundles,

$$T^*(X) \cong T^*(X_0)^{1,0}.$$

But $\{dz_1, \ldots, dz_n\}$ is a local frame for $T^*(X)$ if $(z_1, \ldots, z_n)$ are local coordinates (recall that $\{dz_1, \ldots, dz_n\}$ are dual to $\{\partial/\partial z_1, \ldots, \partial/\partial z_n\}$). We set

$$\frac{\partial}{\partial z_j} = \frac{1}{2}\left(\frac{\partial}{\partial x_j} - i\frac{\partial}{\partial y_j}\right), \quad j = 1, \ldots, n$$

$$\frac{\partial}{\partial \bar{z}_j} = \frac{1}{2}\left(\frac{\partial}{\partial x_j} + i\frac{\partial}{\partial y_j}\right), \quad j = 1, \ldots, n,$$

where $\{\partial/\partial x_1, \ldots, \partial/\partial x_n, \partial/\partial y_1, \ldots, \partial/\partial y_n\}$ is a local frame for $T(X_0)_c$ and $\{\partial/\partial z_1, \ldots, \partial/\partial z_n\}$ is a local frame for $T(X)$ (cf. Examples 2.4 and 2.5). We observe that $\partial/\partial z_j$ so defined *is* the complex (partial) derivative of a holomorphic function, and thus the assertion that these derivatives form a local frame for $T(X)$ is valid. From the above relationships, it follows that

$$dz_j = dx_j + i dy_j$$

$$d\bar{z}_j = dx_j - i dy_j, \quad j = 1, \ldots, n,$$

which gives

$$dx_j = \frac{1}{2}(dz_j + d\bar{z}_j)$$

$$dy_j = \frac{1}{2i}(dz_j - d\bar{z}_j), \quad j = 1, \cdots, n.$$

This in turn implies that for $s \in \mathcal{E}^{p,q}(X)$

$$s = \sum_{I,J}{}' a_{IJ} dz^I \wedge d\bar{z}^J.$$

We have

$$ds = \sum_{j=1}^{n} \sum_{I,J}{}' \left(\frac{\partial a_{IJ}}{\partial x_j} dx_j + \frac{\partial a_{IJ}}{\partial y_j} dy_j \right) \wedge dz^I \wedge d\bar{z}^J$$

$$= \sum_{j=1}^{n} \sum_{I,J}{}' \frac{\partial a_{IJ}}{\partial z_j} dz_j \wedge dz^I \wedge d\bar{z}^J$$

$$+ \sum_{j=1}^{n} \sum_{I,J}{}' \frac{\partial a_{IJ}}{\partial \bar{z}_J} d\bar{z}_J \wedge dz^J \wedge d\bar{z}^J.$$

The first term is of type $(p+1, q)$, and so

$$\partial = \sum_{j=1}^{n} \frac{\partial}{\partial z_j} dz_j,$$

and similarly

$$\bar{\partial} = \sum_{j=1}^{n} \frac{\partial}{\partial \bar{z}_j} d\bar{z}_j,$$

and hence $d = \partial + \bar{\partial}$. Thus the almost complex structure induced by the complex structure of X is integrable.

Q.E.D.

The converse of this theorem is a deep result due to Newlander and Nirenberg [1], whose proof has been simplified in recent years (see, e.g., Kohn [1], Hörmander [2]).

Theorem 3.8 (Newlander-Nirenberg): Let (X, J) be an integrable almost complex manifold. Then there exists a unique complex structure $\mathcal{O}_X$ on X which induces the almost complex structure J.

We shall not prove this theorem, and instead refer the reader to Hörmander [2]. We shall mention, however, that it can easily be reduced to a local problem—and, indeed, to solving particular partial differential equations (namely the inhomogeneous Cauchy-Riemann equations) with estimates. In the case where (X, J) is a real-analytic almost complex manifold, there are simpler proofs (see e.g., Kobayashi-Nomizu, Vol. II [1]). We shall not need this theorem, but we shall mention that it plays an important role in the study of deformations of complex structures on a fixed differentiable manifold, a topic we shall discuss in Chap. V.

CHAPTER II

SHEAF THEORY

Sheaves were introduced some 20 years ago by Jean Leray and have had a profound effect on several mathematical disciplines. Their major virtue is that they unify and give a mechanism for dealing with many problems concerned with passage from local information to global information. This is very useful when dealing with, say, differentiable manifolds, since locally these look like Euclidean space, and hence localized problems can be dealt with by means of all the tools of classical analysis. Piecing together "solutions" of such local problems in a coherent manner to describe, e.g., global invariants, is most easily accomplished via sheaf theory and its associated cohomology theory. The major virtue of sheaf theory is information-theoretic in nature. Most problems could be phrased and perhaps solved without sheaf theory, but the notation would be enormously more complicated and difficult to comprehend.

In Sec. 1 we shall give the basic definition of presheaves and sheaves, including a variety of examples. In Sec. 2 we shall develop one of the basic computational tools associated with a sheaf, namely a resolution, and again there are more examples. Section 3 contains an introduction to cohomology theory via abstract (canonical) soft (or flabby) resolutions, and we shall prove some basic isomorphism theorems which give us an explicit version of de Rham's theorem, for instance. In Sec. 4 we give a brief summary of Čech cohomology theory, an alternative and equally useful method for computing cohomology. General references for this chapter include Bredon [1], Godement [1], and selected chapters in Gunning and Rossi [1] and Hirzebruch [1].

1. Presheaves and Sheaves

In this section we shall introduce the basic concepts of presheaves and sheaves, giving various examples to illustrate the main ideas. We shall start with some formal definitions.

Definition 1.1: A *presheaf* $\mathcal{F}$ over a topological space X is

(a) An assignment to each nonempty open set $U \subset X$ of a set $\mathcal{F}(U)$.
(b) A collection of mappings (called restriction homomorphisms)

$$r_V^U : \mathcal{F}(U) \longrightarrow \mathcal{F}(V)$$

for each pair of open sets U and V such that $V \subset U$, satisfying

(1) $r_U^U =$ identity on $U(= 1_U)$.
(2) For $U \supset V \supset W, r_W^U = r_W^V \circ r_V^U$.

If $\mathcal{F}$ and $\mathcal{G}$ are presheaves over X, then a *morphism* (of presheaves)

$$h : \mathcal{F} \longrightarrow \mathcal{G}$$

is a collection of maps

$$h_U : \mathcal{F}(U) \longrightarrow \mathcal{G}(U)$$

for each open set U in X such that the following diagram commutes:

$$\begin{array}{ccc} \mathcal{F}(U) & \longrightarrow & \mathcal{G}(U) \\ \downarrow r_V^U & & \downarrow r_V^U \\ \mathcal{F}(V) & \longrightarrow & \mathcal{G}(V), \end{array} \qquad V \subset U \subset X.$$

$\mathcal{F}$ is said to be a *subpresheaf* of $\mathcal{G}$ if the maps h_U above are inclusions.

Remark: We shall be dealing primarily with presheaves, $\mathcal{F}$, where $\mathcal{F}(U)$ has some algebraic structure (e.g., abelian groups). In this case we also require that the subpresheaves have the induced substructure (e.g., subgroups) and that restriction homomorphisms and morphisms preserve the algebraic structure (e.g., r_V^U and h_U are group homomorphisms). Moreover, we shall call the elements of $\mathcal{F}(U)$ *sections* of $\mathcal{F}$ over U for reasons which will become apparent later.

Definition 1.2: A presheaf $\mathcal{F}$ is called a *sheaf* if for every collection U_i of open subsets of X with $U = \cup U_i$ then $\mathcal{F}$ satisfies

Axiom S_1: If $s, t \in \mathcal{F}(U)$ and $r_{U_i}^U(s) = r_{U_i}^U(t)$ for all i, then $s = t$.
Axiom S_2: If $s_i \in \mathcal{F}(U_i)$ and if for $U_i \cap U_j \neq \emptyset$ we have

$$r_{U_i \cap U_j}^{U_i}(s_i) = r_{U_i \cap U_j}^{U_j}(s_j)$$

for all i, then there exists an $s \in \mathcal{F}(U)$ such that $r_{U_i}^U(s) = s_i$ for all i.

Morphisms of sheaves (or *sheaf mappings*) are simply morphisms of the underlying presheaf. Moreover, when a subpresheaf of a sheaf $\mathcal{F}$ is also a sheaf, then it will be called a *subsheaf* of $\mathcal{F}$. An *isomorphism* of sheaves (or presheaves) is defined in the obvious way, namely h_U is an isomorphism in the category under consideration for each open set U. Note that Axiom S_1 for a sheaf says that data defined on large open sets U can be determined uniquely by looking at it locally, and Axiom S_2 asserts that local data of a given kind (in a given presheaf) can be pieced together to give global data of the same kind (in the same presheaf).

We would now like to give some examples of presheaves and sheaves.

Example 1.3: Let X and Y be topological spaces and let $\mathcal{C}_{X,Y}$ be the presheaf over X defined by

(a) $\mathcal{C}_{X,Y}(U) := \{f : U \to Y : f \text{ is continuous}\}$.

(b) For $f \in \mathcal{C}_{X,Y}(U)$, $r_V^U(f) := f|_V$, the natural restriction as a function. It is easy to see that this presheaf satisfies Axioms S_1 and S_2 and hence is a sheaf.

Example 1.4: Let X be a topological space and let K be $\mathbf{R}$ or $\mathbf{C}$. Let $\mathcal{C}_X = \mathcal{C}_{X,K}$, as in the above example. This is a sheaf of K-algebras; i.e., $\mathcal{C}_X(U)$ is a K-algebra under pointwise addition, multiplication, and scalar multiplication of functions.

Example 1.5: Let X be an $\mathcal{S}$-manifold (as in Definition 1.1 in Chap. I). Then we see that the assignment $\mathcal{S}_X$ given by

$$\mathcal{S}_X(U) := \mathcal{S}(U) = \text{the } \mathcal{S}\text{-functions on } U$$

defines a subsheaf of $\mathcal{C}_X$. This sheaf is called the *structure sheaf* of the manifold X. In particular, we shall be dealing with $\mathcal{E}_X$, $\mathcal{A}_X$, and $\mathcal{O}_X$, the sheaves of *differentiable, real-analytic*, and *holomorphic functions* on a manifold X.

Example 1.6: Let X be a topological space and let G be an abelian group. The assignment $U \to G$, for U connected, determines a sheaf, called the *constant sheaf* (with coefficients in G). This sheaf will often be denoted simply by the same symbol G when there is no chance of confusion.

We want to give at least one example of a presheaf which is not a sheaf, although our primary interest later on will be sheaves of the type mentioned above.

Example 1.7: Let X be the complex plane, and define the presheaf $\mathcal{B}$ by letting $\mathcal{B}(U)$ be the algebra of bounded holomorphic functions in the open set U. Let $U_i = \{z : |z| < i\}$, and then $\mathbf{C} = \cup\, U_i$. Let $f_i \in \mathcal{B}(U_i)$ be defined by setting $f_i(z) = z$. Then it is quite clear that there is no $f \in \mathcal{B}(\mathbf{C})$ with the property that $f|_{U_i} = f_i$. In fact, by Liouville's theorem, $\mathcal{B}(\mathbf{C}) = \mathbf{C}$. Consequently, $\mathcal{B}$ is not a sheaf, since it violates Axiom S_2.

We see in the above example that the basic reason $\mathcal{B}$ was not a sheaf was that it was not defined by a local property (such as holomorphicity, differentiability, or continuity).

Remark: A presheaf that violates Axiom S_1 can be obtained by taking the sections of $\mathcal{C}_{X,K}$ with X a two point discrete space but letting all proper restrictions be zero.

A natural structure on presheaves which occurs quite often is that of a module.

Definition 1.8: Let $\mathcal{R}$ be a presheaf of commutative rings and let $\mathfrak{M}$ be a presheaf of abelian groups, both over a topological space X. Suppose that for any open set $U \subset X, \mathfrak{M}(U)$ can be given the structure of an $\mathcal{R}(U)$-module such that if $\alpha \in \mathcal{R}(U)$ and $f \in \mathfrak{M}(U)$, then

$$r_V^U(\alpha f) = \rho_V^U(\alpha) r_V^U(f)$$

for $V \subset U$, where r_V^U is the $\mathfrak{M}$-restriction homomorphism and ρ_V^U is the $\mathcal{R}$-restriction homomorphism. Then $\mathfrak{M}$ is called a *presheaf of* $\mathcal{R}$*-modules.* Moreover, if $\mathfrak{M}$ is a sheaf, then $\mathfrak{M}$ will be a *sheaf of* $\mathcal{R}$*-modules.*

Example 1.9: Let $E \to X$ be an $\mathcal{S}$-bundle. Then define a presheaf $\mathcal{S}(E)$ $(= \mathcal{S}_X(E))$† by setting $\mathcal{S}(E)(U) = \mathcal{S}(U, E)$, for U open in X, together with the natural restrictions. Then $\mathcal{S}(E)$ is, in fact, a subsheaf of $\mathcal{C}_{X,E}$ and is called the *sheaf of* $\mathcal{S}$*-sections of the vector bundle* E. As special cases, we have the sheaves of differential forms $\mathcal{E}_X^*$ on a differentiable manifold, or the sheaf of differential forms of type (p, q), $\mathcal{E}_X^{p,q}$, on a complex manifold X. These sheaves are examples of sheaves of $\mathcal{E}_X$-modules, and, more generally, $\mathcal{S}(E)$ is a sheaf of $\mathcal{S}_X$-modules for an $\mathcal{S}$-bundle $E \to X$.

Example 1.10: Let $\mathcal{O}_{\mathbf{C}}$ denote the sheaf of holomorphic functions in the complex plane $\mathbf{C}$ and let $\mathcal{J}$ denote the sheaf defined by the presheaf

$$\begin{cases} U \longrightarrow \mathcal{O}(U), & \text{if } 0 \notin U \\ U \longrightarrow \{f \in \mathcal{O}(U) : f(0) = 0\}, & \text{if } 0 \in U. \end{cases}$$

Then, clearly, this presheaf is a sheaf, and it is also a sheaf of modules over the sheaf of commutative rings $\mathcal{O}_{\mathbf{C}}$ (in fact, it is a sheaf of ideals in the sheaf of rings, going one step further).

The most commonly occurring sheaves of modules in complex analysis have names.

Definition 1.11: Let X be a complex manifold. Then a sheaf of modules over the structure sheaf $\mathcal{O}_X$ of X is called an *analytic sheaf.*

As one knows from algebra, the simplest type of modules are the free modules. We have a corresponding definition for sheaves. First, we note that there is a natural (and obvious) notion of *restriction* of a sheaf (or presheaf) $\mathcal{F}$ on X to a sheaf (or presheaf) on an open subset U of X, to be denoted by $\mathcal{F}|_U$.

Definition 1.12: Let $\mathcal{R}$ be a sheaf of commutative rings over a topological space X.

†$\mathcal{S}_X(E)$ is not to be confused with $\mathcal{S}_E(E)$, which are the global $\mathcal{S}$-functions defined on the manifold E. In context it will be clear which is meant.

(a) Define $\mathcal{R}^p$, for $p \geq 0$, by the presheaf

$$U \longrightarrow \mathcal{R}^p(U) := \underbrace{\mathcal{R}(U) \oplus \cdots \oplus \mathcal{R}(U)}_{p \text{ terms}}.$$

$\mathcal{R}^p$, so defined, is clearly a sheaf of $\mathcal{R}$-modules and is called the *direct sum* of $\mathcal{R}$ (p times; $p = 0$ corresponds to the 0-module).

(b) If $\mathfrak{M}$ is a sheaf of $\mathcal{R}$-modules such that $\mathfrak{M} \cong \mathcal{R}^p$ for some $p \geq 0$, then $\mathfrak{M}$ is said to be a *free sheaf* of modules.

(c) If $\mathfrak{M}$ is a sheaf of $\mathcal{R}$-modules such that each $x \in X$ has a neighborhood U such that $\mathfrak{M}|_U$ is free, then $\mathfrak{M}$ is said to be *locally free.*

The following theorem demonstrates the relationship between vector bundles and locally free sheaves.

Theorem 1.13: Let $X = (X, \mathcal{S})$ be a connected $\mathcal{S}$-manifold. Then there is a one-to-one correspondence between (isomorphism classes of) $\mathcal{S}$-bundles over X and (isomorphism classes of) locally free sheaves of $\mathcal{S}$-modules over X.

Proof: The correspondence is provided by

$$E \longrightarrow \mathcal{S}(E)$$

and it is easy to see that $\mathcal{S}(E)$ is a locally free sheaf of $\mathcal{S}$-modules. Namely, by local triviality, for some neighborhood U of a point $x \in X$, we have $E|_U \cong U \times K^r$, where r is the rank of the vector bundle E. It follows that $\mathcal{S}(E)|_U \cong \mathcal{S}(U \times K^r)$. We claim that

$$\mathcal{S}(U \times K^r) \cong \mathcal{S}|_U \oplus \cdots \oplus \mathcal{S}|_U.$$

From the definition of a section, it follows that $f \in \mathcal{S}(U \times K^r)(V)$ (for V open in U) if and only if $f(x) = (x, g(x))$, where $g : V \to K^r$ and g is an $\mathcal{S}$-morphism (cf. Example I.2.12). Therefore $g = (g_1, \ldots, g_r)$, $g_j \in \mathcal{S}(V)$, and the correspondence above is given by

$$f \longrightarrow (g_1, \ldots, g_r) \in \mathcal{S}_U(V) \oplus \cdots \oplus \mathcal{S}_U(V),$$

which is clearly an isomorphism of sheaves. Therefore $\mathcal{S}(E)$ is a locally free $\mathcal{S}_X$-module.

We shall now show how to construct a vector bundle from a locally free sheaf, which inverts the above construction. Suppose that $\mathcal{L}$ is a locally free sheaf of $\mathcal{S}$-modules. Then we can find an open covering $\{U_\alpha\}$ of X such that

$$g_\alpha : \mathcal{L}|_{U_\alpha} \xrightarrow{\sim} \mathcal{S}^r|_{U_\alpha}$$

for some $r > 0$ (excluding the trivial case); note that r does not depend on α, since X is connected. Then define

$$g_{\alpha\beta} : \mathcal{S}^r|_{U_\alpha \cap U_\beta} \xrightarrow{\sim} \mathcal{S}^r|_{U_\alpha \cap U_\beta}$$

by setting $g_{\alpha\beta} = g_\alpha \circ g_\beta^{-1}$. Now $g_{\alpha\beta}$ is a sheaf mapping, so in particular (when acting on the open set $U_\alpha \cap U_\beta$) it determines an invertible mapping of vector-valued functions $(g_{\alpha\beta})_{U_\alpha \cap U_\beta}$, which we write as

$$g_{\alpha\beta} : \mathcal{S}(U_\alpha \cap U_\beta)^r \longrightarrow \mathcal{S}(U_\alpha \cap U_\beta)^r,$$

which is then a nonsingular $r \times r$ matrix of functions in $\mathcal{S}(U_\alpha \cap U_\beta)$, i.e., $g_{\alpha\beta} : U_\alpha \cap U_\beta \to GL(r, K)$, and hence determines transition functions for a vector bundle E, since the compatibility conditions $g_{\alpha\beta} \cdot g_{\beta\gamma} = g_{\alpha\gamma}$ are trivially satisfied. Thus a vector bundle E can be defined by letting

$$\tilde{E} = \bigcup_\alpha U_\alpha \times K^r \qquad \text{(disjoint union)}$$

and making the identification

$$(x, \xi) \sim (x, g_{\alpha\beta}(x)\xi), \quad \text{if } x \in U_\alpha \cap \bar{U}_\beta \neq \varnothing.$$

(Cf. the remark after Definition I.2.2.)

We leave it to the reader to verify that isomorphism classes are preserved under this correspondence.

Q.E.D.

Remark: Most of the sheaves we shall be dealing with will be locally free sheaves arising from vector bundles; however, there is a generalization which is of great importance for the study of function theory on complex manifolds and, more generally, complex manifolds with singularities—complex spaces. An analytic sheaf $\mathcal{F}$ on a complex manifold X is said to be *coherent* if for each $x \in X$ there is a neighborhood U of x such that there is an exact sequence of sheaves over U,

$$\mathcal{O}^p|_U \longrightarrow \mathcal{O}^q|_U \longrightarrow \mathcal{F}|_U \longrightarrow 0,$$

for some p and q. For a complete discussion of coherent analytic sheaves on complex spaces, see Gunning and Rossi [1]. For instance, let V be a subvariety of $\mathbf{C}^n$; i.e., V is defined as a closed subset in $\mathbf{C}^n$, which is locally given as the set of zeros of a finite number of holomorphic functions. Let $\mathcal{J}_V$ be the subsheaf of $\mathcal{O}$ defined by sections that vanish on V. Therefore $\mathcal{J}_V$ is an ideal sheaf in the sheaf of rings $\mathcal{O}$. Then $\mathcal{J}_V$ is a coherent analytic sheaf (by results of Oka and Cartan; see Gunning and Rossi [1]) but not necessarily locally free. A simple example of this situation is the case where V is simply the origin in $\mathbf{C}^2$; then we see that $\mathcal{J}_V = \mathcal{J}_{\{0\}}$ is similar to Example 1.10. Moreover, $\mathcal{J}_{\{0\}}$ is coherent because of the following exact sequence,

$$0 \longrightarrow \mathcal{O} \xrightarrow{\nu} \mathcal{O}^2 \xrightarrow{\mu} \mathcal{J}_{\{0\}} \longrightarrow 0 \qquad \text{(Koszul complex)},$$

where

$$\mu(f_1, f_2) = z_1 f_1 - z_2 f_2$$

$$\nu(f) = (z_2 f, z_1 f).$$

One can easily check that this is exact (by expanding the functions in power series at the origin and determining the relations between the coefficients).

2. Resolutions of Sheaves

A sheaf $\mathcal{F}$ on a space X is a carrier of localized information about the space X. To get global information about X from $\mathcal{F}$, we want to apply the techniques of homological algebra to sheaves. For this we want to consider exact sequences, quotients, etc. To do this, however, we have to look at another, more localized, model of a sheaf. In fact, we shall make a sheaf into a topological space of a particular type.

Definition 2.1: (a) An *étalé space* over a topological space X is a topological space Y together with a continuous surjective mapping $\pi : Y \longrightarrow X$ such that π is a local homeomorphism.

(b) A *section* of an étalé space $Y \xrightarrow{\pi} X$ over an open set $U \subset X$ is a continuous map $f : U \longrightarrow Y$ such that $\pi \circ f = 1_U$. The set of sections over U is denoted by $\mathbf{\Gamma}(U, Y)$.

It is clear that the sections of an étalé space form a subsheaf of $\mathcal{C}_{X,Y}$. We are going to associate to any presheaf $\mathcal{F}$ over X an étalé space $\tilde{\mathcal{F}} \longrightarrow X$ such that the sheaf of sections of $\tilde{\mathcal{F}}$ gives another model for $\mathcal{F}$ if $\mathcal{F}$ happens to be a sheaf. The reasons for this construction will become clear as we go along.

Consider a presheaf $\mathcal{F}$ over X, and let

$$\mathcal{F}_x := \varinjlim_{x \in U} \mathcal{F}(U)$$

be the direct limit of the sets $\mathcal{F}(U)$ with respect to the restriction maps $\{r_V^U\}$ of $\mathcal{F}$. If $\mathcal{F}$ has an algebraic structure which is preserved under direct limits, then $\mathcal{F}_x$, called the *stalk* of $\mathcal{F}$ at x, will inherit that structure. For instance, this is the case if $\mathcal{F}$ is a presheaf of abelian groups or commutative rings.

There is a natural map

$$r_x^U : \mathcal{F}(U) \longrightarrow \mathcal{F}_x, \qquad x \in U,$$

given by taking an element in $\mathcal{F}(U)$ into its equivalence class in the direct limit. If $s \in \mathcal{F}(U)$, then $s_x := r_x^U(s)$ is called the *germ* of s at x, and s is called a *representative* for the germ s_x. Let

$$\tilde{\mathcal{F}} = \bigcup_{x \in X} \mathcal{F}_x$$

and let $\pi : \tilde{\mathcal{F}} \longrightarrow X$ be the natural projection taking points in $\mathcal{F}_x$ to x. We want to make $\tilde{\mathcal{F}}$ into an étalé space, and all that remains is to give $\tilde{\mathcal{F}}$ a topology. For each $s \in \mathcal{F}(U)$ define the set function

$$\tilde{s} : U \longrightarrow \tilde{\mathcal{F}}$$

by letting $\tilde{s}(x) = s_x$ for each $x \in U$. Note that $\pi \circ \tilde{s} = 1_U$. Let

$$\{\tilde{s}(U)\} \text{ where } U \text{ is open in } X, \mathrm{s} \in \mathcal{F}(U)$$

be a basis for the topology of $\tilde{\mathcal{F}}$. Then all the functions $\tilde{s}$ are continuous. Moreover, it is easy to check that π is continuous and indeed a local homeomorphism ($\tilde{s}$ provides a local inverse at s_x for π for a given representative s of $s_x \in \tilde{\mathfrak{F}}$).

Thus we have associated to each presheaf $\mathfrak{F}$ over X an étalé space. Moreover, if the presheaf has algebraic properties preserved by direct limits, then the étalé space $\tilde{\mathfrak{F}}$ inherits these properties. For example, suppose that $\mathfrak{F}$ is a presheaf of abelian groups. Then $\tilde{\mathfrak{F}}$ has the following properties:

(a) Each stalk is an abelian group.
(b) If $\tilde{\mathfrak{F}} \circ \tilde{\mathfrak{F}} := \{(s,t) \in \tilde{\mathfrak{F}} \times \tilde{\mathfrak{F}} : \pi(s) = \pi(t)\}$, then the map

$$\mu : \tilde{\mathfrak{F}} \circ \tilde{\mathfrak{F}} \longrightarrow \tilde{\mathfrak{F}}$$

given by $(s_x, t_x) \longrightarrow s_x - t_x$ is continuous. This is true since if $(s-t)^{\sim}(U)$ is a basic open set of $s_x - t_x$ for U open in X and $s, t \in \mathfrak{F}(U)$, then the inverse image of $(s-t)^{\sim}(U)$ by the above map is just $\tilde{s}(U) \circ \tilde{t}(U)$, which is a basic open set in $\tilde{\mathfrak{F}} \circ \tilde{\mathfrak{F}}$.

(c) For U open in X, the set of sections of $\tilde{\mathfrak{F}}$ over $U, \mathbf{\Gamma}(U, \tilde{\mathfrak{F}})$ is an abelian group under pointwise addition, i.e., for $s, t \in \mathbf{\Gamma}(U, \tilde{\mathfrak{F}})$

$$(s-t)(x) = s(x) - t(x) \quad \text{for all } x \in U.$$

We see that $s - t$ is continuous since it is given by the following composition of continuous maps:

$$U \xrightarrow{(s,t)} \tilde{\mathfrak{F}} \circ \tilde{\mathfrak{F}} \xrightarrow{\mu} \tilde{\mathfrak{F}}.$$

In associating an étalé space $\tilde{\mathfrak{F}}$ to a presheaf $\mathfrak{F}$, we have also associated a sheaf to $\mathfrak{F}$, namely the sheaf of sections of $\tilde{\mathfrak{F}}$. We call this sheaf the *sheaf generated by* $\mathfrak{F}$. We would now like to look more closely at the relationship between the presheaf, $\mathfrak{F}$, and the sheaf of sections of $\tilde{\mathfrak{F}}$ which we shall call $\bar{\mathfrak{F}}$ for the time being. We have already used the fact that there is a presheaf morphism, which we now denote by

$$\tau : \mathfrak{F} \longrightarrow \bar{\mathfrak{F}},$$

namely $\tau_U : \mathfrak{F}(U) \longrightarrow \bar{\mathfrak{F}}(U)[:= \mathbf{\Gamma}(U, \tilde{\mathfrak{F}})]$ is given by $\tau_U(s) = \tilde{s}$. Recall that $\tilde{s}(x) = r_x^U(s)$ for all $x \in U$. In the case that $\mathfrak{F}$ is a sheaf, we have the following basic result. Its proof will illustrate the use of the sheaf axioms in an abstract setting.

Theorem 2.2: If $\mathfrak{F}$ is a sheaf, then

$$\tau : \mathfrak{F} \longrightarrow \bar{\mathfrak{F}}$$

is a sheaf isomorphism.

Proof: It suffices to show that τ_U is bijective for each U.

(a) τ_U is injective: Suppose that $s', s'' \in \mathfrak{F}(U)$ and $\tau_U(s') = \tau_U(s'')$. Then

$$[\tau_U(s')](x) = [\tau_U(s'')](x) \qquad \text{for all } x \in U;$$

i.e., $r_x^U(s') = r_x^U(s'')$ for all $x \in U$. But when $r_x^U(s') = r_x^U(s'')$ for some $x \in U$, the definition of direct limit implies that there is a neighborhood V of x such that $r_V^U(s') = r_V^U(s'')$. Since this is true for each $x \in U$, we can cover U with open sets U_i such that

$$r_{U_i}^U(s') = r_{U_i}^U(s'')$$

for all i. So since $\mathcal{F}$ is a sheaf, we have, by Axiom S_1, $s' = s''$.

(b) τ_U is surjective: Suppose $\sigma \in \mathbf{\Gamma}(U, \tilde{\mathcal{F}})$. Then for $x \in U$ there is a neighborhood V of x and $s \in \mathcal{F}(V)$ such that

$$\sigma(x) = S_x = [\tau_V(S)](x).$$

Since sections of an étalé space are local inverses for π, any two sections which agree at a point agree in some neighborhood of that point. Hence we have for some V^* a neighborhood of x:

$$\sigma|_{V^*} = \tau_V(s)|_{V^*} = \tau_{V^*}(r_{V^*}^V(s)).$$

Since this is true for any $x \in U$, we can cover U with neighborhoods U_i such that there exists $s_i \in \mathcal{F}(U_i)$ and

$$\tau_{U_i}(s_i) = \sigma|_{U_i}.$$

Moreover, we have

$$\tau_{U_i}(s_i) = \tau_{U_j}(s_j) \quad \text{on} \quad U_i \cap U_j,$$

so by part (a)

$$r_{U_i \cap U_j}^{U_i}(s_i) = r_{U_i \cap U_j}^{U_j}(s_j).$$

Since $\mathcal{F}$ is a sheaf and $U = \cup_i U_i$, there exists $s \in \mathcal{F}(U)$ such that

$$r_{U_i}^U(s) = s_i.$$

Thus

$$\tau_U(s)|_{U_i} = \tau_{U_i}(r_{U_i}^U(s)) = \tau_{U_i}(s_i) = \sigma|_{U_i},$$

and finally $\tau_U(s) = \sigma$.

Q.E.D.

The content of this theorem is that to each sheaf $\mathcal{F}$ one can associate an étalé space $\tilde{\mathcal{F}}$ whose sheaf of sections is the original $\mathcal{F}$; i.e., $\tilde{\mathcal{F}}$ contains the same amount of information as $\mathcal{F}$, and for this reason, a sheaf is very often defined to be an étalé space with algebraic structure along its fibres, as discussed above (see, e.g., Bredon [1] and Gunning and Rossi [1]). For doing analysis, however, the principal object is the presheaf, with its axioms (since most sheaves occur naturally in this form), and the associated étalé space is an auxiliary construction which is useful in constructing the homological machinery which makes sheaves useful objects. One way, in particular, that the étalé space is useful is to pass from a presheaf to a sheaf.

Definition 2.3: Let $\mathcal{F}$ be a presheaf over a topological space X and let $\bar{\mathcal{F}}$ be the sheaf of sections of the étalé space $\tilde{\mathcal{F}}$ associated with $\mathcal{F}$. Then $\bar{\mathcal{F}}$ is the *sheaf generated by* $\mathcal{F}$.

By Theorem 2.2 above, we see that a presheaf, which is a sheaf, generates itself; i.e., $\mathcal{F} = \bar{\mathcal{F}}$. Moreover, we shall use both notations $\mathcal{F}(U)$ and $\mathbf{\Gamma}(U, \mathcal{F})$ to denote the set (or group or module) of sections of $\mathcal{F}$ over U, depending on the context (the word section, of course, coming from the étalé space picture of a sheaf).

We now want to study the elementary homological algebra of sheaves of abelian groups; all the concepts we shall encounter generalize in a natural manner to sheaves of modules.

Definition 2.4: Suppose that $\mathcal{F}$ and $\mathcal{G}$ are sheaves of abelian groups over a space X with $\mathcal{G}$ a subsheaf of $\mathcal{F}$, and let $\mathcal{Q}$ be the sheaf generated by the presheaf $U \to \mathcal{F}(U)/\mathcal{G}(U)$. Then $\mathcal{Q}$ is called the *quotient sheaf of* $\mathcal{F}$ by $\mathcal{G}$ and is denoted by $\mathcal{F}/\mathcal{G}$.

The quotient mapping on presheaves above induces a natural sheaf surjection $\mathcal{F} \longrightarrow \mathcal{F}/\mathcal{G}$ by going to the direct limit, inducing a continuous mapping of étalé spaces, and then considering the induced map on continuous sections. This is then the desired sheaf mapping onto the quotient sheaf.

One of the fundamental concepts of homological algebra is that of exactness.

Definition 2.5: If $\mathcal{A}$, $\mathcal{B}$, and $\mathcal{C}$ are sheaves of abelian groups over X and

$$\mathcal{A} \xrightarrow{g} \mathcal{B} \xrightarrow{h} \mathcal{C}$$

is a sequence of sheaf morphisms, then this sequence is *exact at* $\mathcal{B}$ if the induced sequence on stalks

$$\mathcal{A}_x \xrightarrow{g_x} \mathcal{B}_x \xrightarrow{h_x} \mathcal{C}_x$$

is exact for all $x \in X$. A *short exact sequence* is a sequence

$$0 \longrightarrow \mathcal{A} \longrightarrow \mathcal{B} \longrightarrow \mathcal{C} \longrightarrow 0,$$

which is exact at $\mathcal{A}$, $\mathcal{B}$, and $\mathcal{C}$, where 0 denotes the (constant) zero sheaf.

Remark: Note that exactness is a *local* property. The sheaves are *not* defined to be exact at the presheaf level [i.e., exactness of

$$\mathcal{A}(U) \longrightarrow \mathcal{B}(U) \longrightarrow \mathcal{C}(U)$$

for each U open in X], which, of course, was possible since homomorphism of sheaves *were* so defined. The usefulness of sheaf theory is precisely in finding and categorizing obstructions to the "global exactness" of sheaves.

We shall now give some examples of short exact sequences of sheaves.

Example 2.6: Let X be a connected complex manifold. Let $\mathcal{O}$ be the sheaf of holomorphic functions on X and let $\mathcal{O}^*$ be the sheaf of nonvanishing holomorphic functions on X which is a sheaf of abelian groups under multiplication. Then we have the following sequence:

$$0 \longrightarrow \mathbf{Z} \xrightarrow{i} \mathcal{O} \xrightarrow{\exp} \mathcal{O}^* \longrightarrow 0 \tag{2.1}$$

where $\mathbf{Z}$ is the constant sheaf of integers, i is the inclusion map, and $\exp : \mathcal{O} \longrightarrow \mathcal{O}^*$ is defined by

$$\exp_U (f)(z) = \exp(2\pi i f(z)).$$

Moreover, for some (sufficiently small) simply-connected neighborhood U of $x \in X$ and for some representative $g \in \mathcal{O}^*(U)$ of a germ g_x at x, we can choose $f_x = ((1/2\pi i)\log g)_x$ for some branch of the logarithm function, and we have $\exp_x(f_x) = g_x$. Also, $\exp_x(f_x) = 0$ implies that†

$$\exp 2\pi i f(z) \equiv 1, \qquad z \in U,$$

for any $f \in \mathcal{O}(U)$ which is a representative of the germ f_x on a connected neighborhood U of x. Therefore f is constant on U and is, in fact, an integer, so that

$$\operatorname{Ker}(\exp_x) = \mathbf{Z},$$

and the sequence (2.1) is exact.

Example 2.7: Let $\mathcal{A}$ be a subsheaf of $\mathcal{B}$. Then

$$0 \longrightarrow \mathcal{A} \xrightarrow{i} \mathcal{B} \xrightarrow{q} \mathcal{B}/\mathcal{A} \longrightarrow 0$$

is an exact sequence of sheaves, where i is the natural inclusion and q is the natural quotient mapping.

Example 2.8: As a special case of Example 2.7, we let $X = \mathbf{C}$ and let $\mathcal{O}$ be the holomorphic functions on $\mathbf{C}$. Let $\mathcal{J}$ be the subsheaf of $\mathcal{O}$ consisting of those holomorphic functions which vanish at $z = 0 \in \mathbf{C}$ (Example 1.10). Then we have the following exact sequence of sheaves:

$$0 \longrightarrow \mathcal{J} \longrightarrow \mathcal{O} \longrightarrow \mathcal{O}/\mathcal{J} \longrightarrow 0.$$

We note that

$$(\mathcal{O}/\mathcal{J})_x \cong \begin{cases} \mathbf{C}, & \text{if } x = 0 \\ 0, & \text{if } x \neq 0. \end{cases}$$

Example 2.9: Let X be a connected Hausdorff space and let a, b be two distinct points in X. Let $\mathbf{Z}$ denote the constant sheaf of integers on X and $\mathcal{J}$ denote the subsheaf of $\mathbf{Z}$ which vanishes at a and b. Then

$$0 \longrightarrow \mathcal{J} \longrightarrow \mathbf{Z} \longrightarrow \mathbf{Z}/\mathcal{J} \longrightarrow 0$$

is exact and

$$(\mathbf{Z}/\mathcal{J})_x \cong \begin{cases} \mathbf{Z}, & \text{if } x = a \text{ or } x = b \\ 0, & \text{if } x \neq a \text{ and } x \neq b. \end{cases}$$

†Note that "0" here is the identity element in an abelian group.

Remark: Example 2.9 shows the necessity of using the generated sheaf for the quotient sheaf in Definition 2.4, since the presheaf of quotients of sections of $\mathbf{Z}$ by sections of $\mathcal{J}$ violates Axiom S_2.

Following the terminology of homological algebra for modules, we make the following definitions where sheaf means sheaf of abelian groups or sheaf of modules. A *graded sheaf* is a family of sheaves indexed by integers, $\mathcal{F}^* = \{\mathcal{F}^\alpha\}_{\alpha \in \mathbf{Z}}$. A *sequence of sheaves* (or sheaf sequence) is a graded sheaf connected by sheaf mappings:

$$\cdots \longrightarrow \mathcal{F}^0 \xrightarrow{\alpha_0} \mathcal{F}^1 \xrightarrow{\alpha_1} \mathcal{F}^2 \xrightarrow{\alpha_2} \mathcal{F}^3 \longrightarrow \cdots . \tag{2.2}$$

A *differential sheaf* is a sequence of sheaves where the composite of any pair of mappings is zero; i.e., $\alpha_j \circ \alpha_{j-1} = 0$ in (2.2). A *resolution* of a sheaf $\mathcal{F}$ is an exact sequence of sheaves of the form

$$0 \longrightarrow \mathcal{F} \longrightarrow \mathcal{F}^0 \longrightarrow \mathcal{F}^1 \longrightarrow \cdots \longrightarrow \mathcal{F}^m \longrightarrow \cdots ,$$

which we also denote symbolically by

$$0 \longrightarrow \mathcal{F} \longrightarrow \mathcal{F}^*,$$

the maps being understood.

We shall see later that various types of information for a given sheaf $\mathcal{F}$ can be obtained from knowledge of a given resolution. We shall close this section with various examples of resolutions of sheaves. Their utility in computing cohomology will be demonstrated in the next section.

Example 2.10: Let X be a differentiable manifold of real dimension m and let $\mathcal{E}_X^p$ be the sheaf of real-valued differential forms of degree p. Then there is a resolution of the constant sheaf $\mathbf{R}$ given by

$$0 \longrightarrow \mathbf{R} \xrightarrow{i} \mathcal{E}_X^0 \xrightarrow{d} \mathcal{E}_X^1 \xrightarrow{d} \cdots \longrightarrow \mathcal{E}_X^m \longrightarrow 0, \tag{2.3}$$

where i is the natural inclusion and d is the exterior differentiation operator. Since $d^2 = 0$, it is clear that the above is a differential sheaf. However, the classical Poincaré lemma (see, e.g., Spivak [1], p. 94) asserts that on a star-shaped domain U in $\mathbf{R}^n$, if $f \in \mathcal{E}^p(U)$ is given such that $df = 0$, then there exists a $u \in \mathcal{E}^{p-1}(U)$ $(p > 0)$ so that $du = f$. Therefore the induced mapping d_x on the stalks at $x \in X$ is exact, since we can find representatives in local coordinates in star-shaped domains. At the term $\mathcal{E}_X^0$, exactness is an elementary result from calculus [i.e., $df \equiv 0$ implies that f is a constant (locally)]. We shall denote this resolution by $0 \longrightarrow \mathbf{R} \longrightarrow \mathcal{E}_X^*$ (or $0 \longrightarrow \mathbf{C} \longrightarrow \mathcal{E}_X^*$ if we are using complex coefficients).

Example 2.11: Let X be a topological manifold. We want to derive a resolution for the constant sheaf G over X, where G is an abelian group (which will hold also for more general spaces). Let $S^p(U, G)$ be the group of singular cochains in U with coefficients in G; i.e., $S^p(U, G) = \operatorname{Hom}_{\mathbf{Z}}(S_p(U, \mathbf{Z}), G)$, where $S_p(U, \mathbf{Z})$ is the abelian group of integral singular chains of degree

p in U with the usual boundary map (see, e.g., MacLane [1] or any standard algebraic topology text). Let δ denote the coboundary operator, $\delta : S^p(U, G) \longrightarrow S^{p+1}(U, G)$, and let $\mathcal{S}^p(G)$ be the sheaf over X generated by the presheaf $U \longrightarrow S^p(U, G)$, with the induced differential mapping $\mathcal{S}^p(G) \xrightarrow{\delta} \mathcal{S}^{p+1}(G)$. Consider the unit ball U in Euclidean space. Then the sequence

$$\cdots \longrightarrow S^{p-1}(U, G) \xrightarrow{\delta} S^p(U, G) \xrightarrow{\delta} S^{p+1}(U, G) \longrightarrow \cdots \tag{2.4}$$

is exact, since Ker δ/Im δ is the classical singular cohomology for the unit ball, which is well known to be zero for $p > 0$ (see MacLane [1], pp. 54–61, for an elementary proof of this fact, using barycentric subdivision). Therefore the sequence

$$0 \longrightarrow G \longrightarrow \mathcal{S}^0(G) \xrightarrow{\delta} \mathcal{S}^1(G) \xrightarrow{\delta} \mathcal{S}^2(G) \longrightarrow \cdots \longrightarrow \mathcal{S}^m(G) \longrightarrow \cdots$$

is a resolution of the constant sheaf G, noting that

$$\text{Ker}(\delta : S^0(U, G) \longrightarrow S^1(U, G)) \cong G.$$

We remark that we could also have considered C^∞ chains if X is a differentiable manifold, i.e. (linear combinations of) maps $f : \mathbf{\Delta}^p \longrightarrow U$, where f is a C^∞ mapping defined in a neighborhood of the standard p-simplex $\mathbf{\Delta}^p$. The corresponding results above still hold [in particular, the elementary proof of the exactness of (2.4) still works in the C^∞ case], and we have a resolution by differentiable cochains with coefficients in G:

$$0 \longrightarrow G \longrightarrow \mathcal{S}^0_\infty(G) \longrightarrow \mathcal{S}^1_\infty(G) \longrightarrow \cdots \longrightarrow \mathcal{S}^m_\infty(G) \longrightarrow \cdots,$$

which we abbreviate by

$$0 \longrightarrow G \longrightarrow \mathcal{S}^*_\infty(G). \tag{2.5}$$

Example 2.12: Let X be a complex manifold of complex dimension n, let $\mathcal{E}^{p,q}$ be the sheaf of (p, q) forms on X, and consider the sequence of sheaves, for $p \geq 0$, fixed,

$$0 \longrightarrow \mathbf{\Omega}^p \xrightarrow{i} \mathcal{E}^{p,0} \xrightarrow{\bar\partial} \mathcal{E}^{p,1} \xrightarrow{\bar\partial} \cdots \longrightarrow \mathcal{E}^{p,n} \longrightarrow 0,$$

where $\mathbf{\Omega}^p$ is defined as the kernel sheaf of the mapping $\mathcal{E}^{p,0} \xrightarrow{\bar\partial} \mathcal{E}^{p,1}$, which is the sheaf of *holomorphic differential forms of type* $(p, 0)$ (and we usually say *holomorphic forms of degree* p); i.e., in local coordinates, $\varphi \in \mathbf{\Omega}^p(U)$ if and only if

$$\varphi = \sum_{|I|=p}{}' \varphi_I dz^I, \quad \varphi_I \in \mathcal{O}(U),$$

and we note that $\mathbf{\Omega}^0 = \mathcal{O}(= \mathcal{O}_X)$. Then for each p we have a differential sheaf

$$0 \longrightarrow \mathbf{\Omega}^p \longrightarrow \mathcal{E}^{p,*}, \tag{2.6}$$

since $\bar\partial^2 = 0$, which is, in fact, a resolution of the sheaf $\mathbf{\Omega}^p$, by virtue of the Grothendieck version of the Poincaré lemma for the $\bar\partial$-operator. Namely, if ω is a (p, q)-form defined in a polydisc $\mathbf{\Delta}$ in $\mathbf{C}^n$, $\mathbf{\Delta} = \{z : |z_i| < r, i = 1,$

$\ldots, n\}$, and $\bar{\partial}\omega = 0$ in $\mathbf{\Delta}$, then there exists a $(p, q-1)$-form u defined in a slightly smaller polydisc $\mathbf{\Delta}' \subset\subset \mathbf{\Delta}$, so that $\bar{\partial}u = \omega$ in $\mathbf{\Delta}'$. See Gunning and Rossi [1], p. 27, for an elementary proof of this result using induction (as in one of the classical proofs of the Poincaré lemma) and the general Cauchy integral formula in the complex plane.†

Example 2.13: Let X be a complex manifold and consider the differential sheaf over X,

$$0 \longrightarrow \mathbf{C} \longrightarrow \mathbf{\Omega}^0 \xrightarrow{\partial} \mathbf{\Omega}^1 \longrightarrow \cdots \xrightarrow{\partial} \mathbf{\Omega}^n \longrightarrow 0,$$

where the $\mathbf{\Omega}^p$ are defined in Example 2.12. Then we claim that this is a resolution of the constant sheaf $\mathbf{C}$. First we note that $\partial = d$, when acting on holomorphic forms of degree p, since $d = \partial + \bar{\partial}$, and $\bar{\partial}(\mathbf{\Omega}^p) = 0$ for $p = 0, \ldots, u$; then exactness at $\mathbf{\Omega}^0$ is immediate. Moreover, one can locally solve the equation $\partial u = \omega$ for u if $\partial\omega = 0$ by the same type of proof as for the operator $\bar{\partial}$ indicated in Example 2.12.

Suppose that $\mathcal{L}^*$ and $\mathfrak{M}^*$ are differential sheaves. Then a *homomorphism* $f : \mathcal{L}^* \to \mathfrak{M}^*$ is a sequence of homomorphism $f_j : \mathcal{L}^j \to \mathfrak{M}^j$ which commutes with the differentials of $\mathcal{L}^*$ and $\mathfrak{M}^*$. Similarly, a *homomorphism of resolutions* of sheaves

$$\begin{array}{ccccc} 0 & \longrightarrow & \mathcal{A} & \longrightarrow & \mathcal{A}^* \\ & & \downarrow & & \downarrow \\ 0 & \longrightarrow & \mathcal{B} & \longrightarrow & \mathcal{B}^* \end{array}$$

is a homomorphism of the underlying differential sheaves.

Example 2.14: Let X be a differentiable manifold and let

$$0 \longrightarrow \mathbf{R} \longrightarrow \mathcal{E}^*$$

$$0 \longrightarrow \mathbf{R} \longrightarrow \mathcal{S}^*_\infty(\mathbf{R})$$

be the resolutions of $\mathbf{R}$ given by Examples 2.10 and 2.11, respectively. Then there is a natural homomorphism of differential sheaves

$$I : \mathcal{E}^* \longrightarrow \mathcal{S}^*_\infty(\mathbf{R})$$

which induces a homomorphism of resolutions in the following manner:

$$\begin{array}{ccc} & & \mathcal{E}^* \\ & \overset{i}{\nearrow} & \\ 0 \longrightarrow \mathbf{R} & & \downarrow I \\ & \underset{i}{\searrow} & \\ & & \mathcal{S}^*_\infty(\mathbf{R}). \end{array}$$

The homomorphism I is given by integration over chains; i.e.,

$$I_U : \mathcal{E}^*(U) \longrightarrow \mathcal{S}^*_\infty(U, \mathbf{R})$$

†The same result holds for $\partial : \mathcal{E}^{p,q} \longrightarrow \mathcal{E}^{p+1,q}$, as one can easily see by conjugation.

is given by

$$I_U(\varphi)(c) = \int_c \varphi,$$

where c is a C^∞ chain (with real coefficients, in this case), and then $I_U(\varphi) \in \mathcal{S}^*_\infty(U, \mathbf{R})$. Moreover, by Stokes' theorem it follows that the mapping I commutes with the differentials.

We shall see in the next section how resolutions can be used to represent the cohomology groups of a space. In particular, we shall see that every sheaf admits a canonical abstract resolution with certain nice (cohomological) properties, and we shall then compare this abstract resolution with our more concrete examples of this section.

At this point we mention an analogue of the classical Poincaré lemma mentioned above, for which we shall have an application later on.

Lemma 2.15: Let $\varphi \in \mathcal{E}^{p,q}(U)$ for U open in $\mathbf{C}^n$ and suppose that $d\varphi = 0$. Then for any point $p \in U$ there is a neighborhood N of p and a differential form $\psi \in \mathcal{E}^{p-1,q-1}(N)$ such that

$$\partial\bar\partial\psi = \varphi \quad \text{in} \quad N.$$

Proof: The proof consists of an application of the Poincaré lemmas for the operators d, ∂, and $\bar\partial$ (see Examples 2.10 and 2.12). Namely, since $d\varphi = 0$, we have that there is a $u \in \mathcal{E}_x^{r-1}$ (using germs at x), so that $du = \varphi$, where $r = p + q$ is the total degree of φ. Thus we see that if we write $u = u^{r-1,0} + \cdots + u^{0,r-1}$, we have

$$du = \bar\partial u^{p,q-1} + \partial u^{p-1,q}$$

$$\bar\partial u^{p-1,q} = \partial u^{p,q-1} = 0,$$

and then there exists (by the $\bar\partial$ and ∂ Poincaré lemmas, Example 2.12) forms $\psi_1 \in \mathcal{E}_x^{p-1,q-1}$ and $\psi_2 \in \mathcal{E}_x^{p-1,q-1}$ so that

$$\partial\psi_1 = u^{p,q-1}$$

$$\bar\partial\psi_2 = u^{p-1,q}$$

which implies that

$$\begin{aligned}\varphi = du &= \bar\partial\partial\psi_1 + \partial\bar\partial\psi_2 \\ &= \partial\bar\partial(\psi_2 - \psi_1).\end{aligned}$$

Q.E.D.

Remark: Let $\mathcal{H} = \operatorname{Ker} \partial\bar\partial : \mathcal{E}^{0,0} \longrightarrow \mathcal{E}^{1,1}$ on a complex manifold X. Then there is a fine resolution (see Definition 3.3)

$$0 \longrightarrow \mathcal{H} \longrightarrow \mathcal{E}^{0,0} \xrightarrow{\partial\bar\partial} \mathcal{E}^{1,1} \xrightarrow{d} \mathcal{E}^{2,1} \oplus \mathcal{E}^{1,2} \longrightarrow \cdots,$$

where $\mathcal{H}$ is the sheaf of *pluriharmonic* functions, Lemma 2.15 showing exactness at the $\mathcal{E}^{1,1}$ term (see Bigolin [1]). This is analogous to the resolution of $\mathcal{O}$ by $\mathcal{E}^{0,*}$ and has a similar usefulness.

3. Cohomology Theory

In this section we want to present a brief development of sheaf cohomology theory. We first consider the problem of "lifting" global sections of sheaves. Consider a short exact sequence of sheaves:

$$0 \longrightarrow \mathcal{A} \longrightarrow \mathcal{B} \longrightarrow \mathcal{C} \longrightarrow 0. \tag{3.1}$$

Then it is easy to verify that the induced sequence

$$0 \longrightarrow \mathcal{A}(X) \longrightarrow \mathcal{B}(X) \longrightarrow \mathcal{C}(X) \longrightarrow 0 \tag{3.2}$$

is exact at $\mathcal{A}(X)$ and $\mathcal{B}(X)$ but not necessarily at $\mathcal{C}(X)$. For instance, in Example 2.6, if we let $X = \mathbf{C} - \{0\}$, the punctured plane, then we see that the mapping $\mathcal{O}(X) \longrightarrow \mathcal{O}^*(X)$ is not surjective. Similarly, in Example 2.9, a section of $\mathbf{Z}$ over X has the *same value* at both points a and b, whereas a section of $\mathbf{Z}/\mathcal{J}$ over X may have different values at points a and b and must be zero elsewhere, and thus the map $\mathbf{\Gamma}(X, \mathbf{Z}) \to \mathbf{\Gamma}(X, \mathbf{Z}/\mathcal{J})$ is not surjective.

Cohomology gives a measure to the amount of inexactness of the sequence (3.2) at $\mathcal{C}(X)$. We need to introduce a class of sheaves for which this lifting problem is always solvable, and cohomology will be defined in terms of such sheaves by means of resolutions. Let $\mathcal{F}$ be a sheaf over a space X and let S be a closed subset of X. Let

$$\mathcal{F}(S) := \varinjlim_{U \supset S} \mathcal{F}(U),$$

where the direct limit runs over all open sets U containing S. From the point of view of étalé spaces $\mathcal{F}(S)$ can be identified with the set of (continuous) sections of $\tilde{\mathcal{F}}|_S$, where $\tilde{\mathcal{F}}|_S := \pi^{-1}(S)$, and $\pi : \tilde{\mathcal{F}} \to X$ is the étalé map. We call $\mathcal{F}(S)$ the set (or abelian group) of sections of $\mathcal{F}$ over S, and we shall often denote $\mathcal{F}(S)$ by $\mathbf{\Gamma}(S, \mathcal{F})$. Moreover, we shall assume from now on for simplicity that we are dealing with *sheaves of abelian groups over a paracompact Hausdorff space* X, this being perfectly adequate for the applications in this book.

Definition 3.1: A sheaf $\mathcal{F}$ over a space X is *soft* if for any closed subset $S \subset X$ the restriction mapping

$$\mathcal{F}(X) \longrightarrow \mathcal{F}(S)$$

is surjective; i.e., any section of $\mathcal{F}$ over S can be extended to a section of $\mathcal{F}$ over X.

There are no obstructions to lifting global sections for soft sheaves, as we see in the following theorem.

Theorem 3.2: If $\mathcal{A}$ is a soft sheaf and

$$0 \longrightarrow \mathcal{A} \xrightarrow{g} \mathcal{B} \xrightarrow{h} \mathcal{C} \longrightarrow 0$$

is a short exact sequence of sheaves, then the induced sequence

$$0 \longrightarrow \mathcal{A}(X) \xrightarrow{g_X} \mathcal{B}(X) \xrightarrow{h_X} \mathcal{C}(X) \longrightarrow 0 \tag{3.3}$$

is exact.

Proof: Let $c \in \mathcal{C}(X)$. Then we want to show that there exists a section $b \in \mathcal{B}(X)$ such that $h_X(b) = c$. Since the sequence of sheaves is exact, it follows that for each $x \in X$ there exists a neighborhood U of x and a $b \in \mathcal{B}(U)$ such that $h_U(b) = c|_U$, where $c|_U$ denotes the presheaf restriction from X to U. Thus we can cover X with a family U_i of open sets such that there exists $b_i \in \mathcal{B}(U_i)$ satisfying $h(b_i) = c|_{U_i}$ (dropping the subscript notation for g and h). The object now is to show that the b_i can be pieced together to form a global section.

Since X is paracompact, there exists a locally finite refinement $\{S_i\}$ of $\{U_i\}$ which is still a covering of X and such that the elements S_i of the cover are closed sets. Consider the set of all pairs (b, S), where S is a union of sets in $\{S_i\}$ and $b \in \mathcal{B}(S)$ satisfies $h(b) = c|_S$. The set of all such pairs is partially ordered by $(b, S) \leq (b', S')$ if $S \subset S'$ and $b'|_S = b$. It follows easily from Axiom S_2 in Definition 1.2 that every linearly ordered chain has a maximal element. Thus, by Zorn's lemma there exists a maximal set S and a section $b \in \mathcal{B}(S)$ such that $h(b) = c|_S$. It suffices now to show that $S = X$.

Suppose the contrary. Then there is a set $S_j \in \{S_i\}$ such that $S_j \not\subset S$. Moreover, $h(b - b_j) = c - c = 0$ on $S \cap S_j$. Therefore, by exactness of (3.3) at $\mathcal{B}(X)$ we see that there exists a section $a \in \mathcal{A}(S \cap S_j)$ such that $g(a) = b - b_j$. Since $\mathcal{A}$ is soft, we can extend a to all of X, and using the same notation for the extension, we now define $\tilde{b} \in \mathcal{B}(S \cup S_j)$ by setting

$$\tilde{b} = \begin{cases} b & \text{on } S \\ b_j + g(a) & \text{on } S_j. \end{cases}$$

If follows that $h(b) = c|_{S \cup S_j}$, and hence S is not maximal. This contradiction then proves the theorem.

Q.E.D.

Before continuing with the consequences of Theorem 3.2, we would like to introduce another class of sheaves, which will give us many examples of soft sheaves.

Definition 3.3: A sheaf of abelian groups $\mathcal{F}$ over a paracompact Hausdorff space X is *fine* if for any locally finite open cover $\{U_i\}$ of X there exists a family of sheaf morphisms

$$\{\eta_i : \mathcal{F} \longrightarrow \mathcal{F}\}$$

such that

(a) $\Sigma \eta_i = 1$.

(b) $\eta_i(\mathcal{F}_x) = 0$ for all x in some neighborhood of the complement of U_i.

The family $\{\eta_i\}$ is called a *partition of unity* of $\mathcal{F}$ subordinate to the covering $\{U_i\}$.

Example 3.4: The following sheaves are fine sheaves:

(a) $\mathcal{C}_X$, for X a paracompact Hausdorff space.
(b) $\mathcal{E}_X$, for X a paracompact differentiable manifold.
(c) $\mathcal{E}_X^{p,q}$, for X a paracompact almost-complex manifold.
(d) A locally free sheaf of $\mathcal{E}_X$-modules, where X is a differentiable manifold.
(e) If $\mathcal{R}$ is a fine sheaf of rings with unit, then any module over $\mathcal{R}$ is a fine sheaf.

The first four examples are fine sheaves because multiplication by a continuous or differentiable globally defined function defines a sheaf homomorphism in a natural way. Hence the usual topological and C^∞ partitions of unity define the required sheaf partitions of unity.

Proposition 3.5: Fine sheaves are soft.

Proof: Let $\mathcal{F}$ be a fine sheaf over X and let S be a closed subset of X. Suppose that $s \in \mathcal{F}(S)$. Then there is a covering of S by open sets $\{U_i\}$ in X, and there are sections $s_i \in \mathcal{F}(U_i)$ such that

$$S_i|_{S\cap U_i} = S|_{S\cap U_i}.$$

Let $U_0 = X - S$ and $s_0 = 0$, so that $\{U_i\}$ extends to an open covering of all of X. Since X is paracompact, we may assume that $\{U_i\}$ is locally finite and hence that there is a sheaf partition of unity $\{\eta_i\}$ subordinate to $\{U_i\}$. Now $\eta_i(s_i)$ is a section on U_i which is identically zero in a neighborhood of the boundary of U_i, so it may be extended to a section on all of X. Thus we can define

$$\tilde{s} = \sum_i \eta_i(s_i)$$

in order to obtain the required extension of s.

Q.E.D.

Example 3.6: Let X be the complex plane and let $\mathcal{O} = \mathcal{O}_X$ be the sheaf of holomorphic functions on X. It is easy to see that $\mathcal{O}$ is not soft and hence cannot be fine (which is also easy to see directly). Namely, let $S = \{|z| \leq \frac{1}{2}\}$, and consider a holomorphic function f defined in the unit disc with the unit circle as natural boundary [e.g., $f(z) = \Sigma z^{n!}$]. Then f defines an element of $\mathcal{O}(S)$ which cannot be extended to all of X, and hence $\mathcal{O}$ is not soft.

Example 3.7: Constant sheaves are neither fine nor soft. Namely, if G is a constant sheaf over X and a and b are two distinct points, then let $s \in G(\{a\} \cup \{b\})$ be defined by setting $s(a) = 0$ and $s(b) \neq 0$. Then it is clear that s cannot be extended to a global section of G over X.

Now that we have some familiar examples of soft sheaves, we return to some consequences of Theorem 3.2.

Corollary 3.8: If $\mathcal{A}$ and $\mathcal{B}$ are soft and

$$0 \longrightarrow \mathcal{A} \longrightarrow \mathcal{B} \longrightarrow \mathcal{C} \longrightarrow 0$$

is exact, then $\mathcal{C}$ is soft.

Proof: Let S be a closed subset of X and restrict the sequence above to the set S. Then Theorem 3.2 applies, and the given section of $\mathcal{C}$ over S to be extended to all of X comes from a section of $\mathcal{B}$ over S, which by softness then extends to all of X. Its image in $\mathcal{C}(X)$ is a suitable extension.

Q.E.D.

Corollary 3.9: If

$$0 \longrightarrow \mathcal{S}_0 \longrightarrow \mathcal{S}_1 \longrightarrow \mathcal{S}_2 \longrightarrow \cdots$$

is an exact sequence of soft sheaves, then the induced section sequence

$$0 \longrightarrow \mathcal{S}_0(X) \longrightarrow \mathcal{S}_1(X) \longrightarrow \mathcal{S}_2(X) \longrightarrow \cdots$$

is also exact.

Proof: Let $\mathcal{K}_i = \operatorname{Ker}(\mathcal{S}_i \to \mathcal{S}_{i+1})$. Then we have short exact sequences

$$0 \longrightarrow \mathcal{K}_i \longrightarrow \mathcal{S}_i \longrightarrow \mathcal{K}_{i+1} \longrightarrow 0.$$

For $i = 0$, $\mathcal{K}_1 = \mathcal{S}_0$, and $\mathcal{S}_0$ is soft. Thus we have the induced short exact sequence

$$0 \longrightarrow \mathcal{K}_1(X) \longrightarrow \mathcal{S}_1(X) \longrightarrow \mathcal{K}_2(X) \longrightarrow 0$$

by Theorem 3.2. An induction using Corollary 3.8 shows that $\mathcal{K}_i$ is soft for all i, and so we obtain short exact sequences:

$$0 \longrightarrow \mathcal{K}_i(X) \longrightarrow \mathcal{S}_i(X) \longrightarrow \mathcal{K}_{i+1}(X) \longrightarrow 0.$$

Splicing these sequences gives the desired result.

Q.E.D.

We are now in a position to construct a canonical soft resolution for any sheaf over a topological space X. Let $\mathcal{S}$ be the given sheaf and let $\tilde{\mathcal{S}} \xrightarrow{\pi} X$ be the étalé space associated to $\mathcal{S}$. Let $\mathcal{C}^0(\mathcal{S})$ be the presheaf defined by

$$\mathcal{C}^0(\mathcal{S})(U) = \{f : U \longrightarrow \tilde{\mathcal{S}} : \pi \circ f = 1_U\}.$$

This presheaf is a sheaf and is called the *sheaf of discontinuous sections of* $\mathcal{S}$ *over* X.† There is clearly a natural injection

$$0 \longrightarrow \mathcal{S} \longrightarrow \mathcal{C}^0(\mathcal{S}).$$

†Recall that *sections* were defined to be continuous in Definition 2.1, so *discontinuous section* is a generalization of the concept of section.

Now let $\mathcal{F}^1(\mathcal{S}) = \mathcal{C}^0(\mathcal{S})/\mathcal{S}$ and define $\mathcal{C}^1(\mathcal{S}) = \mathcal{C}^0(\mathcal{F}^1(\mathcal{S}))$. By induction we define

$$\mathcal{F}^i(\mathcal{S}) = \mathcal{C}^{i-1}(\mathcal{S})/\mathcal{F}^{i-1}(\mathcal{S})$$

and

$$\mathcal{C}^i(\mathcal{S}) = \mathcal{C}^0(\mathcal{F}^i(\mathcal{S})).$$

We then have the following short exact sequences of sheaves:

$$0 \longrightarrow \mathcal{S} \longrightarrow \mathcal{C}^0(\mathcal{S}) \longrightarrow \mathcal{F}^1(\mathcal{S}) \longrightarrow 0$$

$$0 \longrightarrow \mathcal{F}^i(\mathcal{S}) \longrightarrow \mathcal{C}^i(\mathcal{S}) \longrightarrow \mathcal{F}^{i+1}(\mathcal{S}) \longrightarrow 0.$$

By splicing these two short exact sequences together, we obtain the long exact sequence

$$0 \longrightarrow \mathcal{S} \longrightarrow \mathcal{C}^0(\mathcal{S}) \longrightarrow \mathcal{C}^1(\mathcal{S}) \longrightarrow \mathcal{C}^2(\mathcal{S}) \longrightarrow \cdots,$$

which we call the *canonical resolution of* $\mathcal{S}$. We abbreviate this by writing

$$0 \longrightarrow \mathcal{S} \longrightarrow \mathcal{C}^*(\mathcal{S}). \tag{3.4}$$

The sheaf of discontinuous sections $\mathcal{C}^0(\mathcal{S})$ is a soft sheaf, for any sheaf $\mathcal{S}$, and for this reason we call the resolution (3.4) the canonical soft resolution of $\mathcal{S}$.

Remark: A sheaf $\mathcal{S}$ is called *flabby* if $\mathcal{S}(X) \to \mathcal{S}(U)$ is surjective for all open sets U in X. It can be shown that a flabby sheaf is soft (see Godement [1]). To avoid the restriction of paracompactness in the above arguments, one must deal with flabby sheaves rather than soft sheaves. However, we note that most of our examples of soft sheaves are *not* flabby.

We are now in a position to give a definition of the cohomology groups of a space with coefficients in a given sheaf. Suppose that $\mathcal{S}$ is a sheaf over a space X and consider the canonical soft resolution given by (3.4). By taking global sections, (3.4) induces a sequence of the form

$$\begin{aligned} 0 \longrightarrow \Gamma(X, \mathcal{S}) \longrightarrow \Gamma(X, \mathcal{C}^0(\mathcal{S})) \longrightarrow \Gamma(X, \mathcal{C}^1(\mathcal{S})) \\ \longrightarrow \cdots \longrightarrow \Gamma(X, \mathcal{C}^q(\mathcal{S})) \longrightarrow, \end{aligned} \tag{3.5}$$

and this sequence of abelian groups forms a cochain complex.† This sequence is exact at $\Gamma(X, \mathcal{C}^0(\mathcal{S}))$, and if $\mathcal{S}$ is soft, it is exact everywhere by Corollary 3.9. Let

$$C^*(X, \mathcal{S}) := \Gamma(X, \mathcal{C}^*(\mathcal{S})),$$

and we rewrite (3.5) in the form

$$0 \longrightarrow \Gamma(X, \mathcal{S}) \longrightarrow C^*(X, \mathcal{S}).$$

†A cochain complex means that the composition of successive maps in the sequence is zero, but the sequence is not necessarily exact. We shall assume some elementary homological algebra, and we refer to, e.g., MacLane [1], Chap. I.

Definition 3.10: Let $\mathcal{S}$ be a sheaf over a space X and let

$$H^q(X, \mathcal{S}) := H^q(C^*(X, \mathcal{S})),$$

where $H^q(C^*(X, \mathcal{S}))$ is the qth derived group of the cochain complex $C^*(X, \mathcal{S})$; i.e.,

$$H^q(C^*) = \frac{\operatorname{Ker}(C^q \longrightarrow C^{q+1})}{\operatorname{Im}(C^{q-1} \longrightarrow C^q)}, \quad \text{where } C^{-1} = 0.$$

The abelian groups $H^q(X, \mathcal{S})$ are defined for $q \geq 0$ and are called the *sheaf cohomology groups of the space X of degree q and with coefficients in* $\mathcal{S}$.

As we shall see later, there are various ways of representing more explicitly such cohomology groups in a given geometric situation. This abstract definition is a convenient way to derive the general functorial properties of cohomology groups, as we shall see in the next theorem.

Theorem 3.11: Let X be a paracompact Hausdorff space. Then

(a) For any sheaf $\mathcal{S}$ over X,
  (1) $H^0(X, \mathcal{S}) = \Gamma(X, \mathcal{S})$ $(= \mathcal{S}(X))$.
  (2) If $\mathcal{S}$ is soft, then $H^q(X, \mathcal{S}) = 0$ for $q > 0$.
(b) For any sheaf morphism

$$h : \mathcal{A} \longrightarrow \mathcal{B}$$

there is, for each $q \geq 0$, a group homomorphism

$$h_q : H^q(X, \mathcal{A}) \longrightarrow H^q(X, \mathcal{B})$$

such that
  (1) $h_0 = h_X : \mathcal{A}(X) \to \mathcal{B}(X)$.
  (2) h_q is the identity map if h is the identity map, $q \geq 0$.
  (3) $g_q \circ h_q = (g \circ h)_q$ for all $q \geq 0$, if g: $\mathscr{B} \to \mathscr{C}$ is a second sheaf morphism.
(c) For each short exact sequence of sheaves

$$0 \longrightarrow \mathcal{A} \longrightarrow \mathcal{B} \longrightarrow \mathcal{C} \longrightarrow 0$$

there is a group homomorphism

$$\delta^q : H^q(X, \mathcal{C}) \longrightarrow H^{q+1}(X, \mathcal{A})$$

for all $q \geq 0$ such that
  (1) The induced sequence

$$0 \longrightarrow H^0(X, \mathcal{A} \longrightarrow H^0(X, \mathcal{B}) \longrightarrow H^0(X, \mathcal{C}) \xrightarrow{\delta^0} H^1(X, \mathcal{A}) \longrightarrow \cdots$$

$$\longrightarrow H^q(X, \mathcal{A}) \longrightarrow H^q(X, \mathcal{B}) \longrightarrow H^q(X, \mathcal{C}) \xrightarrow{\delta^q} H^{q+1}(X, \mathcal{A}) \longrightarrow$$

is exact.
  (2) A commutative diagram

$$\begin{array}{ccccccccc} 0 & \longrightarrow & \mathcal{A} & \longrightarrow & \mathcal{B} & \longrightarrow & \mathcal{C} & \longrightarrow & 0 \\ & & \downarrow & & \downarrow & & \downarrow & & \\ 0 & \longrightarrow & \mathcal{A}' & \longrightarrow & \mathcal{B}' & \longrightarrow & \mathcal{C}' & \longrightarrow & 0 \end{array}$$

induces a commutative diagram

$$\begin{array}{ccccccccc} 0 \longrightarrow & H^0(X,\mathcal{A}) & \longrightarrow & H^0(X,\mathcal{B}) & \longrightarrow & H^0(X,\mathcal{C}) & \longrightarrow & H^1(X,\mathcal{A}) & \longrightarrow \cdots \\ & \downarrow & & \downarrow & & \downarrow & & \downarrow & \\ 0 \longrightarrow & H^0(X,\mathcal{A}') & \longrightarrow & H^0(X,\mathcal{B}') & \longrightarrow & H^0(X,\mathcal{C}') & \longrightarrow & H^1(X,\mathcal{A}') & \longrightarrow \cdots . \end{array}$$

Proof:

(a), (1) We have that the resolution

$$0 \longrightarrow \Gamma(X, \mathcal{S}) \longrightarrow C^0(X, \mathcal{S}) \longrightarrow C^1(X, \mathcal{S}) \longrightarrow \cdots$$

is exact at $C^0(X, \mathcal{S})$, and so

$$\Gamma(X, \mathcal{S}) = \operatorname{Ker}(C^0(X, \mathcal{S}) \longrightarrow C^1(X, \mathcal{S})) = H^0(X, \mathcal{S}).$$

(a), (2) This follows easily from Corollary 3.9.

For the proof of (b) and (c) we shall show first that

$$h : \mathcal{A} \longrightarrow \mathcal{B}$$

induces naturally a cochain complex map†

$$h^* : C^*(\mathcal{A}) \longrightarrow C^*(\mathcal{B}). \tag{3.6}$$

First we define

$$h^0 : \mathcal{C}^0(\mathcal{A}) \longrightarrow \mathcal{C}^0(\mathcal{B})$$

by letting $h^0(s_x) = (h \circ s)_x$, where s is a discontinuous section of $\mathcal{A}$. Now h^0 induces a quotient map

$$\begin{array}{ccc} \tilde{h}^0 : \mathcal{C}^0(\mathcal{A})/\mathcal{A} & \longrightarrow & \mathcal{C}^0(\mathcal{B})/\mathcal{B} \\ \| & & \| \\ \mathcal{F}^1(\mathcal{A}) & & \mathcal{F}^1(\mathcal{B}), \end{array}$$

and, as above, $\tilde{h}^0$ induces

$$\begin{array}{ccc} \tilde{h}^1 : \mathcal{C}^0(\mathcal{F}^1(\mathcal{A})) & \longrightarrow & \mathcal{C}^0(\mathcal{F}^1(\mathcal{B})) \\ \| & & \| \\ \mathcal{C}^1(\mathcal{A}) & & \mathcal{C}^1(\mathcal{B}). \end{array}$$

Repeating the above procedure, we obtain, for each $q \geq 0$,

$$h^q : \mathcal{C}^q(\mathcal{A}) \longrightarrow \mathcal{C}^q(\mathcal{B}).$$

The induced section maps give the required complex map (3.6). It is clear that h^* is functorial [i.e., satisfies compatibility conditions similar to those in (b), (1)–(3)]. Moreover, if

$$0 \longrightarrow \mathcal{A} \longrightarrow \mathcal{B} \longrightarrow \mathcal{C} \longrightarrow 0$$

is exact, then this implies that

$$0 \longrightarrow \mathcal{C}^*(\mathcal{A}) \longrightarrow \mathcal{C}^*(\mathcal{B}) \longrightarrow \mathcal{C}^*(\mathcal{C}) \longrightarrow 0$$

†Letting $C^*(\mathcal{A}) = C^*(X, \mathcal{A})$, etc.

is an exact sequence of complexes of sheaves. However, the sheaves in these complexes are all soft, and hence it follows that

$$0 \longrightarrow C^*(\mathcal{A}) \longrightarrow C^*(\mathcal{B}) \longrightarrow C^*(\mathcal{C}) \longrightarrow 0$$

is an exact sequence of cochain complexes of abelian groups. It now follows from elementary homological algebra that there is a long exact sequence for the derived cohomology groups

$$\begin{aligned} &\longrightarrow H^q(C^*(\mathcal{A})) \longrightarrow H^q(C^*(\mathcal{B})) \longrightarrow H^q(C^*(\mathcal{C})) \\ &\xrightarrow{\delta^q} H^{q+1}(C^*(\mathcal{A})) \longrightarrow, \end{aligned} \tag{3.7}$$

where the mapping δ^q is defined in the following manner. Consider the following commutative diagram of exact sequences:

$$\begin{array}{ccccccccc} 0 & \longrightarrow & C^{q+1}(\mathcal{A}) & \xrightarrow{\mu'} & C^{q+1}(\mathcal{B}) & \xrightarrow{\nu'} & C^{q+1}(\mathcal{C}) & \longrightarrow & 0 \\ & & \uparrow \alpha & & \uparrow \beta & & \uparrow \gamma & & \\ 0 & \longrightarrow & C^q(\mathcal{A}) & \xrightarrow{\mu} & C^q(\mathcal{B}) & \xrightarrow{\nu} & C^q(\mathcal{C}) & \longrightarrow & 0. \end{array}$$

Suppose that $c \in \operatorname{Ker} \gamma$. Then by exactness, $c = \nu(b)$. Consider the element $\beta(b)$. Then $\nu'(\beta(b)) = \gamma(\nu(b))$, by commutativity, and hence $\beta(b) = \mu'(a)$ for some $a \in C^{q+1}(\mathcal{A})$. It is easy to check that (1) a is a closed element of $C^{q+1}(\mathcal{A})$, (2) the cohomology class of a in $H^{q+1}(\mathcal{A})$ is independent of the various choices made, and (3) the induced mapping $\delta^q : H^q(X, \mathcal{C}) \to H^{q+1}(X, \mathcal{A})$ makes the sequence (3.7) exact (the operator δ^q is often called the Bockstein operator). From these constructions it is not difficult to verify the assertions in (b) and (c).

Q.E.D.

Remark: The assertions (a), (b), and (c) in the above theorem can be used as axioms for cohomology theory, and one can prove existence and uniqueness for such an axiomatic theory. What we have in the theorem is the existence proof; see, e.g., Gunning and Rossi [1] for the additional uniqueness. There are other existence proofs; e.g., Čech theory is a popular one (cf. Hirzebruch [1]). In Sec. 4 we shall give a short summary of Čech theory.

We now want to give the proof of an important theorem which will give us a means of computing the abstract sheaf cohomology in given geometric situations. First we have the following definition.

Definition 3.12: A resolution of a sheaf $\mathcal{S}$ over a space X

$$0 \longrightarrow \mathcal{S} \longrightarrow \mathcal{A}^*$$

is called *acyclic* if $H^q(X, \mathcal{A}^p) = 0$ for all $q > 0$ and $p \geq 0$.

Note that a fine or soft resolution of a sheaf is necessarily acyclic (Theorem 3.11). Acyclic resolutions of sheaves give us one way of computing the cohomology groups of a sheaf, because of the following theorem (sometimes called the *abstract de Rham theorem*).

Theorem 3.13: Let $\mathcal{S}$ be a sheaf over a space X and let

$$0 \longrightarrow \mathcal{S} \longrightarrow \mathcal{A}^*$$

be a resolution of $\mathcal{S}$. Then there is a natural homomorphism

$$\gamma^p : H^p(\boldsymbol{\Gamma}(X, \mathcal{A}^*)) \longrightarrow H^p(X, \mathcal{S}),$$

where $H^p(\boldsymbol{\Gamma}(X, \mathcal{A}^*))$ is the pth derived group of the cochain complex $\boldsymbol{\Gamma}(X, \mathcal{A}^*)$. Moreover, if

$$0 \longrightarrow \mathcal{S} \longrightarrow \mathcal{A}^*$$

is acyclic, γ^p is an isomorphism.

Proof: Let $\mathcal{K}^p = \mathrm{Ker}(\mathcal{A}^p \to \mathcal{A}^{p+1}) = \mathrm{Im}(\mathcal{A}^{p-1} \to \mathcal{A}^p)$ so that $\mathcal{K}^\circ = \mathcal{S}$. We have short exact sequences

$$0 \longrightarrow \mathcal{K}^{p-1} \longrightarrow \mathcal{A}^{p-1} \longrightarrow \mathcal{K}^p \longrightarrow 0,$$

and this induces, by Theorem 3.11,

$$0 \longrightarrow \boldsymbol{\Gamma}(X, \mathcal{K}^{p-1}) \longrightarrow \boldsymbol{\Gamma}(X, \mathcal{A}^{p-1}) \longrightarrow \boldsymbol{\Gamma}(X, \mathcal{K}^p) \longrightarrow H^1(X, \mathcal{K}^{p-1})$$
$$\longrightarrow H^1(X, \mathcal{A}^{p-1}) \longrightarrow \cdots .$$

We also notice that

$$\mathrm{Ker}(\boldsymbol{\Gamma}(X, \mathcal{A}^p) \longrightarrow \boldsymbol{\Gamma}(X, \mathcal{A}^{p+1})) \cong \boldsymbol{\Gamma}(X, \mathcal{K}^p),$$

so that

$$H^p(\boldsymbol{\Gamma}(X, \mathcal{A}^*)) \cong \boldsymbol{\Gamma}(X, \mathcal{K}^p)/\mathrm{Im}(\boldsymbol{\Gamma}(X, \mathcal{A}^{p-1}) \longrightarrow \boldsymbol{\Gamma}(X, \mathcal{K}^p)).$$

Therefore, using the exact sequence above, we have defined

$$\gamma_1^p : H^p(\boldsymbol{\Gamma}(X, \mathcal{A}^*)) \longrightarrow H^1(X, \mathcal{K}^{p-1}),$$

and γ_1^p is injective. Moreover, if the resolution is acyclic,

$$H^1(X, \mathcal{A}^{p-1}) = 0$$

and γ_1^p is an isomorphism.

We now consider the exact sequences of the form

$$0 \longrightarrow \mathcal{K}^{p-r} \longrightarrow \mathcal{A}^{p-r} \longrightarrow \mathcal{K}^{p-r+1} \longrightarrow 0$$

for $2 \le r \le p$, and we obtain from the induced long exact sequences

$$\gamma_r^p : H^{r-1}(X, \mathcal{K}^{p-r+1}) \longrightarrow H^r(X, \mathcal{K}^{p-r}),$$

and again γ_r^p is an isomorphism if the resolution is acyclic. Therefore we define

$$\gamma_p = \gamma_p^p \circ \gamma_{p-1}^p \cdot \cdots \cdot \gamma_2^p \circ \gamma_1^p,$$

i.e.,

$$H^p(\Gamma(X, \mathcal{A}^*)) \xrightarrow{\gamma_1{}^p} H^1(X, \mathcal{K}^{p-1}) \xrightarrow{\gamma_2{}^p} H^2(X, \mathcal{K}^{p-2}) \xrightarrow{\gamma_3{}^p}$$

$$\cdots \xrightarrow{\gamma_p{}^p} H^p(X, \mathcal{K}^\circ) = H^p(X, \mathcal{S}),$$

and γ^p is an isomorphism if the resolution is acyclic.

The assertion that γ^p is *natural* in Theorem 3.13 means that if

$$\begin{array}{ccccc} 0 & \longrightarrow & \mathcal{S} & \longrightarrow & \mathcal{A}^* \\ & & \downarrow f & & \downarrow g \\ 0 & \longrightarrow & \mathcal{J} & \longrightarrow & \mathcal{B}^* \end{array}$$

is a homomorphism of resolutions, then

$$\begin{array}{ccc} H^p(\mathbf{\Gamma}(X, \mathcal{A}^*)) & \xrightarrow{\gamma^p} & H^p(X, \mathcal{S}) \\ \downarrow g_p & & \downarrow f_p \\ H^p(\mathbf{\Gamma}(X, \mathcal{B}^*)) & \xrightarrow{\gamma^p} & H^p(X, \mathcal{J}) \end{array}$$

is also commutative, where g_p is the induced map on the cohomology of the complexes. This is not difficult to check and follows from the naturality assertions in Theorem 3.11.

Q.E.D.

Remark: Note that in the proof of the previous theorem we did not use the definition of sheaf cohomology, but only the formal properties of cohomology as given in Theorem 3.11 (i.e., the same result holds for any other definition of cohomology which satisfies the properties of Theorem 3.11).

Corollary 3.14: Suppose that

$$\begin{array}{ccccc} 0 & \longrightarrow & \mathcal{S} & \longrightarrow & \mathcal{A}^* \\ & & \downarrow f & & \downarrow g \\ 0 & \longrightarrow & \mathcal{J} & \longrightarrow & \mathcal{B}^* \end{array}$$

is a homomorphism of resolutions of sheaves. Then there is an induced homomorphism

$$H^p(\mathbf{\Gamma}(X, \mathcal{A}^*)) \xrightarrow{g_p} H^p(\mathbf{\Gamma}(X, \mathcal{B}^*)),$$

which is, moreover, an isomorphism if f is an isomorphism of sheaves and the resolutions are both acyclic.

As a consequence of this corollary, we easily obtain de Rham's theorem (see Example 2.14 for the notation).

Theorem 3.15 (de Rham): Let X be a differentiable manifold. Then the natural mapping

$$I : H^p(\mathcal{E}^*(X)) \longrightarrow H^p(\mathcal{S}^*_\infty(X, \mathbf{R}))$$

induced by integration of differential forms over $\mathbf{C}^\infty$ singular chains with real coefficients is an isomorphism.

Proof: As in Example 2.14, consider the resolutions of $\mathbf{R}$ given by

$$0 \longrightarrow \mathbf{R} \begin{array}{l} \overset{i}{\nearrow} \mathcal{E}^* \\ \underset{i}{\searrow} \mathcal{S}^*_\infty(\mathbf{R}) = \mathcal{S}^*_\infty. \end{array}$$

Then the sheaves $\mathcal{E}^*$ and $\mathcal{S}^*_\infty$ are both soft. Since $\mathcal{E}^*$ is fine, it remains only to show that the sheaves $\mathcal{S}^p_\infty$ are soft. First we note that the sheaf $\mathcal{S}^p_\infty$ is an $\mathcal{S}^0_\infty$-module (given by cup product on open sets). Then we claim that $\mathcal{S}^0_\infty$ is soft. This follows from the observation that $\mathcal{S}^0_\infty = \mathcal{S}^0 = \mathcal{C}^0(X, \mathbf{R})$; i.e., for each point of X (a singular 0-cochain), we assign a value of $\mathbf{R}$. We now need the following simple lemma, which asserts that $\mathcal{S}^p$ is soft, which concludes the proof, in view of Corollary 3.14.

Q.E.D.

Lemma 3.16: If $\mathfrak{M}$ is a sheaf of modules over a soft sheaf of rings $\mathcal{R}$, then $\mathfrak{M}$ is a soft sheaf.

Proof: Let $s \in \Gamma(K, \mathfrak{M})$ for K a closed subset of X. Then s extends to some open neighborhood U of K. Let $\rho \in \Gamma(K \cup (X - U), \mathcal{R})$ be defined by

$$\rho \equiv \begin{cases} 1 & \text{on } K \\ 0 & \text{on } X - U. \end{cases}$$

Then, since $\mathcal{R}$ is soft, ρ extends to a section over X, and $\rho \cdot s$ is the desired extension of s.

Q.E.D.

We now have an analogue of de Rham's theorem for complex manifolds, due to Dolbeault [1].

Theorem 3.17 (Dolbeault): Let X be a complex manifold. Then

$$H^q(X, \mathbf{\Omega}^p) \cong \frac{\operatorname{Ker}(\mathcal{E}^{p,q}(X) \xrightarrow{\bar\partial} \mathcal{E}^{p,q+1}(X))}{\operatorname{Im}(\mathcal{E}^{p,q-1}(X) \xrightarrow{\bar\partial} \mathcal{E}^{p,q}(X))}.$$

Proof: The resolution given in Example 2.12 is a fine resolution, and we can apply Theorem 3.13.

Q.E.D.

We want to consider a generalization of Theorem 3.17, and for this we need to introduce the tensor product of sheaves of modules.

Definition 3.18: Let $\mathfrak{M}$ and $\mathfrak{N}$ be sheaves of modules over a sheaf of commutative rings, $\mathcal{R}$. Then $\mathfrak{M} \otimes_{\mathcal{R}} \mathfrak{N}$, the *tensor product of* $\mathfrak{M}$ *and* $\mathfrak{N}$ is the sheaf generated by the presheaf

$$U \longrightarrow \mathfrak{M}(U) \otimes_{\mathcal{R}(U)} \mathfrak{N}(U).$$

Remark: The necessity of using the generated sheaf is demonstrated by considering the presheaf

$$U \longrightarrow \mathcal{O}(E)(U) \otimes_{\mathcal{O}(U)} \mathcal{E}(U),$$

where $E \to X$ is a holomorphic vector bundle with no non-trivial global holomorphic sections (see Example 2.13). The presheaf does not satisfy Axiom $\mathbf{S}_2$, since $\mathcal{O}(E)(X) \otimes_{\mathcal{O}(X)} \mathcal{E}(X) = 0$, but $\mathcal{O}(E)(U_j) \otimes_{\mathcal{O}(U_j)} \mathcal{E}(U_j) \cong \mathcal{E}(E)(U_j) \neq 0$ for the sets of any trivializing cover $\{U_j\}$ of X.

It follows from Definition 3.18 that

$$(\mathfrak{M} \otimes_{\mathcal{R}} \mathfrak{N})_x = \mathfrak{M}_x \otimes_{\mathcal{R}_x} \mathfrak{N}_x.$$

This easily implies the following lemma.

Lemma 3.19: If $\mathcal{I}$ is a locally free sheaf of $\mathcal{R}$-modules and

$$0 \longrightarrow \mathcal{A}' \longrightarrow \mathcal{A} \longrightarrow \mathcal{A}'' \longrightarrow 0$$

is a short exact sequence of $\mathcal{R}$-modules, then

$$0 \longrightarrow \mathcal{A}' \otimes_{\mathcal{R}} \mathcal{I} \longrightarrow \mathcal{A} \otimes_{\mathcal{R}} \mathcal{I} \longrightarrow \mathcal{A}'' \otimes_{\mathcal{R}} \mathcal{I} \longrightarrow 0$$

is also exact.

Recalling Example 2.12, we have a resolution of sheaves of $\mathcal{O}$-modules over a complex manifold X:

$$0 \longrightarrow \mathbf{\Omega}^p \longrightarrow \mathcal{E}^{p,0} \xrightarrow{\bar\partial} \mathcal{E}^{p,1} \xrightarrow{\bar\partial} \cdots \longrightarrow \mathcal{E}^{p,n} \longrightarrow 0.$$

Moreover, if E is a holomorphic vector bundle, then $\mathcal{O}(E)$ is a locally free sheaf, and so, using Lemma 3.19, we have the following resolution:

$$0 \longrightarrow \mathbf{\Omega}^p \otimes_{\mathcal{O}} \mathcal{O}(E) \longrightarrow \mathcal{E}^{p,0} \otimes_{\mathcal{O}} \mathcal{O}(E) \xrightarrow{\bar\partial\otimes 1} \cdots \xrightarrow{\bar\partial\otimes 1} \mathcal{E}^{p,n} \otimes_{\mathcal{O}} \mathcal{O}(E) \longrightarrow 0. \tag{3.8}$$

We also notice that

$$\mathbf{\Omega}^p \otimes_{\mathcal{O}} \mathcal{O}(E) \cong \mathcal{O}(\wedge^p T^*(X) \otimes_{\mathbb{C}} E)$$

and that

$$\begin{aligned}\mathcal{E}^{p,q} \otimes_{\mathcal{O}} \mathcal{O}(E) &\cong \mathcal{E}^{p,q} \otimes_{\mathcal{E}} \mathcal{E}(E)\\ &\cong \mathcal{E}(\wedge^{p,q} T^*(X) \otimes_{\mathbb{C}} E),\end{aligned}$$

where $\mathcal{E}(E)$ is the sheaf of differentiable sections of the differentiable bundle E. This follows from the fact that

$$\mathcal{O}(E) \otimes_{\mathcal{O}} \mathcal{E} = \mathcal{E}(E),$$

since $\mathcal{E}^{p,q}$ is also an $\mathcal{E}$-module.

We call $\mathcal{O}(X, \wedge^p T^*(X) \otimes_{\mathbf{C}} E)$ the (global) *holomorphic p-forms on X with coefficients* in E, which we shall denote for simplicity by $\mathbf{\Omega}^p(X, E)$, and we shall denote the sheaf of holomorphic p-forms with coefficients in E by $\mathbf{\Omega}^p(E)$. Analogously, we let

$$\mathcal{E}^{p,q}(X, E) := \mathcal{E}(X, \wedge^{p,q} T^*(X) \otimes_{\mathbf{C}} E) \tag{3.9}$$

be the differentiable (p, q)-forms on X with coefficients in E. Therefore the resolution (3.8) can be written in the form (letting $\bar{\partial}_E = \bar{\partial} \otimes 1$)

$$0 \longrightarrow \mathbf{\Omega}^p(E) \longrightarrow \mathcal{E}^{p,0}(E) \xrightarrow{\bar{\partial}_E} \mathcal{E}^{p,1}(E) \longrightarrow \cdots \xrightarrow{\bar{\partial}_E} \mathcal{E}^{p,n}(E) \longrightarrow 0, \tag{3.10}$$

and since it is a fine resolution, we have the following generalization of Dolbeault's theorem.

Theorem 3.20: Let X be a complex manifold and let $E \longrightarrow X$ be a holomorphic vector bundle. Then

$$H^q(X, \mathbf{\Omega}^p(E)) \cong \frac{\operatorname{Ker}(\mathcal{E}^{p,q}(X, E) \xrightarrow{\bar{\partial}_E} \mathcal{E}^{p,q+1}(X, E))}{\operatorname{Im}(\mathcal{E}^{p,q-1}(X, E) \xrightarrow{\bar{\partial}_E} \mathcal{E}^{p,q}(X, E))}.$$

4. Čech Cohomology with Coefficients in a Sheaf

Suppose that X is a topological space and that $\mathcal{S}$ is a sheaf of abelian groups on X. Let $\mathfrak{U} = \{U_\alpha\}$ be a covering of X by open sets. A *q-simplex*, σ, is an ordered collection of $q + 1$ sets of the covering $\mathfrak{U}$ with nonempty intersection; i.e.,

$$\sigma = (U_0, \ldots, U_q)$$

and $\bigcap_{i=0}^q U_i \neq \emptyset$. The set $\bigcap_{U_i \in \sigma} U_i$ is called the *support* of the simplex σ, denoted $|\sigma|$. A *q-cochain* of $\mathfrak{U}$ with coefficients in $\mathcal{S}$ is a mapping f which associates to each q-simplex, σ,

$$f(\sigma) \in \mathcal{S}(|\sigma|).$$

The set of q-cochains will be denoted by $C^q(\mathfrak{U}, \mathcal{S})$ and is an abelian group (by pointwise addition).

We define a *coboundary operator*

$$\delta : C^q(\mathfrak{U}, \mathcal{S}) \to C^{q+1}(\mathfrak{U}, \mathcal{S})$$

as follows. If $f \in C^q(\mathfrak{U}, \mathcal{J})$ and $\sigma = (U_0, \ldots, U_{q+1})$, define

$$\delta f(\sigma) = \sum_{i=0}^{q+1} (-1)^i \, r_{|\sigma|}^{|\sigma_i|} \, f(\sigma_i),$$

where $\sigma_i = (U_0, \ldots, U_{i-1}, U_{i+1}, \ldots U_{q+1})$ and $r_{|\sigma|}^{|\sigma_i|}$ is the sheaf restriction mapping. It is clear that δ is a group homomorphism and that $\delta^2 = 0$. Thus we have a cochain complex:

$$C^*(\mathfrak{U}, \mathcal{S}) := C^0(\mathfrak{U}, \mathcal{S}) \longrightarrow \cdots \longrightarrow C^q(\mathfrak{U}, \mathcal{S}) \xrightarrow{\delta} C^{q+1}(\mathfrak{U}, \mathcal{S})$$
$$\longrightarrow C^{q+2}(\mathfrak{U}, \mathcal{S}) \longrightarrow \cdots$$

The cohomology of this cochain complex is the *Čech cohomology of* $\mathfrak{U}$ *with coefficients in* $\mathcal{S}$; i.e., letting

$$Z^q(\mathfrak{U}, \mathcal{S}) = \text{Ker } \delta : C^q(\mathfrak{U}, \mathcal{S}) \longrightarrow C^{q+1}(\mathfrak{U}, \mathcal{S}),$$
$$B^q(\mathfrak{U}, \mathcal{S}) = \text{Im } \delta : C^{q-1}(\mathfrak{U}, \mathcal{S}) \longrightarrow C^q(\mathfrak{U}, \mathcal{S})$$

we define

$$H^q(\mathfrak{U}, \mathcal{S}) := H^q(C^*(\mathfrak{U}, \mathcal{S})) = Z^q(\mathfrak{U}, \mathcal{S})/B^q(\mathfrak{U}, \mathcal{S}).$$

We shall now summarize the properties of the Čech cohomology. For proofs, see the references listed below.

(a) If $\mathfrak{W}$ is a refinement of $\mathfrak{U}$, there is a natural group homomorphism

$$\mu^{\mathfrak{U}}_{\mathfrak{W}} : H^q(\mathfrak{U}, \mathcal{S}) \longrightarrow H^q(\mathfrak{W}, \mathcal{S})$$

and

$$\varinjlim_{\mathfrak{U}} H^q(\mathfrak{U}, \mathcal{S}) \cong H^q(X, \mathcal{S}),$$

where $H^q(X, \mathcal{S})$ is the cohomology defined in Definition 3.10.

(b) If $\mathfrak{U}$ is a covering such that

$$H^q(|\sigma|, \mathcal{S}) = 0$$

for $q \geq 1$ and all simplices σ in $\mathfrak{U}$, then

$$H^q(X, \mathcal{S}) \cong H^q(\mathfrak{U}, \mathcal{S})$$

for all $q \geq 0$ ($\mathfrak{U}$ is called a *Leray* cover).

(c) If X is paracompact and $\mathfrak{U}$ is a locally finite covering of X, then

$$H^q(\mathfrak{U}, \mathcal{S}) = 0$$

for $q > 0$ and $\mathcal{S}$ a fine sheaf over X.

We shall most often use resolutions of particular sheaves in order to represent cohomology, principally because the techniques we develop are derived from the theory of partial differential equations and are applied to differential forms and their generalizations. Čech theory, on the other hand, is very important in complex analysis and arises very naturally in such problems as Cousin I and II and their generalizations, being the general theory of Stein manifolds. See, e.g., Gunning and Rossi [1] and Gunning [1]. More generally, see Bredon [1], Godement [1], or Hirzebruch [1].

CHAPTER III

DIFFERENTIAL GEOMETRY

This chapter is an exposition of some of the basic ideas of Hermitian differential geometry, with applications to Chern classes and holomorphic line bundles. In Sec. 1 we shall give the basic definitions of the Hermitian analogues of the classical concepts of (Riemannian) metric, connection, and curvature. This is carried out in the context of differentiable **C**-vector bundles over a differentiable manifold X. More specific formulas are obtained in the case of holomorphic vector bundles (in Sec. 2) and holomorphic line bundles (in Sec. 4). In Sec. 3 is presented a development of Chern classes from the differential-geometric viewpoint. In Sec. 4 this approach to characteristic class theory is compared with the classifying space approach and with the sheaf-theoretic approach (in the case of line bundles). We prove that the Chern classes are primary obstructions to finding trivial subbundles of a given vector bundle, and, in particular, to the given vector bundle being itself trivial. In the case of line bundles, we give a useful characterization of which cohomology classes in $H^2(X, \mathbf{Z})$ are the first Chern class of a line bundle. Additional references for the material covered here are Chern [2], Griffiths [2], and Kobayashi and Nomizu [1].

1. Hermitian Differential Geometry

In this section we want to develop some of the basic differential-geometric concepts in the context of holomorphic vector bundles and, more generally, differentiable **C**-vector bundles. The basic purpose is to develop certain concepts such as metrics, connections, and curvatures which will have various applications in later sections. We do not relate these concepts in detail to their more classical counterparts in real differential geometry, as there are recent texts which do this quite well (e.g., Helgason [1] and Kobayashi and Nomizu [1]). We shall give more specific references as we go along.

In this section we shall denote by the term *vector bundle* a differentiable **C**-vector bundle over a differentiable manifold, $E \to X$. An analogous treatment can be given for **R**-vector bundles, but our applications are primarily to Chern class theory and holomorphic vector bundles, both of which require complex-linear fibres.

Suppose that $E \to X$ is a vector bundle of rank r and that $f = (e_1, \dots, e_r)$ is a frame at $x \in X$; i.e., there is a neighborhood U of x and sections $\{e_1, \dots, e_r\}$, $e_j \in \mathcal{E}(U, E)$, which are linearly independent at each point of U. If we want to indicate the dependence of the frame f on the domain of definition U, we write f_U, although normally this will be understood to be some local neighborhood of a given point. Suppose that $f = f_U$ is a given frame and that $g : U \to GL(r, \mathbf{C})$ is a differentiable mapping. Then there is an action of g on the set of all frames on the open set U defined by

$$f \longrightarrow fg,$$

where

$$(fg)(x) = \left(\sum_{\rho=1}^{r} g_{\rho 1}(x) e_\rho(x), \dots, \sum_{\rho=1}^{r} g_{\rho r}(x) e_\rho(x) \right) \quad x \in U,$$

is a new frame, i.e., $fg(x) = f(x)g(x)$, and we have the usual matrix product. Clearly, fg is a new frame defined on U, and we call such a mapping g a *change of frame*. Moreover, given any two frames f and f' over U, we see that there exists a change of frame g defined over U such that $f' = fg$.†

Using frames, we shall find local representations for all the differential geometric objects that we are going to define. We start by giving a local representation for sections of a vector bundle. Let $E \to X$ be a vector bundle, and suppose that $\xi \in \mathcal{E}(U, E)$ for U open in X. Let $f = (e_1, \dots, e_r)$ be a frame over U for E (which does not always exist, but will if U is a sufficiently small neighborhood of a given point). Then

$$\xi = \sum_{\rho=1}^{r} \xi^\rho(f) e_\rho \tag{1.1}$$

where $\xi^\rho(f) \in \mathcal{E}(U)$ are uniquely determined smooth functions on U. This induces a mapping

$$\mathcal{E}(U, E) \xrightarrow{\ell_f} \mathcal{E}(U)^r \cong \mathcal{E}(U, U \times \mathbf{C}^r), \tag{1.2}$$

which we write as

$$\xi \longrightarrow \xi(f) = \begin{bmatrix} \xi^1(f) \\ \cdot \\ \cdot \\ \cdot \\ \xi^r(f) \end{bmatrix},$$

†The set of all frames over open sets in X is the sheaf of sections of the *principal bundle* $P(E)$ associated with E, often called the *frame bundle* of E, a concept we shall not need; see, e.g., Kobayashi and Nomizu [1], or Steenrod [1]. Namely, the principal bundle $P(E)$ has fibres isomorphic to $GL(r, C)$, with the same transition functions as the vector bundle $E \to X$.

where $\xi^\rho(f)$ are defined by (1.1). Suppose that g is a change of frame over U. Then we compute that

$$\xi^\rho(fg) = \sum_{\sigma=1}^{r} g_{\rho\sigma}^{-1}\xi^\sigma(f),$$

which implies that

$$\xi(fg) = g^{-1}\xi(f)$$

or

$$g\xi(fg) = \xi(f), \tag{1.3}$$

all products being matrix multiplication at a given point $x \in U$. Therefore (1.1) gives a vector representation for sections $\xi \in \mathcal{E}(U, E)$, and (1.3) shows how the vector is transformed under a change of frame for the vector bundle E. Moreover, if E is a holomorphic vector bundle, then we shall also have *holomorphic frames*, i.e., $f = (e_1, \ldots, e_r)$, $e_j \in \mathcal{O}(U, E)$, and $e_1 \wedge \cdots \wedge e_r(x) \neq 0$, for $x \in U$; and *holomorphic changes of frame*, i.e., holomorphic mappings $g : U \to GL(r, \mathbf{C})$. Then with respect to a holomorphic frame we have the vector representation

$$\mathcal{O}(U, E) \xrightarrow{\ell_f} \mathcal{O}(U)^r, \tag{1.4}$$

given by $\xi \to \xi(f)$ as before, and the transformation rule for a holomorphic change of frame is still given by (1.3).

Our object now is to give definitions of three fundamental differential-geometric concepts: metric, connection, and curvature. We shall then give some examples in the next section to illustrate the definitions.

Definition 1.1: Let $E \to X$ be a vector bundle. A *Hermitian metric* h on E is an assignment of a Hermitian inner product $\langle , \rangle_x$ to each fibre E_x of E such that for any open set $U \subset X$ and $\xi, \eta \in \mathcal{E}(U, E)$ the function

$$\langle \xi, \eta \rangle : U \longrightarrow \mathbf{C}$$

given by

$$\langle \xi, \eta \rangle(x) = \langle \xi(x), \eta(x) \rangle_x$$

is C^∞.

A vector bundle E equipped with a Hermitian metric h is called a *Hermitian vector bundle*. Suppose that E is a Hermitian vector bundle and that $f = (e_1, \ldots, e_r)$ is a frame for E over some open set U. Then define

$$h(f)_{\rho\sigma} = \langle e_\sigma, e_\rho \rangle, \tag{1.5}$$

and let $h(f) = [h(f)_{\rho\sigma}]$ be the $r \times r$ matrix of the C^∞ functions $\{h(f)_{\rho\sigma}\}$, where $r = \operatorname{rank} E$. Thus $h(f)$ is a positive definite Hermitian symmetric matrix and is a (local) representative for the Hermitian metric h with respect to the frame f. For any $\xi, \eta \in \mathcal{E}(U, E)$, we write

$$\langle \xi, \eta \rangle = \Big\langle \sum_\rho \xi^\rho(f) e_\rho, \sum_\sigma \eta^\sigma(f) e_\sigma \Big\rangle$$

$$= \sum_{\rho,\sigma} \overline{\eta^\sigma(f)} h_{\sigma\rho}(f) \xi^\rho(f)$$

(1.6) $$\langle \xi, \eta \rangle = {}^t\overline{\eta(f)} h(f) \xi(f),$$

where the last product is matrix multiplication and ${}^tA$ denotes the transpose of the matrix A. Moreover, if g is a change of frame over U, it is easy to check that

(1.7) $$h(fg) = {}^t\bar{g} h(f) g,$$

which is the transformation law for local representations of the Hermitian metric.

Theorem 1.2: Every vector bundle $E \to X$ admits a Hermitian metric.

Proof: There exists a locally finite covering $\{U_\alpha\}$ of X and frames f_α defined on U_α. Define a Hermitian metric h_α on $E|_{U\alpha}$ by setting, for $\xi, \eta \in E_x$, $x \in U_\alpha$,

$$\langle \xi, \eta \rangle_x^\alpha = {}^t\overline{\eta(f_\alpha)}(x) \cdot \xi(f_\alpha)(x).$$

Now let $\{\rho_\alpha\}$ be a C^∞ partition of unity subordinate to the covering $\{U_\alpha\}$ and let, for $\xi, \eta \in E_x$,

$$\langle \xi, \eta \rangle_x = \sum \rho_\alpha(x) \langle \xi, \eta \rangle_x^\alpha.$$

We can now verify that $\langle\, ,\, \rangle$ so defined gives a Hermitian metric for $E \to X$. First, it is clear that if $\xi, \eta \in \mathcal{E}(U, E)$, then the function

$$x \longrightarrow \langle \xi(x), \eta(x) \rangle_x = \sum_\alpha \rho_\alpha(x) \langle \xi(x), \eta(x) \rangle_x^\alpha$$

$$= \sum_\alpha \rho_\alpha(x) {}^t\overline{\eta(f_\alpha)}(x) \cdot \xi(f_\alpha)(x)$$

is a C^∞ function on U. It is easy to verify that h is indeed a Hermitian inner product on each fibre of E, and we leave this verification to the reader.

Q.E.D.

We now want to consider differential forms with vector bundle coefficients. Suppose that $E \to X$ is a vector bundle. Then we let

$$\mathcal{E}^p(X, E) = \mathcal{E}(X, \wedge^p T^*(X) \otimes_{\mathbb{C}} E)$$

be the *differential forms of degree p on X with coefficients in* E (cf. the discussion following Lemma II.3.19). We want to relate this definition to one involving tensor products over the structure sheaf.

Lemma 1.3: Let E and E' be vector bundles over X. Then there is an isomorphism

$$\tau : \mathcal{E}(E) \otimes_{\mathcal{E}} \mathcal{E}(E') \xrightarrow{\cong} \mathcal{E}(E \otimes E').$$

Proof: We shall define the mapping τ on presheaves generating the above sheaves

$$\tau_U : \mathcal{E}(U, E) \otimes_{\mathcal{E}(U)} \mathcal{E}(U, E') \longrightarrow \mathcal{E}(E \otimes E')(U)$$

by setting

$$\tau_U(\xi \otimes \eta)(x) = \xi(x) \otimes \eta(x) \in E_x \otimes E'_x.$$

If $f = (e_1, \ldots, e_r)$ and $f' = (e'_1, \ldots, e'_r)$ are frames for E and E' over an open set U, then we see that for any $\gamma \in \mathcal{E}(U, E \otimes E')$ we can write

$$\gamma(x) = \sum_{\alpha,\beta} \gamma_{\alpha\beta}(x) e_\alpha(x) \otimes e'_\beta(x), \quad \gamma_{\alpha\beta} \in \mathcal{E}(U).$$

But this shows that

$$\gamma \in \mathcal{E}(U, E) \otimes_{\mathcal{E}(U)} \mathcal{E}(U, E'),$$

and this implies easily that $\{\tau_U\}$ defines a sheaf isomorphism when we pass to the sheaves generated by these presheaves.

Q.E.D.

Corollary 1.4: Let E be a vector bundle over X. Then

$$\mathcal{E}^p \otimes_{\mathcal{E}} \mathcal{E}(E) \cong \mathcal{E}^p(E).$$

We denote the image of $\varphi \otimes \xi$ under the isomorphism in Corollary 1.4 by $\varphi \cdot \xi \in \mathcal{E}^p(X, E)$, where $\varphi \in \mathcal{E}^p(X)$ and $\xi \in \mathcal{E}(X, E)$. Suppose that f is a frame for E over U. Then we have a local representation for $\xi \in \mathcal{E}^p(U, E)$ similar to (1.2) given by

$$\mathcal{E}^p(U, E) \xrightarrow{l_f} [\mathcal{E}^p(U)]^r$$

(1.8)
$$\xi \longrightarrow \begin{bmatrix} \xi^1(f) \\ \cdot \\ \cdot \\ \cdot \\ \xi^r(f) \end{bmatrix},$$

defined by the relation

(1.1′)
$$\xi = \sum_{\rho=1}^{r} \xi^\rho(f) \cdot e_\rho.$$

Namely, let $x \in U$ and let $(\omega_1, \ldots, \omega_s)$ be a frame for $\wedge^p T^*(X) \otimes \mathbf{C}$ at x. Then we can write

$$\xi(x) = \sum_{\rho,k} \varphi_{k\rho}(x) \omega_k(x) \otimes e_\rho(x).$$

where the $\varphi_{k\rho}$ are uniquely determined C^∞ functions defined near x. Let

$$\xi^\rho = \sum_k \varphi_{k\rho} \omega_k,$$

and it is easy to check that the differential form ξ^ρ so determined is independent of the choice of frame $(\omega_1, \ldots, \omega_s)$. Since x was an arbitrary point of U, the differential forms $\{\xi^\rho\}$ are defined in all of U, and thus the mapping (1.8) (local representation of vector-valued differential forms) is well defined and, indeed, is an isomorphism. Moreover, we have the transformation law for a change of frame

$$\xi(fg) = g^{-1}\xi(f), \quad \xi \in \mathcal{E}^p(X, E) \tag{1.3'}$$

exactly as in (1.3) for sections. We now make the following definition.

Definition 1.5: Let $E \to X$ be a vector bundle. Then a *connection D* on $E \to X$ is a **C**-linear mapping

$$D : \mathcal{E}(X, E) \longrightarrow \mathcal{E}^1(X, E),$$

which satisfies

$$D(\varphi\xi) = d\varphi \cdot \xi + \varphi D\xi, \tag{1.9}$$

where $\varphi \in \mathcal{E}(X)$ and $\xi \in \mathcal{E}(X, E)$.

Remarks: (a) Relation (1.9) implies that D is a first-order differential operator (cf. Sec. 2 in Chap. IV) mapping $\mathcal{E}(X, E)$ to $\mathcal{E}(X, T^*(X) \otimes E)$, as we shall see below.

(b) In the case where $E = X \times \mathbf{C}$, the trivial line bundle, we see that we may take ordinary exterior differentiation

$$d : \mathcal{E}(X) \longrightarrow \mathcal{E}^1(X)$$

as a connection on E. Thus a connection is a generalization of exterior differentiation to vector-valued differential forms, and we shall later extend the definition of D to higher-order forms.

We now want to give a local description of a connection. Let f be a frame over U for a vector bundle $E \to X$, equipped with a connection D. Then we define the *connection matrix* $\theta(D, f)$ associated with the connection D and the frame f by setting

$$\theta(D, f) = [\theta_{\rho\sigma}(D, f)], \quad \theta_{\rho\sigma}(D, f) \in \mathcal{E}^1(U),$$

where

$$De_\sigma = \sum_{\rho=1}^{r} \theta_{\rho\sigma}(D, f) \cdot e_\rho. \tag{1.10}$$

We shall denote the matrix $\theta(D, f)$ by $\theta(f)$ (for a fixed connection) or often simply by θ (for a fixed frame in a given computation). We can use the connection matrix to explicitly represent the action of D on sections of E. Namely, if $\xi \in \mathcal{E}(U, E)$, then, for a given frame f,

$$\begin{aligned} D\xi &= D\Big(\sum_\rho \xi^\rho(f)e_\rho\Big) \\ &= \sum_\sigma d\xi^\sigma(f) \cdot e_\sigma + \sum_\rho \xi^\rho(f)De_\rho \\ &= \sum_\sigma \Big[d\xi^\sigma(f) + \sum_\rho \xi^\rho(f)\theta_{\sigma\rho}(f)\Big] \cdot e_\sigma \end{aligned}$$

$$D\xi = \sum_\sigma [d\xi(f) + \theta(f)\xi(f)] \cdot e_\sigma, \tag{1.11}$$

where we have set

$$d\xi(f) = \begin{bmatrix} d\xi^1(f) \\ \cdot \\ \cdot \\ \cdot \\ d\xi^r(f) \end{bmatrix},$$

and the wedge product inside the brackets in (1.11) is ordinary matrix multiplication of matrices with differential form coefficients. Thus we see that

$$\begin{aligned} D\xi(f) &= d\xi(f) + \theta(f)\xi(f) \\ &= [d + \theta(f)]\xi(f) \end{aligned}$$

thinking of $d+\theta(f)$ as being an operator acting on vector-valued functions.

Remark: If we let $E = T(X)$, then the real analogue of a connection in the differential operator sense as defined above defines an affine connection in the usual sense (cf. Helgason [1], Nomizu [1], Sternberg [1], and Kobayashi and Nomizu [1]). If $\omega = (\omega_1, \ldots, \omega_n)$ is a frame for $T^*(X)$ over U, then

$$\theta_{\rho\sigma} = \sum_{k=1}^{n} \mathbf{\Gamma}^{\rho}_{\sigma k}\omega_k, \quad \mathbf{\Gamma}^{\rho}_{\sigma k} \in \mathcal{E}(U).$$

In the classical case these are the Schwarz-Christoffel symbols associated with (or defining) a given connection.

Suppose that $E \to X$ is a vector bundle equipped with a connection D (as we shall see below, every vector bundle admits a connection). Let Hom (E, E) be the vector bundle whose fibres are $\mathrm{Hom}(E_x, E_x)$. We want to show that the connection D on E induces in a natural manner an element

$$\mathbf{\Theta}_E(D) \in \mathcal{E}^2(X, \mathrm{Hom}(E, E)),$$

to be called the *curvature tensor*.

First we want to give a local description of an arbitrary element $\chi \in \mathcal{E}^p(X, \mathrm{Hom}(E, E))$. Let f be a frame for E over U in X. Then $f = (e_1, \ldots, e_r)$ becomes a basis for the free $\mathcal{E}^p(U)$-module

$$\mathcal{E}^p(U, \mathrm{Hom}(E, E)) \cong \mathcal{E}^p(U) \otimes_{\mathcal{E}(U)} \mathcal{E}(U, \mathrm{Hom}(E, E)).$$

Since $E|_U \cong U \times \mathbf{C}^r$, by using f to effect a trivialization, we see that

$$\mathcal{E}(U, \mathrm{Hom}(E, E)) \cong \mathfrak{M}_r(U) = \mathfrak{M}_r \otimes_{\mathbf{C}} \mathcal{E}(U),$$

where $\mathfrak{M}_r$ is the vector space of $r \times r$ matrices, and thus $\mathfrak{M}_r(U)$ is the $\mathcal{E}(U)$-module of $r \times r$ matrices with coefficients in $\mathcal{E}(U)$. Therefore there is associated with χ under the above isomorphisms, an $r \times r$ matrix

$$\chi(f) = [\chi(f)_{\rho\sigma}], \quad \chi(f)_{\rho\sigma} \in \mathcal{E}^p(U). \tag{1.12}$$

Moreover, we see easily that χ determines a global homomorphism of vector bundles

$$\chi : \mathcal{E}(X, E) \longrightarrow \mathcal{E}^p(X, E),$$

defined fibrewise in the natural manner. The frame f gives local representations for elements in $\mathcal{E}(X, E)$ and $\mathcal{E}^p(X, E)$ and the matrix (1.12) is chosen so that the following diagram commutes,

$$\begin{array}{ccc} \mathcal{E}(U, E) & \xrightarrow{\chi} & \mathcal{E}^p(U, E) \\ \wr\| & & \wr\| \\ \mathcal{E}(U)^r & \xrightarrow{\chi(f)} & [\mathcal{E}^p(U)]^r \\ \xi(f) & \longrightarrow & \chi(f)\xi(f) = \eta(f), \end{array}$$

where

$$\eta^\rho(f) = \sum_\sigma \chi(f)_{\rho\sigma}\xi^\sigma(f)$$

is matrix multiplication and the vertical isomorphisms are given by (1.2) and (1.8), respectively. Under this convention it is easy to compute how the local representation for χ behaves under a change of frame; namely, if

$$\eta(fg) = \chi(fg)\xi(fg),$$

then we see that

$$g^{-1}\eta(f) = \chi(fg)g^{-1}\xi(f),$$

which implies that

$$\chi(fg) = g^{-1}\chi(f)g; \tag{1.13}$$

i.e., χ transforms by a similarity transformation. Conversely, any assignment of a matrix of p-forms $\chi(f)$ to a given frame f which is defined for all frames and satisfies (1.13) defines an element $\chi \in \mathcal{E}^p(X, \operatorname{Hom}(E, E))$, as is easy to verify.

Returning to the problem of defining the curvature, let $E \longrightarrow X$ be a vector bundle with a connection D and let $\theta(f) = \theta(D, f)$ be the associated connection matrix. We define

$$\Theta(D, f) = d\theta(f) + \theta(f) \wedge \theta(f), \tag{1.14}$$

which is an $r \times r$ matrix of 2-forms; i.e.,

$$\Theta_{\rho\sigma} = d\theta_{\rho\sigma} + \sum \theta_{\rho k} \wedge \theta_{k\sigma}.$$

We call $\Theta(D, f)$ the *curvature matrix* associated with the connection matrix $\theta(f)$. We have the following two simple propositions, the first showing how $\theta(f)$ and $\Theta(f)$ transform, and the second relating $\Theta(f)$ to the operator $d + \theta(f)$.

Lemma 1.6: Let g be a change of frame and define $\theta(f)$ and $\Theta(f)$ as above. Then

(a) $dg + \theta(f)g = g\theta(fg)$,
(b) $\Theta(fg) = g^{-1}\Theta(f)g$.

Proof:

(a) If

$$fg = \left(\sum g_{\rho 1}e_\rho, \ldots, \sum g_{\rho r}e_\rho\right) = (e_1', \ldots, e_r'),$$

then

$$D(e'_\sigma) = \sum_\nu \theta_{\nu\sigma}(fg)e'_\nu$$

$$= \sum_{\nu,\rho} \theta_{\nu\sigma}(fg)g_{\rho\nu}e_\rho,$$

and, on the other hand,

$$D\Big(\sum_\rho g_{\rho\sigma}e_\rho\Big) = \sum_\rho dg_{\rho\sigma}e_\rho + \sum_{\rho,\tau} g_{\rho\sigma}\theta_{\tau\rho}e_\tau.$$

By comparing coefficients, we obtain

$$g\theta(fg) = dg + \theta(f)g. \tag{1.15}$$

(b) Take the exterior derivative of the matrix equation (1.15), obtaining

$$d\theta(f) \cdot g - \theta(f) \cdot dg = dg \cdot \theta(fg) + g \cdot d\theta(fg). \tag{1.16}$$

Also,

$$\theta(fg) = g^{-1}dg + g^{-1}\theta(f)g, \tag{1.17}$$

and thus we obtain by substituting (1.17) into (1.16) an algebraic expression for $gd\theta(fg)$ in terms of the quantities $d\theta(f), \theta(f), dg, g$, and g^{-1}. Then we can write

$$g[d\theta(fg) + \theta(fg) \wedge \theta(fg)] \tag{1.18}$$

in terms of these same quantities. Writing this out and simplifying, we find that (1.18) is the same as

$$[d\theta(f) + \theta(f) \wedge \theta(f)]g,$$

which proves part (b).

Q.E.D.

Lemma 1.7: $[d + \theta(f)][d + \theta(f)]\xi(f) = \boldsymbol{\Theta}(f)\xi(f).$

Proof: By straightforward computation we have (deleting the notational dependence on f)

$$\begin{aligned}(d+\theta)(d+\theta)\xi &= d^2\xi + \theta \cdot d\xi + d(\theta \cdot \xi) + \theta \wedge \theta \cdot \xi \\ &= \theta \cdot d\xi + d\theta \cdot \xi - \theta \cdot d\xi + \theta \wedge \theta \cdot \xi \\ &= d\theta \cdot \xi + \theta \wedge \theta \cdot \xi \\ &= \boldsymbol{\Theta} \cdot \xi.\end{aligned}$$

Q.E.D.

The proof of the above lemma illustrates why we have taken care to see that the abstract operations and equations at the section level correspond, with respect to a local frame, to matrix operations and equations.

We now make the following definition.

Definition 1.8: Let D be a connection in a vector bundle $E \longrightarrow X$. Then the *curvature* $\boldsymbol{\Theta}_E(D)$ is defined to be that element $\boldsymbol{\Theta} \in \mathcal{E}^2(X, \mathrm{Hom}(E, E))$ such that the **C**-linear mapping

$$\boldsymbol{\Theta} : \mathcal{E}(X, E) \longrightarrow \mathcal{E}^2(X, E)$$

has the representation with respect to a frame

$$\boldsymbol{\Theta}(f) = \boldsymbol{\Theta}(D, f) = d\theta(f) + \theta(f) \wedge \theta(f).$$

We see by Lemma 1.6(b) that $\boldsymbol{\Theta}_E(D)$ is well defined, since $\boldsymbol{\Theta}(D, f)$ satisfies the transformation property (1.13), which ensures that $\boldsymbol{\Theta}(D, f)$ determines a global element in $\mathcal{E}^2(X, \mathrm{Hom}(E, E))$.

Remark: It follows from the local definition of $\boldsymbol{\Theta}_E(D)$ that the curvature is an $\mathcal{E}(X)$-linear mapping

$$\boldsymbol{\Theta} : \mathcal{E}(X, E) \longrightarrow \mathcal{E}^2(X, E),$$

and it is this linearity property that makes $\boldsymbol{\Theta}$ into a tensor in the classical sense. Note that the transformation formula for $\theta(f)$ involves derivatives of the change of frames and that of course the connection D is not $\mathcal{E}(X)$-linear. If we denote by $D_z\xi$ the natural contraction of $Z \otimes D\xi$ for $Z \in T(X)$ and $\xi \in \mathcal{E}(X, E)$, then the classical curvature tensor $R(Z, W) = D_Z D_W - D_W D_Z - D_{[Z,W]}$ defined from this affine connection agrees with $\boldsymbol{\Theta}(Z, W) \in \mathcal{E}(X, \mathrm{Hom}(E, E))$. This follows by an exterior algebra computation and (1.14), since for a frame f over U, $D\xi(f) = d\xi(f) + \theta(f) \wedge \xi(f)$ implies

$$D_Z\xi(f) = Z\xi(f) + \theta(f)(Z)\xi(f).$$

We can now define the action of D on higher-order differential forms by setting

$$D\xi(f) = d\xi(f) + \theta(f) \wedge \xi(f),$$

where $\xi \in \mathcal{E}^p(X, E)$. Thus

$$D : \mathcal{E}^p(X, E) \longrightarrow \mathcal{E}^{p+1}(X, E)$$

if it is well defined. But we only have to check whether the image satisfies the transformation law (1.3′) in order to see that the image of D is a well-defined E-valued $(p + 1)$-form. To check this, we see that

$$g[d\xi(fg) + \theta(fg)\xi(fg)] = d(g\xi(fg)) - dg \cdot \xi(fg) + [dg + \theta(f)g] \wedge g^{-1}\xi(f)$$

from (1.3) and Lemma 1.6(a), which reduces to

$$d\xi(f) + \theta(f) \wedge \xi(f).$$

Thus we have the extension of D to differential forms (E-valued) of higher order. This extension is known as *covariant differentiation*, and we have proved the following.

Proposition 1.9: $D^2 = \boldsymbol{\Theta}$, as an operator mapping

$$\mathcal{E}^p(X, E) \longrightarrow \mathcal{E}^{p+2}(X, E), \text{ where } D^2 = D \circ D.$$

The only unproved part is for $p > 0$, but we observe that Lemma 1.7 is still valid in this case. Then the curvature is the obstruction to $D^2 = 0$ and is therefore the obstruction that the sequence

$$\mathcal{E}(X, E) \xrightarrow{D} \mathcal{E}^1(X, E) \xrightarrow{D} \mathcal{E}^2(X, E) \longrightarrow \cdots \longrightarrow$$

be a *complex* (cf. Sec. 5 in Chap. IV).

The differential forms $\mathcal{E}^p(X, \mathrm{Hom}(E, E))$ are locally matrices of p-forms. We want to use this fact to define a Lie product on the algebra

$$\mathcal{E}^*(X, \mathrm{Hom}(E, E)) = \sum_p \mathcal{E}^p(X, \mathrm{Hom}(E, E)).$$

We proceed as follows. If $\chi \in \mathcal{E}^p(X, \mathrm{Hom}(E, E))$ and f is a frame for E over the open set U, then we have seen before that

$$\chi(f) \in \mathfrak{M}_r \otimes_{\mathbf{C}} \mathcal{E}^p(U),$$

and thus if $\psi \in \mathcal{E}^q(X, \mathrm{Hom}(E, E))$, we define

$$(1.19) \qquad [\chi(f), \psi(f)] = \chi(f) \wedge \psi(f) - (-1)^{pq}\psi(f) \wedge \chi(f),$$

where the right-hand side is matrix multiplication. If g is a change of frame, then by (1.13) we have

$$\chi(fg) = g^{-1}\chi(f)g$$

$$\psi(fg) = g^{-1}\psi(f)g,$$

and thus

$$[\chi(fg), \psi(fg)] = g^{-1}[\chi(f), \psi(f)]g$$

by a straightforward substitution. Therefore the Lie bracket is well defined on $\mathcal{E}^*(\chi, \mathrm{Hom}(E, E))$ and satisfies the Jacobi identity, making $\mathcal{E}^*(X, \mathrm{Hom}(E, E))$ into a Lie algebra (cf., e.g., Helgason [1]).

Suppose that E is equipped with a connection D and that we let $\theta(f), \mathbf{\Theta}(f)$ be the local connection and curvature forms with respect to some frame f. Then we can prove a version of the *Bianchi identity* in this context, for which we shall have use later.

Proposition 1.10: $d\mathbf{\Theta}(f) = [\mathbf{\Theta}(f), \theta(f)]$.

Proof: Letting $\theta = \theta(f)$ and $\mathbf{\Theta} = \mathbf{\Theta}(f)$, we have

$$\mathbf{\Theta} = d\theta + \theta \wedge \theta,$$

and thus

$$\begin{aligned} d\mathbf{\Theta} &= d^2\theta + d\theta \wedge \theta - \theta \wedge d\theta \\ &= d\theta \wedge \theta - \theta \wedge d\theta. \end{aligned}$$

But

$$\begin{aligned} [\mathbf{\Theta}, \theta] &= [d\theta + \theta \wedge \theta, \theta] \\ &= d\theta \wedge \theta + \theta \wedge \theta \wedge \theta \\ &\quad - (-1)^{2\cdot 1}(\theta \wedge d\theta + \theta \wedge \theta \wedge \theta) \\ &= d\theta \wedge \theta - \theta \wedge d\theta. \end{aligned}$$

Q.E.D.

We now want to show that any differentiable vector bundle admits a connection. In the next section we shall see some examples when we look at the special case of holomorphic vector bundles. Assume that E is a Hermitian vector bundle over X. Then we can extend the metric h on E in a natural manner to act on E-valued covectors. Namely, set

$$\langle \omega \otimes \xi, \omega' \otimes \xi' \rangle_x = \omega \wedge \bar{\omega}' \langle \xi, \xi' \rangle_x \tag{1.20}$$

for $\omega \in \wedge^p T_x^*(X), \omega' \in \wedge^q T_x^*(X)$, and $\xi, \xi' \in E$, for $x \in X$. Thus the extension of the inner product to differential forms induces a mapping

$$h : \mathcal{E}^p(X, E) \otimes \mathcal{E}^q(X, E) \longrightarrow \mathcal{E}^{p+q}(X).$$

A connection D on E is said to be *compatible* with a Hermitian metric h on E if

$$d\langle \xi, \eta \rangle = \langle D\xi, \eta \rangle + \langle \xi, D\eta \rangle. \tag{1.21}$$

Suppose that $f = (e_1, \ldots, e_r)$ is any frame and that D is a connection compatible with a Hermitian metric on E. Then we see that [letting $h(f) = h$, $\theta(f) = \theta$]

$$\begin{aligned} dh_{\rho\sigma} &= d\langle e_\sigma, e_\rho \rangle = \langle De_\sigma, e_\rho \rangle + \langle e_\sigma, De_\rho \rangle \\ &= \Big\langle \sum_\tau \theta_{\tau\sigma} e_\tau, e_\rho \Big\rangle + \Big\langle e_\sigma, \sum_\mu \theta_{\mu\rho} e_\mu \Big\rangle \\ &= \sum_\tau \theta_{\tau\sigma} h_{\rho\tau} + \sum_\mu \bar{\theta}_{\mu\rho} h_{\mu\sigma} \\ &= (h\theta)_{\rho\sigma} + ({}^t\bar{\theta} h)_{\rho\sigma}, \end{aligned}$$

and thus

$$dh = h\theta + {}^t\bar{\theta} h \tag{1.22}$$

is a necessary condition that h and the connection D be compatible. Moreover, it is sufficient. Namely, suppose that (1.22) is satisfied for all frames. Then one obtains immediately

$$d\langle \xi, \eta \rangle = d({}^t\bar{\eta} h \xi) = {}^t(d\bar{\eta}) h \xi + {}^t\bar{\eta}(dh)\xi + {}^t\bar{\eta} h d\xi$$

in terms of a local frame. Substituting (1.22) into the above equation, we get four terms which group together as

$${}^t(d\bar{\eta} + \overline{\theta\eta}) h \xi + {}^t\bar{\eta} h (d\xi + \theta\xi) = \langle \xi, D\eta \rangle + \langle D\xi, \eta \rangle.$$

Proposition 1.11: Let $E \longrightarrow X$ be a Hermitian vector bundle. Then there exists a connection D on E compatible with the Hermitian metric on E.

Proof: A unitary frame f has the property that $h(f) = I$. Such frames always exist near a given point x_0, since the Gram-Schmidt orthogonalization process allows one to find r local sections which form an orthonormal basis for E_x at all points x near x_0. In particular, we can find a locally finite covering U_α and unitary frames f_α defined in U_α. The condition (1.21) reduces to

$$0 = \theta + {}^t\bar{\theta}$$

for a unitary frame; i.e., θ is to be skew-Hermitian. In each U_α we can choose the trivial skew-Hermitian matrix of the form $\theta_\alpha = 0$; i.e., $\theta(f_\alpha) = 0$. If we make a change of frame in U_α, then we see that we require that

$$\theta(f_\alpha g) = g^{-1}dg + 0 \tag{1.23}$$

by Lemma 1.6(a). Therefore, define $\theta(f_\alpha g)$ by (1.23), and noting that $h(f_\alpha g) = {}^t\bar{g}h(f)g = {}^t\bar{g}g$, we obtain

$$\begin{aligned} dh(f_\alpha g) &= d({}^t\bar{g} \cdot g) \\ &= d{}^t\bar{g} \cdot g + {}^t\bar{g} \cdot dg \\ &= d{}^t\bar{g}({}^t\bar{g})^{-1} \cdot {}^t\bar{g} \cdot g + {}^t\bar{g} \cdot g \cdot g^{-1} \cdot dg \\ &= {}^t\bar{\theta}(f_\alpha g)h(f_\alpha g) + h(f_\alpha g)\theta(f_\alpha g), \end{aligned}$$

which verifies the compatibility. Let $\{\varphi_\alpha\}$ be a partition of unity subordinate to $\{U_\alpha\}$ and let D_α be the connection in $E|_{U_\alpha}$ defined by

$$(D_\alpha \xi)(f_\alpha) = d\xi(f_\alpha).$$

D_α is defined with respect to other frames over U_α by formula (1.23) and is compatible with the Hermitian metric on $E|_{U_\alpha}$, by construction. Then we let $D = \sum_\alpha \varphi_\alpha D_\alpha$, which is a well-defined (first-order partial-differential) operator

$$D : \mathcal{E}(X, E) \longrightarrow \mathcal{E}^1(X, E).$$

Moreover, D is compatible with the metric h on E since

$$\begin{aligned} \langle D\xi, \eta\rangle + \langle \xi, D\eta\rangle &= \sum_\alpha \varphi_\alpha[\langle D_\alpha\xi, \eta\rangle + \langle \xi, D_\alpha\eta\rangle] \\ &= \sum_\alpha \varphi_\alpha d\langle \xi, \eta\rangle = d\langle \xi, \eta\rangle. \end{aligned}$$

Q.E.D.

Remark: It is clear by the construction in the proof of Proposition 1.11 that a connection compatible with a metric is by no means unique because of the various choices made along the way. In the holomorphic category, we shall obtain a unique connection satisfying an additional restriction on the type of θ.

2. The Canonical Connection and Curvature of a Hermitian Holomorphic Vector Bundle

Suppose now that $E \longrightarrow X$ is a holomorphic vector bundle over a complex manifold X. If E, as a differentiable bundle, is equipped with a differentiable Hermitian metric, h, we shall refer to it as a *Hermitian holomorphic vector bundle*.

Recall that since X is a complex manifold,

$$\mathcal{E}^*(E) = \sum_r \mathcal{E}^r(E) = \sum_{p,q} \mathcal{E}^{p,q}(E),$$

where

$$\mathcal{E}^{p,q}(E) = \mathcal{E}_X^{p,q} \otimes_{\mathcal{E}_X} \mathcal{E}(E).$$

Suppose then that we have a connection on E

$$D : \mathcal{E}(X, E) \longrightarrow \mathcal{E}^1(X, E) = \mathcal{E}^{1,0}(X, E) \oplus \mathcal{E}^{0,1}(X, E).$$

Then D splits naturally into $D = D' + D''$, where

$$D' : \mathcal{E}(X, E) \longrightarrow \mathcal{E}^{1,0}(X, E)$$

$$D'' : \mathcal{E}(X, E) \longrightarrow \mathcal{E}^{0,1}(X, E).$$

Theorem 2.1: If h is a Hermitian metric on a holomorphic vector bundle $E \longrightarrow X$, then h induces canonically a connection, $D(h)$, on E which satisfies, for W an open set in X,

(a) For $\xi, \eta \in \mathcal{E}(W, E)$

$$d\langle \xi, \eta \rangle = \langle D\xi, \eta \rangle + \langle \xi, D\eta \rangle;$$

i.e., D is *compatible* with the metric h.

(b) If $\xi \in \mathcal{O}(W, E)$, i.e., is a holomorphic section of E, then $D''\xi = 0$.

Proof: First, we point out that (b) is equivalent to the fact that the connection matrix $\theta(f)$ is of type (1,0) for a holomorphic frame f. This follows, since for $\xi \in \mathcal{O}(W, E)$ and f a holomorphic frame, we have

$$\begin{aligned} D\xi(f) &= (d + \theta(f))\xi(f) \\ &= (\partial + \theta^{(1,0)}(f))\xi(f) + (\bar{\partial} + \theta^{(0,1)}(f))\xi(f), \end{aligned}$$

where $\theta = \theta^{(1,0)} + \theta^{(0,1)}$ is the natural decomposition. Therefore

$$D'\xi(f) = (\partial + \theta^{(1,0)}(f))\xi(f)$$

and

$$D''\xi(f) = (\bar{\partial} + \theta^{(0,1)}(f))\xi(f).$$

But $\bar{\partial}\xi(f) = 0$ since ξ and f are holomorphic. Thus

$$D''\xi(f) = \theta^{(0,1)}(f)\xi(f).$$

Suppose now that we have a connection D satisfying (a) and (b). Then let $f = (e_1, \ldots, e_r)$ be a holomorphic frame over $U \subset X$ and θ the associated connection matrix. Since D is compatible with the metric h, we have, by (1.22),

$$dh = h\theta + {}^t\bar{\theta}h.$$

Since, in addition, D satisfies (b), we have seen that θ is of type (1, 0). Thus, by examining types we see that

$$\partial h = h\theta$$

and

$$\bar{\partial} h = {}^t\bar{\theta}h,$$

from which it follows that

$$\theta = h^{-1}\partial h. \tag{2.1}$$

We can then define θ by (2.1). Such a connection matrix clearly satisfies (a) and (b). Moreover, if $f' = fg$ is another holomorphic frame, we have

$$h(fg) = {}^t\bar{g}h(f)g,$$

so that

$$h^{-1}(fg) = g^{-1}h(f)^{-1}[{}^t\bar{g}]^{-1}$$

and

$$\begin{aligned} g\theta(fg) &= g[h^{-1}(fg)\partial h(fg)] \\ &= h(f)^{-1}[{}^t\bar{g}]^{-1}\partial[{}^t\bar{g}h(f)g] \\ &= h(f)^{-1}[{}^t\bar{g}]^{-1}[{}^t\bar{g}\partial h(f)g + {}^t\bar{g}h(f)\partial g + \partial{}^t\bar{g}h(f)g]. \end{aligned}$$

But g is a holomorphic change of frame, from which it follows that

$$\partial{}^t\bar{g} = \overline{\bar{\partial}{}^t g} = 0 \quad \text{and} \quad \partial g = dg.$$

Thus

$$\begin{aligned} g\theta(fg) &= h(f)^{-1}\partial h(f)g + dg \\ &= \theta(f)g + dg. \end{aligned}$$

Recalling Lemma 1.6(a), we see that this is the necessary transformation formula for θ to define a global connection.

Q.E.D.

This theorem gives a simple formula for the canonical connection in terms of the metric h; namely,

$$\theta(f) = h(f)^{-1}\partial h(f) \tag{2.2}$$

for a holomorphic frame f. Moreover, $D = D' + D''$ has the following representation with respect to a holomorphic frame f:

$$\begin{aligned} D' &= \partial + \theta(f) \\ D'' &= \bar{\partial}. \end{aligned} \tag{2.3}$$

Thus we have the following proposition.

Proposition 2.2: Let D be the canonical connection of a Hermitian holomorphic vector bundle $E \longrightarrow X$, with Hermitian metric h. Let $\theta(f)$ and $\mathbf{\Theta}(f)$ be the connection and curvature matrices defined by D with respect to a holomorphic frame f. Then

(a) $\theta(f)$ is of type (1, 0), and $\partial\theta(f) = -\theta(f) \wedge \theta(f)$.
(b) $\mathbf{\Theta}(f) = \bar{\partial}\theta(f)$, and $\mathbf{\Theta}(f)$ is of type (1, 1).
(c) $\bar{\partial}\mathbf{\Theta}(f) = 0$, and $\partial\mathbf{\Theta}(f) = [\mathbf{\Theta}(f), \theta(f)]$.

Proof: Let $h = h(f)$, $\theta = \theta(f)$, and $\mathbf{\Theta} = \mathbf{\Theta}(f)$. Then we first note that θ is of type (1, 0) by (2.2). Then by using

$$\partial h^{-1} = -h^{-1} \cdot \partial h \cdot h^{-1}$$

$$\partial^2 = 0,$$

we see that

$$\partial\theta = \partial(h^{-1}\partial h) = -h^{-1}\cdot\partial h\cdot h^{-1}\wedge\partial h$$
$$= -(h^{-1}\partial h)\wedge(h^{-1}\partial h) = -\theta\wedge\theta,$$

which gives us part (a). Part (b) is a simple computation, namely,

$$\mathbf{\Theta} = d\theta + \theta\wedge\theta = \partial\theta + \theta\wedge\theta + \bar\partial\theta$$
$$= \bar\partial\theta,$$

by using part (a). Part (c) then follows from

$$\bar\partial\mathbf{\Theta} = \bar\partial^2\theta = 0$$

and Proposition 1.10.

Q.E.D.

Let $E \longrightarrow X$ be a holomorphic vector bundle and let $f = (e_1, \ldots, e_r)$ be a frame for E defined near a point $p \in X$. Choose local coordinates $z = (z_1, \ldots, z_n)$ near p so that p is given by $z = 0$. Then we can write

$$f(z) = (e_1(z), \ldots, e_r(z))$$

to denote the dependence on the variable z near $z = 0$. Suppose that h is a Hermitian metric on $E \to X$ and that $f(z)$ is the above frame. Then we write

$$h(z) = h(f(z))$$

near $z = 0$. The next lemma tells us that we may always find a local frame f near p such that $h(f(z))$ has a very nice form. Let $\mathbf{\Theta}(z) = \mathbf{\Theta}(f(z))$.

Lemma 2.3: There exists a holomorphic frame f such that

(a) $h(z) = I + O(|z|^2)$.
(b) $\mathbf{\Theta}(0) = \bar\partial\partial h(0)$.

Proof: Suppose that (a) holds. Then it follows that

$$h^{-1}(z) = I + O(|z|^2),$$

from which we see that

$$\mathbf{\Theta}(z) = \bar\partial\partial h(z) + O(|z|),$$

and hence (b) follows.

To show (a), we shall make two changes of frame. First we note that $h(0)$ is a positive definite Hermitian matrix, and thus there exists a nonsingular matrix $g \in GL(r, \mathrm{C})$ such that

$$g^*h(0)g = I,$$

where for any matrix M we let $M^* = {}^t\bar M$. The matrix g induces a change of frame $f \longrightarrow \tilde f = f\cdot g$, and we see that

$$\tilde h(z) = h(\tilde f(z)) = h(fg)$$
$$= g^*hg$$

$$\tilde h(z) = I + O(|z|). \tag{2.4}$$

Assume now that $h(z)$ is given satisfying (2.4). We want to consider a change of frame of the form

$$g = I + A(z),$$

where $A(z) = (\sum_j A^j_{\rho\sigma} z_j)$ is a matrix of linear holomorphic functions of z. Since $A(0) = 0$, this change of frame will preserve (2.4). By choosing $A(z)$ such that

$$\tilde{h}(z) = g(z)^* h(z) g(z) = I + O(|z|^2), \tag{2.5}$$

we will have proved (a). But (2.5) is equivalent, by Taylor's theorem, to the vanishing of the first derivatives of (2.5) at $z = 0$; i.e., $d\tilde{h}(0) = 0$. Thus we compute

$$\begin{aligned} d\tilde{h}(z) &= dh(z) + dA^*(z) \cdot h(z) \\ &\quad + h(z) dA(z) + O(|z|). \end{aligned}$$

Therefore

$$d\tilde{h}(0) = \partial h(0) + dA(0) + \bar{\partial} h(0) + dA^*(0).$$

Suppose that we let

$$A^i_{\rho\sigma} = \frac{-\partial h_{\rho\sigma}(0)}{\partial z_i}, \tag{2.6}$$

Then we see that

$$dA(0) = -\partial h(0),$$

which implies that

$$dA^*(0) = -\bar{\partial} h(0).$$

Then the choice of $A(z)$ given by (2.6), depending on the derivatives of the metric h, ensures that (2.5) holds.

Q.E.D.

This lemma allows us to compute the curvature $\mathbf{\Theta}$ at a particular point without having to compute the inverse of the local representation for the metric, provided that we have the right frame.

We want to give one principal example concerning the computation of connections and curvatures. Further examples of specific Hermitian metrics on tangent bundles are found in Chap. VI, where we shall discuss Kähler manifolds. In Sec. 4 we shall look at the special case of line bundles in more detail.

Example 2.4: Let $U_{r,n} \longrightarrow G_{r,n}$ be the universal bundle over the Grassmannian manifold $G_{r,n}$ (Example 1.2.6). We see that a frame $f = (e_1, \ldots, e_r)$ for $U_{r,n} \longrightarrow G_{r,n}$ consists of an open set $U \subset G_{r,n}$ and smooth functions

$$e_j : U \longrightarrow \mathbf{C}^n,$$

so that $e_1 \wedge \ldots \wedge e_r \neq 0$. Thus $f = (e_1, \ldots, e_r)$ can be thought of as an $n \times r$ matrix with coefficients being smooth functions in U and whose columns are the vectors $\{e_j\}$, and the matrix f is of maximal rank at each point $z \in U$. A holomorphic frame will simply have holomorphic coefficients. We define a metric on $U_{r,n}$ by letting

$$h(f) = {}^t\bar{f}f \tag{2.7}$$

for any frame f for $U_{r,n}$. This metric results from considering $U_{r,n} \subset G_{r,n} \times \mathbf{C}^n$ and restricting the standard Hermitian metric on $\mathbf{C}^n$ to the fibres of $U_{r,n} \longrightarrow G_{r,n}$. First we note that $h(f)$ is positive definite since (recall that f has maximal rank)

$$ {}^t\bar{z}h(f)z = {}^t(\overline{fz})(fz) = |fz|^2 > 0 \quad \text{if } z \neq 0.$$

Moreover, if g is a change of frame, then we compute that

$$h(fg) = {}^t(\overline{fg})(fg) = {}^t\bar{g}{}^t\bar{f}fg = {}^t\bar{g}h(f)g,$$

so that (1.7) is satisfied, and thus we see that h defined by (2.7) on frames gives a well-defined Hermitian metric on $U_{r,n}$, since the frame representation transforms correctly.

We can now compute the canonical connection and curvature for $U_{r,n}$ with respect to this natural metric. If f is any holomorphic frame for $U_{r,n}$, then by (2.2) and Proposition 2.2, we see that

$$\theta(f) = h^{-1}(f)\partial h(f)$$
$$\mathbf{\Theta}(f) = \bar{\partial}(h^{-1}(f)\partial h(f)).$$

We obtain, letting $\theta = \theta(f)$, etc., as before,

$$\mathbf{\Theta} = h^{-1} \cdot {}^t\overline{df} \wedge df - h^{-1} \cdot {}^t\overline{df} \cdot f \cdot h^{-1} \wedge {}^t\bar{f} \cdot df, \tag{2.8}$$

where $h^{-1} = [{}^t\bar{f}f]^{-1}$. In the case $r = 1$ (projective space), we can obtain a more explicit formula. If $\varphi \in [\mathcal{E}^p(W)]^n$, $\psi \in [\mathcal{E}^q(W)]^n$, for W an open subset of $\mathbf{C}^n$, we set

$$\langle \varphi, \psi \rangle = (-1)^{pq}{}^t\bar{\psi} \wedge \varphi,$$

which generalizes the usual Hermitian inner product on vectors in $\mathbf{C}^n$ [note that this is compatible with (1.19), where we have the usual inner product on $E = W \times \mathbf{C}^n$ given by $\langle u, v \rangle = {}^t\bar{v}u, u, v \in \mathbf{C}^n$]. Then the curvature form for $U_{1,n}$ becomes

$$\mathbf{\Theta}(f) = -\frac{\langle f, f \rangle \langle df, df \rangle - \langle df, f \rangle \wedge \langle f, df \rangle}{\langle f, f \rangle^2}, \tag{2.9}$$

where f is a holomorphic frame for $U_{1,n}$. If we choose f to be of the form

$$f = \begin{bmatrix} \xi_1 \\ \cdot \\ \cdot \\ \cdot \\ \xi_n \end{bmatrix},$$

where $\xi_j \in \mathcal{O}(U)$ and $\sum |\xi_j|^2 = |f|^2 \neq 0$, then

$$df = \begin{bmatrix} d\xi_1 \\ \cdot \\ \cdot \\ \cdot \\ d\xi_n \end{bmatrix}, \qquad \begin{aligned} {}^t\overline{df} &= (\overline{d\xi}_1, \ldots, \overline{d\xi}_n) \\ &= (d\bar{\xi}_1, \ldots, d\bar{\xi}_n), \end{aligned}$$

and we obtain

$$\Theta(f) = -\frac{|f|^2 \sum_{i=1}^{n} d\xi_i \wedge d\bar{\xi}_i - \sum_{i,j=1}^{n} \bar{\xi}_i \xi_j d\xi_i \wedge d\bar{\xi}_j}{|f|^4}. \tag{2.10}$$

Recall that the functions $\xi_1, \ldots, \xi_n$ are functions of the local coordinates on $G_{1,n} = \mathbf{P}_{n-1}$, and that, in particular, $\Theta(f)$ is a well-defined 2-form on $U \subset \mathbf{P}_{n-1}$. Alternatively, we can think of $(\xi_1, \ldots, \xi_n)$ as being homogeneous coordinates for $\mathbf{P}_{n-1}$, and by the homogeneity of (2.9), we see that the expression in (2.9) induces a well-defined 2-form on all of $\mathbf{P}_{n-1}$, which agrees with the 2-form on U mentioned above. We shall see this differential form again when we study Kähler metrics in Chap. V.

Returning to the general case of $U_{r,n} \to G_{r,n}$, we have seen in Lemma 2.3 that, by a proper choice of holomorphic frame for $U_{r,n}$ and a proper choice of local holomorphic coordinates near some fixed point, we can find a very simple expression for the curvature. We shall now see an example of this. Let

$$f_0 = \begin{bmatrix} I_r \\ 0 \end{bmatrix}$$

be a frame for $U_{r,n}$ at the point $x_0 \in G_{r,n}$ defined by

$$x_0 = \langle \begin{bmatrix} I_r \\ 0 \end{bmatrix} \rangle,$$

where $\langle\, \rangle$ denotes the span of the columns of the frame matrix inside, which is a subspace of $\mathbf{C}^n$ and thus a point in $G_{r,n}$. Letting

$$B_\epsilon = \{Z \in \mathfrak{M}_{n-r,r} \cong \mathbf{C}^{(n-r)r} : |Z| < \epsilon\},$$

the mapping

$$B_\epsilon \longrightarrow G_{r,n}$$

given by

$$Z \longrightarrow \langle \begin{bmatrix} I_r \\ Z \end{bmatrix} \rangle$$

is a coordinate system for $G_{r,n}$ near x_0, with the property that x_0 corresponds to $Z = 0$. There is a natural action of $GL(n, \mathbf{C})$ on $G_{r,n}$ given by left multiplication of frames (i.e., left multiplication of homogeneous coordinates). Namely, if $f \in M_{n,r}$, $x = \langle f \rangle$, and $u \in GL(n, \mathbf{C})$, then set

$$u(x) = \langle u \cdot f \rangle.$$

Moreover, $U(n)$, the unitary group, is transitive on $G_{r,n}$ under this action (a well-known fact of linear algebra).

Therefore if y_0 is any point in $G_{r,n}$ and $y_0 = u(x_0)$ for a unitary matrix u, then the mapping

$$Z \longrightarrow u\begin{bmatrix} I \\ Z \end{bmatrix}$$

gives local coordinates at $y_0 \in G_{r,n}$. The metric at y_0 has the form, with respect to the frame

$$f(z) = u\begin{bmatrix} I \\ Z \end{bmatrix},$$

$$\begin{aligned} h(Z) = h(f(Z)) &= \left(u\begin{bmatrix} I \\ Z \end{bmatrix}\right)^* \left(u\begin{bmatrix} I \\ Z \end{bmatrix}\right) \\ &= I + Z^* Z \\ &= I + O(|Z|^2), \end{aligned}$$

which is the form occurring in Lemma 2.3(a) (note that the dependence on u disappears completely). Thus we see that

$$\Theta(y_0) = \Theta(0) = \bar{\partial}\partial(I + Z^* Z)(0)$$

$$\Theta(y_0) = dZ^* \wedge dZ(0) \tag{2.11}$$

which is the same for all points of $G_{r,n}$ with respect to these *particular* systems of local coordinates. We shall use this expression for the curvature to compute certain Chern classes of this vector bundle in the next section.

3. Chern Classes of Differentiable Vector Bundles

Our object in this section is to give a differential-geometric derivation of the Chern classes of a differentiable **C**-vector bundle $E \to X$. The Chern classes will turn out to be the primary obstruction to admitting global frames, or, more generally, admitting k global sections $\xi_1, \ldots, \xi_k$, $1 \le k \le \operatorname{rank}_{\mathbf{C}} E$, such that $\xi_1 \wedge \ldots \wedge \xi_k \neq 0$, at each point of E (i.e., they are to be obstructions to E or some nonzero subbundle of E being trivial). Classically, the Chern classes are related to the Euler characteristic of a compact manifold X, which for oriented 2-manifolds, for instance, decides completely whether or not there are nonvanishing vector fields on X. More specifically, if E is a **C**-vector bundle of rank r, then the Chern classes $c_j(E)$, $j = 1, \ldots, r$, will be elements of the de Rham group $H^{2j}(X, \mathbf{R})$ having certain functorial properties. As we shall see, they can be defined in terms of the curvature of E with respect to a connection. Our approach here follows the exposition of Bott and Chern [1], based on the original ideas of Chern and Weil.

To begin, we need some multilinear algebra. Recall that $\mathfrak{M}_r$, is the set of $r \times r$ matrices with complex entries. A k-linear form

$$\tilde{\varphi} : \mathfrak{M}_r \times \cdots \times \mathfrak{M}_r \longrightarrow \mathbf{C}$$

is said to be *invariant* if

$$\tilde{\varphi}(gA_1g^{-1}, \ldots, gA_kg^{-1}) = \tilde{\varphi}(A_1, \ldots, A_k)$$

for $g \in GL(r, \mathbf{C})$, $A_i \in \mathfrak{M}_r$. Let $\tilde{I}_k(\mathfrak{M}_r)$ be the C-vector space of all invariant k-linear forms on $\mathfrak{M}_r$.

Suppose that $\tilde{\varphi} \in \tilde{I}_k(\mathfrak{M}_r)$. Then $\tilde{\varphi}$ induces

$$\varphi : \mathfrak{M}_r \longrightarrow \mathbf{C}$$

by setting

$$\varphi(A) = \tilde{\varphi}(A, \ldots, A).$$

It is clear then that φ is a homogeneous polynomial of degree k in the entries of A. Moreover, for $g \in GL(r, \mathbf{C})$,

$$\varphi(gAg^{-1}) = \varphi(A),$$

and we say then that φ is *invariant*. Let $I_k(\mathfrak{M}_r)$ be the set of invariant homogeneous polynomials of degree k as above. Since the isomorphism of the symmetric tensor algebra $S(\mathfrak{M}_r^*)$ and the polynomials on $\mathfrak{M}_r$ preserves degrees (see Sternberg [1]), one obtains† from $\varphi \in I_k(\mathfrak{M}_r)$ an element $\tilde{\varphi} \in \tilde{I}_k(\mathfrak{M}_r)$ such that

$$\tilde{\varphi}(A, \ldots, A) = \varphi(A).$$

We shall omit the tilde and use the same symbol for the multilinear form and its restriction to the diagonal.

Example 3.1: The usual determinant of an $r \times r$ matrix is a mapping

$$\det : \mathfrak{M}_r \longrightarrow \mathbf{C},$$

which is clearly a member of $I_r(\mathfrak{M}_r)$. Moreover, for $A \in \mathfrak{M}_r$ and I, the identity in $\mathfrak{M}_r$, we see that

$$\det(I + A) = \sum_{k=0}^{r} \mathbf{\Phi}_k(A),$$

where each $\mathbf{\Phi}_k \in I_k(\mathfrak{M}_r)$. Note that $\mathbf{\Phi}_k, k = 0, \ldots, r$, so defined is a *real* mapping; i.e., if M has real entries, then $\mathbf{\Phi}_k(M)$ is real.

We would like to extend the action of $\varphi \in \tilde{I}_k(\mathfrak{M}_r)$ to $\mathcal{E}^*(\mathrm{Hom}(E, E))$. First, we define the extension to $\mathfrak{M}_r \otimes_{\mathcal{E}} \mathcal{E}^p$. If U is open in X and $A_i \cdot w_i \in \mathfrak{M}_r(U) \otimes_{\mathcal{E}(U)} \mathcal{E}^p(U)$, then set

$$\varphi_U(A_1 \cdot w_1, \ldots, A_k \cdot w_k) = w_1 \wedge \cdots \wedge w_k \varphi(A_1, \ldots, A_k).$$

By linearity φ becomes a well-defined k-linear form on $\mathfrak{M}_r \otimes_{\mathcal{E}} \mathcal{E}^p$. If $\xi_j \in \mathcal{E}^p(U, \mathrm{Hom}(E, E))$, $j = 1, \ldots, k$, then set

$$\varphi_U(\xi_1, \ldots, \xi_k) = \varphi_U(\xi_1(f), \ldots, \xi_k(f)).$$

† This process is called *polarization* and a specific formula for $\tilde{\varphi}$ is

$$\tilde{\varphi}(A_1, \ldots, A_k) = \frac{(-1)^k}{k!} \sum_{j=1}^{k} \sum_{i_1 < \cdots < i_j} (-1)^j \varphi(A_{i_1} + \cdots + A_{i_j}).$$

This shows that the invariance of $\tilde{\varphi}$ follows from that of φ.

We can check that this definition is independent of the choice of frame. Namely, if g is a change of frame, then by (1.13)

$$\varphi_U(\xi_1(fg), \ldots, \xi_k(fg)) = \varphi_U(g^{-1}\xi_1(f)g, \ldots, g^{-1}\xi_k(f)g)$$
$$= \varphi_U(\xi_1(f), \ldots, \xi_k(f)),$$

by the invariance of φ and the induced invariance of φ when acting on matrices with differential form coefficients. Thus we get an extension of φ to all of X,

$$\varphi_x : \mathcal{E}^p(X, \mathrm{Hom}(E, E)) \times \cdots \times \mathcal{E}^p(X, \mathrm{Hom}(E, E)) \to \mathcal{E}^{pk}(X),$$

which when restricted to the diagonal induces the action of the invariant polynomial $\varphi \in I_k(\mathfrak{M}_r)$ on $\mathcal{E}^p(X, \mathrm{Hom}(E, E))$, which we denote by

$$\varphi_X : \mathcal{E}^p(X, \mathrm{Hom}(E, E)) \longrightarrow \mathcal{E}^{pk}(X).$$

Now suppose that we have a connection

$$D : \mathcal{E}(X, E) \longrightarrow \mathcal{E}^1(X, E)$$

defined on $E \to X$. Then we have the curvature $\boldsymbol{\Theta}_E(D)$, as defined in Definition 1.8. So if $\varphi \in I_k(\mathfrak{M}_r)$, $\varphi_x(\boldsymbol{\Theta}_E(D))$ is a global $2k$-form on X. We can now state the following basic result due to A. Weil (cf. Bott and Chern [1]).

Theorem 3.2: Let $E \to X$ be a differentiable **C**-vector bundle, let D be a connection on E, and suppose that $\varphi \in I_k(\mathfrak{M}_r)$. Then

(a) $\varphi_X(\boldsymbol{\Theta}_E(D))$ is closed.

(b) The image of $\varphi_X(\boldsymbol{\Theta}_E(D))$ in the de Rham group $H^{2k}(X, \mathbf{C})$ is independent of the connection D.

Proof: To prove (a), we shall show that for $\varphi \in I_k(\mathfrak{M}_r)$, the associated invariant k-linear form φ satisfying

$$\varphi(gA_1g^{-1}, \ldots, gA_kg^{-1}) = \varphi(A_1, \ldots, A_k)$$

for all $g \in GL(r, \mathbf{C})$ satisfies

(3.1) $$\sum_j \varphi(A_1, \ldots, [A_j, B], \ldots, A_k) = 0$$

for all $A_j, B \in \mathfrak{M}_r$.

Assuming (3.1), we shall first see that (a) holds. Recalling the definition of the Lie product on $\mathfrak{M}_r \otimes \mathcal{E}^*$ preceding Proposition 1.10, equation (3.1) gives, for U open in X,†

(3.2) $$\sum_\alpha (-1)^{f(\alpha)} \varphi_U(A_1, \ldots, [A_\alpha, B], \ldots, A_k) = 0$$

for all $A_\alpha \in \mathfrak{M}_r \otimes \mathcal{E}^{p_\alpha}(U)$ and $B \in \mathfrak{M}_r \otimes \mathcal{E}^q(U)$, where $f(\alpha) = \deg B \sum_{\beta \leq \alpha} \deg A_\beta$. Moreover, it follows from the definition of a k-linear form that

(3.3) $$d\varphi_U(A_1, \ldots, A_k) = \sum_\alpha (-1)^{g(\alpha)} \varphi_U(A_1, \ldots, dA_\alpha, \ldots, A_k)$$

†We have previously defined the action of φ only on $\mathfrak{M}_r \otimes \mathcal{E}^p$, but this clearly extends to an action on $\mathfrak{M}_r \otimes \mathcal{E}^*$.

for $A_\alpha \in \mathfrak{M}_r \otimes \mathcal{E}^{p_\alpha}(U)$, where $g(\alpha) = \sum_{\beta<\alpha} \deg A_\beta$. We want to show that $d\varphi_x(\boldsymbol{\Theta}) = 0$, and it suffices to show that for a frame f over U,

$$d\varphi_U(\boldsymbol{\Theta}(f)) = 0.$$

But from equation (3.3) we have [letting $\boldsymbol{\Theta}(f) = \boldsymbol{\Theta}$]

$$d\varphi_U(\boldsymbol{\Theta}) = d\varphi_U(\boldsymbol{\Theta}, \ldots, \boldsymbol{\Theta}) = \sum \varphi_U(\boldsymbol{\Theta}, \ldots, d\boldsymbol{\Theta}, \ldots, \boldsymbol{\Theta}),$$

noting that deg $\boldsymbol{\Theta}$ is even. From Proposition 1.10 we have that

$$d\varphi_U(\boldsymbol{\Theta}) = \sum \varphi_U(\boldsymbol{\Theta}, \ldots, [\boldsymbol{\Theta}, \theta], \ldots, \boldsymbol{\Theta}),$$

but this vanishes by equation (3.2), and thus $\varphi_X(\boldsymbol{\Theta}_E)$ is a closed form.

Now all that remains is to show that the invariance of φ implies equation (3.1). First, if $f(t)$ and $g(t)$ are power series with matrix coefficients which converge for all $t \in \mathbf{C}$, i.e.,

$$f(t) = \sum_n A_n t^n \quad \text{and} \quad g(t) = \sum_n B_n t^n,$$

then

$$f(t)g(t) = A_0 B_0 + (A_1 B_0 + A_0 B_1)t + O(|t|^2),$$

and if φ is a linear functional on $\mathfrak{M}_r$, then

$$\varphi(f(t)) = \sum_n \varphi(A_n)t^n.$$

Now for $A, B \in \mathfrak{M}_r$ it follows from the above remarks that

$$e^{-tB} A e^{tB} - A = t[A, B] + O(|t|^2). \tag{3.4}$$

We now want to show that (3.1) holds. We consider, for simplicity, the case $k = 2$, the general case being an immediate generalization. Thus, if $\varphi \in I_2(\mathfrak{M}_r)$, by the invariance of the associated bilinear form we obtain

$$\varphi(e^{-tB} A_1 e^{tB}, e^{-tB} A_2 e^{tB}) - \varphi(A_1, A_2) = 0$$

for all $t \in \mathbf{C}$ and $A_1, A_2, B \in \mathfrak{M}_r$, since $e^{-tB} \cdot e^{tB} = I$. By adding and subtracting $\varphi(e^{-tB} A_1 e^{tB}, A_2)$ to/from the above identity, we obtain

$$\varphi(e^{-tB} A_1 e^{tB}, e^{-tB} A_2 e^{tB}) - \varphi(e^{-tB} A_1 e^{tB}, A_2) + \varphi(e^{-tB} A_1 e^{tB}, A_2) - \varphi(A_1, A_2) \equiv 0.$$

Applying (3.4) to each of the differences above, we find that

$$\varphi(e^{-tB} A_1 e^{tB}, t[A_2, B]) + O(|t|^2) + \varphi(t[A_1, B] + O(|t|^2), A_2) = t\{\varphi(A_1, [A_2, B]) + \varphi([A_1, B], A_2)\} + O(|t|^2) \equiv 0.$$

Thus the coefficient of t must also vanish identically, and this proves (3.1) in the case $k = 2$. It is now clear that the general case is obtained in the same way by adding and subtracting the appropriate $k - 1$ terms to/from the difference

$$\varphi(e^{-tB} A_1 e^{tB}, \ldots, e^{-tB} A_k e^{tB}) - \varphi(A_1, \ldots, A_k),$$

and we omit further details.

Now that $\varphi_X(\mathbf{\Theta}_E(D))$ is closed, it makes sense to consider its image in the de Rham group $H^{2k}(X, \mathbf{C})$. To prove part (b), we shall show that for two connections D_1, D_2 on $E \longrightarrow X$ there is a differential form α so that

$$\varphi(\mathbf{\Theta}_E(D_1)) - \varphi(\mathbf{\Theta}_E(D_2)) = d\alpha. \tag{3.5}$$

To do this, we need to consider one-parameter families of differential forms on X and one-parameter families of connections on $E \to X$, and to point out some of their properties.

Let $\alpha(t)$ be a C^∞ one-parameter family of differential forms on $X, t \in \mathbf{R}$; i.e., α has the local representation

$$\alpha(t) = \sum a_I(x, t) dx_I$$

for $t \in \mathbf{R}$ and a_I is C^∞ in x and t (cf. Sec. 2 in Chap. I). Define locally

$$\dot{\alpha}(t) = \frac{\partial \alpha(t)}{\partial t} = \sum \frac{\partial a_I}{\partial t} dx_I$$

$$\int_a^b \alpha(t)dt = \sum \left(\int_a^b a_I(x, t)dt \right) dx_I$$

It is easy to check that these definitions are independent of the local coordinates used and that $\dot{\alpha}(t)$ and $\int_a^b \alpha(t)dt$ are well-defined global differential forms. Also,

$$\frac{\partial}{\partial t}(\alpha(t) \wedge \beta(t)) = \frac{\partial \alpha}{\partial t}(t) \wedge \beta(t) + \alpha(t) \wedge \frac{\partial \beta}{\partial t}(t)$$

and

$$\int_a^b \dot{\alpha}(t)dt = \alpha(t)|_a^b = \alpha(b) - \alpha(a).$$

For a differentiable vector bundle $E \to X$, we define a C^∞ one-parameter family of connections on E to be a family of connections $\{D_t\}_{t \in \mathbf{R}}$ such that for a C^∞ frame f over U open in X the connection matrix $\theta_t(f) := \theta(D_t, f)$ has coefficients which are C^∞ one-parameter families of differential forms on E.† Suppose that D_t is such a family of connections. Then for a C^∞ frame f over U and $\xi \in \mathcal{E}(U, E)$ we have

$$\begin{aligned} \frac{\partial}{\partial t} D_t \xi(f) &= \frac{\partial}{\partial t}(d\xi(f) + \theta_t(f)\xi(f)) \\ &= \left(\frac{\partial}{\partial t} \theta_t(f) \right) \xi(f). \end{aligned}$$

Moveover, since a change of frame is independent of t, this clearly defines for each $t_0 \in \mathbf{R}$ a mapping

$$\dot{D}_{t_0} : \mathcal{E}(X, E) \longrightarrow \mathcal{E}^1(X, E)$$

†We shall need only C^1 families of connections in the applications, which have the analogous definition.

by

$$\dot{D}_{t0}(\xi) = \frac{\partial}{\partial t} D_t \xi|_{t0}.$$

Moreover, this mapping is $\mathcal{E}_X$-linear. Therefore $\dot{D}_{t0}$ defines an element of $\mathcal{E}^1(X, \mathrm{Hom}(E, E))$ which we also call $\dot{D}_{t0}$. As we pointed out above, $\dot{D}_{t0}$ has a local representation

$$\dot{\theta}_{t0}(f) := \dot{D}_{t0}(f) = \frac{\partial}{\partial t}\theta_t(f)|_{t0}.$$

We can now reduce the proof of part (b) to the following lemma, which will be proved below.

Lemma 3.3: Let D_t be a C^∞ one-parameter family of connections, and for each $t \in \mathbf{R}$, let $\mathbf{\Theta}_t$ be the induced curvature. Then for any $\varphi \in I_k(\mathfrak{M}_r)$,

$$\varphi_X(\mathbf{\Theta}_b) - \varphi_X(\mathbf{\Theta}_a) = d\left(\int_a^b \varphi'(\mathbf{\Theta}_t; \dot{D}_t)dt\right),$$

where

$$\varphi'(\xi; \eta) = \sum_\alpha \varphi(\xi, \xi, \ldots, \xi, \underset{(\alpha)}{\eta}, \xi, \ldots, \xi),$$

(α) denotes the αth argument, and $\xi, \eta \in \mathcal{E}^*(X, \mathrm{Hom}(E, E))$.

Namely, if D_1 and D_2 are two given connections, for $E \longrightarrow X$, then let

$$D_t = tD_1 + (1-t)D_2,$$

which is clearly a C^∞ one-parameter family of connections on E. Thus, by Lemma 3.3, we see that

$$\varphi_X(\mathbf{\Theta}_E(D_1)) - \varphi_X(\mathbf{\Theta}_E(D_2)) = \varphi_X(\mathbf{\Theta}_1) - \varphi_X(\mathbf{\Theta}_2) = d\alpha,$$

where

$$\alpha = \int_0^1 \varphi'(\mathbf{\Theta}_t; \dot{D}_t)\, dt.$$

Q.E.D.

Proof of Lemma 3.3: It suffices to show that, for a frame f over U, we have

(3.6) $$\dot{\varphi}_U(\mathbf{\Theta}) = d\varphi'_U(\mathbf{\Theta}; \dot{\theta}),$$

where $\mathbf{\Theta} = \mathbf{\Theta}_E(D_t, f)$, $\theta = \theta(D_t, f)$, and the dot denotes differentiation with respect to the parameter t, as above. Here we use the simple fact that exterior differentiation commutes with integration with respect to the parameter t. We proceed by computing

$$\begin{aligned} d\varphi'_U(\mathbf{\Theta}; \dot{\theta}) &= d\Big(\sum_\alpha \varphi_U\big(\mathbf{\Theta}, \ldots, \underset{(\alpha)}{\dot{\theta}}, \ldots, \mathbf{\Theta}\big)\Big) \\ &= \sum_\alpha \Big\{\sum_{i<\alpha} \varphi_U\big(\mathbf{\Theta}, \ldots, \underset{(i)}{d\mathbf{\Theta}}, \ldots, \underset{(\alpha)}{\dot{\theta}}, \ldots, \mathbf{\Theta}\big) \\ &\quad + \varphi_U\big(\mathbf{\Theta}, \ldots, \underset{(\alpha)}{d\dot{\theta}}, \ldots, \mathbf{\Theta}\big) \\ &\quad - \sum_{i>\alpha} \varphi_U\big(\mathbf{\Theta}, \ldots, \underset{(\alpha)}{\dot{\theta}}, \ldots, \underset{(i)}{d\mathbf{\Theta}}, \ldots, \mathbf{\Theta}\big)\Big\}. \end{aligned}$$

By adding and subtracting

$$\sum_{\alpha} \varphi_U(\mathbf{\Theta}, \ldots, [\underset{(\alpha)}{\dot{\theta}}, \theta], \ldots, \mathbf{\Theta})$$

to/from the above equation and noting that

$$\dot{\mathbf{\Theta}} = d\dot{\theta} + [\dot{\theta}, \theta] \qquad \text{(differentiation of (1.14))}$$

$$d\mathbf{\Theta} = [\mathbf{\Theta}, \theta] \qquad \text{(Bianchi identity, Proposition 1.10)},$$

we obtain the equation

$$\begin{aligned} d\varphi'_U(\mathbf{\Theta}; \dot{\theta}) = &\sum_{\alpha} \varphi_U(\mathbf{\Theta}, \ldots, \underset{(\alpha)}{\dot{\mathbf{\Theta}}}, \ldots, \mathbf{\Theta}) \\ &+ \sum_{\alpha} \{ \sum_{i<\alpha} \varphi_U(\mathbf{\Theta}, \ldots, [\mathbf{\Theta}, \underset{(i)}{\theta}], \ldots, \underset{(\alpha)}{\dot{\theta}}, \ldots, \mathbf{\Theta}) \\ &- \varphi_U(\mathbf{\Theta}, \ldots, [\dot{\theta}, \underset{(\alpha)}{\theta}], \ldots, \mathbf{\Theta}) \\ &- \sum_{i>\alpha} \varphi_U(\mathbf{\Theta}, \ldots, \underset{(\alpha)}{\dot{\theta}}, \ldots, [\mathbf{\Theta}, \underset{(i)}{\theta}], \ldots, \mathbf{\Theta}) \}. \end{aligned}$$

By (3.2), we see that the second sum over α vanishes, and we are left with

$$\begin{aligned} d\varphi'_U(\mathbf{\Theta}; \dot{\theta}) &= \sum_{\alpha} \varphi_U(\mathbf{\Theta}, \ldots, \underset{(\alpha)}{\dot{\mathbf{\Theta}}}, \ldots, \mathbf{\Theta}) \\ &= \dot{\varphi}_U(\mathbf{\Theta}), \end{aligned}$$

which is (3.6).

Q.E.D.

We are now in a position to define Chern classes of a differentiable vector bundle. From Example 3.1 we consider the invariant polynomials $\mathbf{\Phi}_k \in \boldsymbol{I}_k(\mathfrak{M}_r)$ defined by the equation

$$\det(I + A) = \sum_k \mathbf{\Phi}_k(A), \qquad A \in \mathfrak{M}_r.$$

Definition 3.4: Let $E \longrightarrow X$ be a differentiable vector bundle equipped with a connection D. Then the kth *Chern form of* E *relative to the connection* D is defined to be

$$c_k(E, D) = (\mathbf{\Phi}_k)_X\Big(\frac{i}{2\pi}\mathbf{\Theta}_E(D)\Big) \in \mathcal{E}^{2k}(X).$$

The (*total*) *Chern form of* E *relative to* D is defined to be

$$c(E, D) = \sum_{k=0}^{r} c_k(E, D), \qquad r = \text{rank } E.$$

The kth *Chern class* of the vector bundle E, denoted by $c_k(E)$, is the cohomology class of $c_k(E, D)$ in the de Rham group $H^{2k}(X, \mathbf{C})$, and the *total Chern class* of E, denoted by $c(E)$, is the cohomology class of $c(E, D)$ in $H^*(X, \mathbf{C})$; i.e., $c(E) = \sum_{k=0}^{r} c_k(E)$.

It follows from Theorem 3.2 that the Chern classes are well defined and independent of the connection D used to define them. Thus the Chern classes are topological cohomology classes in the base space of the vector bundle

E. We shall see shortly that they are indeed obstructions to finding, e.g., global frames. First we want to show that the Chern classes are real cohomology classes.

Proposition 3.5: Let D be a connection on a Hermitian vector bundle E compatible with the Hermitian metric h. Then the Chern form $c(E, D)$ is a real differential form, and it follows that $c(E) \in H^*(X, \mathbf{R})$, under the canonical inclusion $H^*(X, \mathbf{R}) \subset H^*(X, \mathbf{C})$.

Proof: It suffices to show that for a local frame f the matrix representation for the Chern form is a real differential form. Therefore let $h = h(f)\boldsymbol{\Theta} = \boldsymbol{\Theta}(D, f)$, as usual, and recall that D being compatible with the metric h was equivalent to the condition (1.22),

$$dh = h\theta + {}^t\bar{\theta}h,$$

whose exterior derivative is given by

$$0 = dh \wedge \theta + hd\theta + d^t\bar{\theta} \cdot h - {}^t\bar{\theta} \wedge dh.$$

By substituting the above expression for dh, we obtain

$$0 = h\boldsymbol{\Theta} + {}^t\bar{\boldsymbol{\Theta}}h. \tag{3.7}$$

In particular, if f is a unitary frame, we note that $\boldsymbol{\Theta}$ is skew-Hermitian. Using (3.7) we can show that if

$$c := c(E, D, f) = \det\left(I + \frac{i}{2\pi}\boldsymbol{\Theta}\right),$$

then $c = \bar{c}$; i.e., c is a real differential form. Namely,

$$\begin{array}{ccc} \det\left(h + \dfrac{i}{2\pi}\boldsymbol{\Theta}h\right) & = & \det\left(I + \dfrac{i}{2\pi}\boldsymbol{\Theta}\right) \cdot \det h \\ \| & & \\ \det\left(h - \dfrac{i}{2\pi}h^t\bar{\boldsymbol{\Theta}}\right) & = & \det\ h \cdot \det\left(I - \dfrac{i}{2\pi}{}^t\bar{\boldsymbol{\Theta}}\right), \end{array}$$

where the vertical equality is given by (3.7). Now it follows that

$$\begin{aligned} c = \det\left(I + \frac{i}{2\pi}\boldsymbol{\Theta}\right) &= \det\left(I - \frac{i}{2\pi}{}^t\bar{\boldsymbol{\Theta}}\right) \\ &= \det\left(I - \frac{i}{2\pi}\bar{\boldsymbol{\Theta}}\right) = \bar{c}. \end{aligned}$$

Q.E.D.

We want to prove some functorial properties of the Chern classes. In doing so we shall see that it is often convenient to choose a particular connection to find a useful representative for the Chern classes. We remark that the de Rham group $H^*(X, \mathbf{R})$ on a differentiable manifold X carries a ring structure induced by wedge products; i.e., if

$$c, c' \in H^*(X, \mathbf{R})$$

and $c = [\varphi]$ and $c' = [\varphi']$, then

$$c \cdot c' = [\varphi \wedge \varphi'],$$

which is easily checked to be well defined.†

Theorem 3.6: Suppose that E and E' are differentiable **C**-vector bundles over a differentiable manifold X. Then

(a) If $\varphi : Y \longrightarrow X$ is a differentiable mapping where Y is a differentiable manifold, then

$$c(\varphi^* E) = \varphi^* c(E),$$

where $\varphi^* E$ is the pullback vector bundle and $\varphi^* c(E)$ is the pullback of the cohomology class $c(E)$.

(b) $c(E \oplus E') = c(E) \cdot c(E')$, where the product is in the de Rham cohomology ring $H^*(X, \mathbf{R})$.

(c) $c(E)$ depends only on the isomorphism class of the vector bundle E.

(d) If E^* is the dual vector bundle to E, then

$$c_j(E^*) = (-1)^j c_j(E).$$

Proof:

(a) Let D be any connection on $E \longrightarrow X$. To prove part (a), it will suffice to define a connection D^* on $\varphi^* E$ so that

$$\varphi^*(\mathbf{\Theta}(D)) = \mathbf{\Theta}(D^*),$$

where φ^* is the induced map on curvature. We proceed as follows. Suppose that $f = (e_1, \ldots, e_r)$ is a frame over U in X. Then $f^* = \left(e_1^*, \ldots, e_r^*\right)$, where $e_i^* = e_i \circ \varphi$, is a frame for $\varphi^* E$ over $\varphi^{-1}(U)$, and frames of the form f^* cover Y. Also, if $g : U \longrightarrow GL(r, \mathbf{C})$ is a change of frame over U, then $g^* = g \circ \varphi$ is a change of frame in $\varphi^* E$ over $\varphi^{-1}(U)$. Now define a connection matrix

$$\theta^*(f^*) := \varphi^* \theta(f) = [\varphi^* \theta_{\rho\sigma}],$$

where $\varphi^* \theta_{\rho\sigma}$ is the induced map on forms. Moreover, it is easy to see that

$$g^* \theta^*(f^* g^*) = \theta^*(f^*) g^* + dg^*$$

so that θ^* defines a global connection on $h^* E$. And, finally, we have

$$\begin{aligned}
\mathbf{\Theta}(D^*, f^*) &= d\theta^*(f^*) + \theta^*(f^*) \wedge \theta^*(f^*) \\
&= d\varphi^* \theta(f) + \varphi^* \theta(f) \wedge \varphi^* \theta(f) \\
&= \varphi^*(d\theta(f) + \theta(f) \wedge \theta(f)) \\
&= \varphi^* \mathbf{\Theta}(D, f),
\end{aligned}$$

which completes the proof of part (a).

†This is a representation for the *cup product* of algebraic topology; see, e.g., Bredon [1] and Greenberg [1].

(b) Given D and D', connections on E and E', respectively, it suffices to find a connection $D^{\oplus}$ on $E \oplus E'$ so that

$$c(E \oplus E', D^{\oplus}) = c(E, D) \wedge c(E', D').$$

Also, as in part (a), it suffices to consider a local argument. Therefore for θ and θ' connection matrices over U on E and E', respectively, it is easy to see that

$$\theta^{\oplus} = \begin{bmatrix} \theta & 0 \\ 0 & \theta' \end{bmatrix}$$

is a connection matrix defining a global connection on $E \oplus E'$ (the details are left to the reader). The associated curvature matrix is given by

$$\mathbf{\Theta}^{\oplus} = \begin{bmatrix} \mathbf{\Theta} & 0 \\ 0 & \mathbf{\Theta}' \end{bmatrix}.$$

Thus

$$\begin{aligned} c(E \oplus E', D^{\oplus})|_U &= \det \begin{bmatrix} I + \dfrac{i}{2\pi}\mathbf{\Theta} & 0 \\ 0 & I' + \dfrac{i}{2\pi}\mathbf{\Theta}' \end{bmatrix} \\ &= \det\left[I + \frac{i}{2\pi}\mathbf{\Theta} \right] \det\left[I' + \frac{i}{2\pi}\mathbf{\Theta}' \right] \\ &= c(E, D)|_U \wedge c(E', D')|_U . \end{aligned}$$

(c) Suppose that $\alpha : E \longrightarrow E'$ is a vector bundle isomorphism. Then we want to show that $c(E) = c(E')$. This is simple, and similar to the argument in part (a). Let D be a connection on E, and define a connection D' on E' by defining the connection matrix for D' by the relation

$$\theta'(f') = \theta(f),$$

where f is a frame for E and $f' = (\alpha(e_1), \ldots, \alpha(e_r))$ is a frame for E'. As in (a), this is a connection for E', and it follows that $\mathbf{\Theta}'(f') = \mathbf{\Theta}(f)$, and hence $c(E) = c(E')$.

(d) Suppose that the duality between E and E^* is represented by $\langle\, , \rangle$ (not to be confused with a metric) and that D is a connection on E. If f and f^* are dual frames over an open set U, i.e., $\langle e_\sigma, e^*_\rho \rangle = \delta_{\sigma\rho}$, then we can define a connection D^* in E^* by setting

$$\theta^* = \theta(D^*, f^*) = -{}^t\theta(D, f). \tag{3.8}$$

We can check that θ^* defined by (3.8) is indeed a connection on E^*. Suppose that g is a change of frame $f \longrightarrow fg$ on E. Then the induced change of frame for the dual frame f^* is given by $f^* \longrightarrow f^{*t}(g^{-1})$, as is easy to verify. Thus, if we let $g^* = ({}^t g)^{-1}$, we have to check that

$$\theta^*(f^* g^*) = (g^*)^{-1} dg^* + (g^*)^{-1} \theta^*(f^*) g^* \tag{3.9}$$

to see that θ^* is a well-defined connection on E^*. But (3.9) holds if and only if

$$-{}^t\theta(fg) = {}^t g d^t(g^{-1}) - {}^t g^t\theta(f)^t(g^{-1}),$$

which simplifies to, after taking transposes and using the fact that $dg^{-1} = -g^{-1}dgg^{-1}$,

$$\theta(fg) = g^{-1}dg + g^{-1}\theta(f)g,$$

which holds, since $\theta(f)$ is a connection matrix. Therefore the curvature for E^* is

$$\begin{aligned}\mathbf{\Theta}^* &= d\theta^* + \theta^* \wedge \theta^* \\ &= -d^t\theta + {}^t\theta \wedge {}^t\theta \\ &= -d^t\theta - {}^t(\theta \wedge \theta) \\ &= -{}^t(d\theta + \theta \wedge \theta) \\ &= -{}^t\mathbf{\Theta}.\end{aligned}$$

Thus the Chern forms restricted to U are related by

$$\begin{aligned}c_k(E^*, D^*) &= \mathbf{\Phi}_k\left(-\frac{i}{2\pi}\mathbf{\Theta}\right) \\ &= (-1)^k\mathbf{\Phi}_k\left(\frac{i}{2\pi}\mathbf{\Theta}\right) \\ &= (-1)^k c_k(E, D),\end{aligned}$$

where we note that the invariant polynomial $\mathbf{\Phi}_k$ is homogeneous of degree k and is invariant with respect to transpose (since det is).

Q.E.D.

Remark: In the case where $E \longrightarrow X$ is a holomorphic vector bundle and h is a Hermitian matrix on E, h^*, the induced metric on E^*, is given by

$$h^*(f^*) = {}^t(h^{-1}(f)),$$

where f and f^* are dual holomorphic frames. From this we see that

$$\begin{aligned}\theta^* &= (h^*)^{-1}\partial h^* \\ &= {}^t h\partial^t(h^{-1}) \\ &= -(\partial^t h)^t(h^{-1}) \\ &= -{}^t(h^{-1}\partial h) = -{}^t\theta\end{aligned}$$

and

$$\mathbf{\Theta}^* = -\bar{\partial}^t\theta = -{}^t\mathbf{\Theta}.$$

We now use the above functorial properties to derive the obstruction-theoretic properties of Chern classes, i.e., the obstructions to finding global sections.

Theorem 3.7: Let $E \longrightarrow X$ be a differentiable vector bundle of rank r. Then

(a) $c_0(E) = 1$.
(b) If $E \cong X \times \mathbf{C}^r$ is trivial, then $c_j(E) = 0$, $j = 1, \ldots, r$; i.e., $c(E) = 1$.
(c) If $E \cong E' \oplus T_s$, where T_s is a trivial vector bundle of rank s, then

$$c_j(E) = 0, \quad j = r - s + 1, \ldots, r.$$

Proof:

(a) This is obvious from the definition of Chern classes.
(b) If $E = X \times \mathbf{C}^r$, then $\mathcal{E}(X, E) \cong (\mathcal{E}(X))^r$, and a connection

$$D : \mathcal{E}(X, E) \longrightarrow \mathcal{E}^1(X, E)$$

can be defined by

$$D\xi = d\xi = d \begin{bmatrix} \xi_1 \\ \cdot \\ \cdot \\ \cdot \\ \xi_r \end{bmatrix},$$

where $\xi_j \in \mathcal{E}(X)$. In this case the connection matrix θ is identically zero. Then the curvature vanishes, and we have

$$c(E, D) = \det(I + 0) = 1,$$

which implies that $c_j(E, D) = 0$, $j > 0$.
(c) We compute

$$\begin{aligned} c(E) &= c(E' \oplus T_s) \\ &= c(E') \cdot c(T_s) \\ &= c(E') \cdot 1 \end{aligned}$$

by Theorem 3.6 and part (b). Moreover, E' is of rank $r - s$, and so we have

$$c(E) = 1 + c_1(E) + \cdots + c_r(E) = 1 + c_1(E') + \cdots + c_{r-s}(E'),$$

from which it follows that

$$c_j(E) = 0, \quad j = r - s + 1, \ldots, r.$$

Q.E.D.

We shall now use Theorem 3.7 to show that some of our examples of vector bundles discussed in Chap. I are indeed nontrivial vector bundles by showing that they have nonvanishing Chern classes.

Example 3.8: Consider $T(\mathbf{P}_1(\mathbf{C}))$, which is R-linear isomorphic to $T(S^2)$, the real tangent bundle to the 2-sphere S^2, and we shall show that it has a nonzero first Chern class. The natural metric on $T(\mathbf{P}_1(\mathbf{C}))$ is the *chordal metric* defined by

$$h(z) = h\left(\frac{\partial}{\partial z}, \frac{\partial}{\partial z}\right) = \frac{1}{(1 + |z|^2)^2}$$

in the z-plane; if $w = 1/z$ is the coordinate system at infinity (from the classical point of view), $h(\partial/\partial w, \partial/\partial w)$ has the same form. We compute

$$\theta(z) = h(z)^{-1}\partial h(z)$$
$$= (1 + |z|^2)^2 \,\partial\left(\frac{1}{(1+|z|^2)^2}\right)$$
$$\theta(z) = -\frac{2\bar{z}}{(1+|z|^2)}dz$$
$$\Theta = \bar{\partial}\theta = \frac{2}{(1+|z|^2)^2}dz \wedge d\bar{z}.$$

Therefore

$$c_1(E, h) = \frac{i}{\pi(1+|z|^2)^2}dz \wedge d\bar{z}$$
$$= \frac{2dx \wedge dy}{\pi(1+|z|^2)^2}.$$

Now

$$\int_{\mathbf{P}_1} c_1(E, h) = \frac{2}{\pi}\int_0^\infty \int_0^{2\pi} \frac{\rho d\rho d\theta}{(1+\rho^2)^2}$$
$$= 4\int_0^\infty \frac{\rho d\rho}{(1+\rho^2)^2}$$
$$= 2\int_1^\infty \frac{du}{u^2}$$
$$= 2.$$

Thus the closed differential form $c_1(E, h)$ cannot be exact, since its integral over the 2-cycle $\mathbf{P}_1$ is nonzero. Therefore $T(\mathbf{P}_1(\mathbf{C}))$ is a nontrivial complex line bundle. Note that the integral of the Chern class over $\mathbf{P}_1$ was in fact 2, which is the Euler characteristic of $\mathbf{P}_1$. This is true in much greater generality. Namely, the classical Gauss-Bonnet theorem asserts that the integral of the Gaussian curvature over a compact 2-manifold is the Euler characteristic (see e.g. Eisenhart [1]). More generally,

$$\int_X c_n(T(X)) = \chi(X)$$

for a compact n-dimensional complex manifold X (see Chern [2]). We shall see the above computation on the 2-sphere in a different context in the next section.

Example 3.9: Consider the universal bundle $E = U_{2,3} \longrightarrow G_{2,3}$, which is a vector bundle with fibres isomorphic to $\mathbf{C}^2$. In Example 2.4 we have computed the curvature in an appropriate coordinate system, and we obtained

$$\Theta(y_0) = dZ^* \wedge dZ(0),$$

using the notation of Example 2.4. Thus we find that $Z = (Z_{11}, Z_{12})$, $Z_{1j} \in \mathbf{C}$, and we have the 2×2 curvature matrix

$$dZ^* \wedge dZ = \begin{bmatrix} d\bar{Z}_{11} \wedge dZ_{11} & d\bar{Z}_{11} \wedge dZ_{12} \\ d\bar{Z}_{12} \wedge dZ_{11} & d\bar{Z}_{12} \wedge dZ_{12} \end{bmatrix}$$

from which we compute

$$\begin{aligned} c_2(E,h)(y_0) &= \det\left(\frac{i}{2\pi} dZ^* \wedge dZ\right) \\ &= \left(-\frac{1}{2\pi^2}\right) dZ_{11} \wedge d\bar{Z}_{11} \wedge dZ_{12} \wedge d\bar{Z}_{12} \\ &= \left(-\frac{1}{2\pi^2}\right) \cdot \left(\frac{2}{i}\right)^2 dX_{11} \wedge dY_{11} \wedge dX_{12} \wedge dY_{12} \\ &= \frac{2}{\pi^2} dX_{11} \wedge dY_{11} \wedge dX_{12} \wedge dY_{12}, \end{aligned}$$

which shows that $c_2(E,h)$ is a volume form for $G_{2,3}$ and, consequently, that

$$\int_{G_{2,3}} c_2(E,h) > 0.$$

This shows that $c_2(E,h) \neq 0$. Thus E has no trivial subbundles and is itself not trivial.

4. Complex Line Bundles

In this section we are going to continue our study of Chern classes of vector bundles by restricting attention to *complex line bundles*, i.e., differentiable or holomorphic **C**-vector bundles of rank 1. In particular, we shall characterize which cohomology classes in $H^2(X, \mathbf{R})$ (for a given differentiable manifold X) are the first Chern class of a complex line bundle over X, a result which has an important application in Chap. VI when we prove Kodaira's fundamental theorem characterizing which abstract compact complex manifolds admit an embedding into complex projective space.

We start with the following two propositions, which are true for vector bundles of any rank.

Proposition 4.1: Let $E \longrightarrow X$ be a differentiable vector bundle. Then there is a finite open covering $\{U_\alpha\}, \alpha = 1, \ldots, N$, of X such that $E|_{U_\alpha}$ is trivial.

Proof: If X is compact, then the result is obvious. By definition we are assuming that X is paracompact (see Chap. I). Now let $\{V_\beta\}$ be an open covering of X such that $E|_{v\beta}$ is trivial. By a standard result in topology, X has topological dimension n implies that there is a refinement $\{U_\alpha\}$ of $\{V_\beta\}$ with the property that the intersection of any $(n+2)$ elements of the covering $\{U_\alpha\}$ is empty, which, in particular means that $\{U_\alpha\}$ is a locally finite covering of X. Let $\{\varphi_\alpha\}$ be a partition of unity subordinate to the covering $\{U_\alpha\}$.

Let A_i be the set of unordered $(i+1)$-tuples of distinct elements of the index set of $\{\varphi_\alpha\}$. Given $a \in A_i, a = \{\alpha_0, \ldots, \alpha_i\}$, let

$$W_{ia} = \{x \in X : \varphi_\alpha(x) < \min[\varphi_{\alpha_0}(x), \ldots, \varphi_{\alpha_i}(x)] \text{ for all } \alpha \neq \alpha_0, \ldots, \alpha_i\}.$$

Then it follows that each W_{ia} is open, and $W_{ia} \cap W_{ib} = \varnothing$ if $a \neq b$. Moreover,

$$W_{ia} \subset \operatorname{supp} \varphi_{\alpha_0} \cap \cdots \cap \operatorname{supp} \varphi_{\alpha_i} \subset U_\alpha$$

for some α, where supp $\varphi_{\alpha j}$ = support of $\varphi_{\alpha j}$. Then we see that if we let

$$X_i = \bigcup_a W_{ia}, \quad i = 0, \ldots, n,$$

then (a) $E|_{X_i}$ is trivial and (b) $\cup X_i = X$. Assertion (a) follows from the fact that $E|_{W_{ia}}$ is trivial, since $W_{ia} \subset U_\alpha$ and $W_{ia} \cap W_{ib} = \varnothing, a \neq b$. If $x \in X$, then x is contained in at most $n+1$ of the sets $\{U_\alpha\}$, and so at most $n+1$ of the functions $\{\varphi_\alpha\}$ are positive at x. Let $a = \{\alpha_0, \ldots, \alpha_i\}$, where $\varphi_{\alpha_0}, \ldots, \varphi_{\alpha_i}$ are the only functions in $\{\varphi_\alpha\}$ which are positive at $x, 0 \leq i \leq n$. Then it follows that

$$0 = \varphi_\alpha(x) < \min\{\varphi_{\alpha 0}(x), \ldots, \varphi_{\alpha i}(x)\}$$

for any $\alpha \neq \alpha_0, \ldots, \alpha_i$, and hence $x \in W_{ia} \subset X_i$. Thus $\{X_i\}$ is a finite open covering of X such that $E|_{X_i}$ is trivial.

Q.E.D.

Proposition 4.2: Let $E \to X$ be a differentiable **C**-vector bundle of rank r. Then there is an integer $N > 0$ and a differentiable mapping $\mathbf{\Phi} : X \to G_{r,N}(\mathbf{C})$ such that $\mathbf{\Phi}^*(U_{r,N}) \cong E$, where $U_{r,N} \to G_{r,N}$ is the universal bundle.

Remark: This is one-half of the classification theorem for vector bundles, theorem I.2.17, discussed in Sec. 2. of Chap. I.

Proof: Consider the dual vector bundle $E^* \to X$. By Proposition 4.1, there exists a finite open cover of $X, \{U_\alpha\}$, and a finite number of frames $f_\alpha = (e_1^\alpha, \ldots, e_r^\alpha), \alpha = 1, \ldots, k$, for the vector bundle E^*. By a simple partition of unity argument, we see that there exists a finite number of global sections of the vector bundle $E^*, \xi_1, \ldots, \xi_N \in \mathcal{E}(X, E^*)$, such that at any point $x \in X$ there are r sections $\{\xi_{\alpha_1}, \ldots, \xi_{\alpha_r}\}$ which are linearly independent at x (and hence in a neighborhood of x). We want to use the sections $\xi_1, \ldots, \xi_N$ to define a mapping

$$\mathbf{\Phi} : X \longrightarrow G_{r,N}.$$

Suppose that f^* is a frame for E^* near $x_0 \in X$. Then

$$M(f^*) = [\xi_1(f^*)(x), \ldots, \xi_N(f^*)(x)] \tag{4.1}$$

is an $r \times N$ matrix of maximal rank, whose coefficients are C^∞ functions defined near x_0. The rows of M span an r-dimensional subspace of $\mathbf{C}^N$, and we denote this subspace by $\mathbf{\Phi}(x)$. A priori, $\mathbf{\Phi}(x)$ depends on the choice of frame, but we see that if g is a change of frame, then

$$\begin{aligned} M(f^*g) &= [\xi_1(f^*g), \ldots, \xi_N(f^*g)] \\ &= [g^{-1}\xi_1(f^*), \ldots, g^{-1}\xi_N(f^*)] \\ &= g^{-1}M(f^*). \end{aligned}$$

Thus the rows of $M(f^*)$ and $M(f^*g)$ span the same subspace, and therefore the mapping

$$\mathbf{\Phi} : X \longrightarrow G_{r,N}$$

is well defined at every point. It follows from the construction, by looking at local coordinates in $G_{r,N}$, that $\mathbf{\Phi}$; is a differentiable mapping. We now claim that $\mathbf{\Phi}^* U_{r,N} \cong E$. To see this, it suffices to define a bundle morphism $\tilde{\mathbf{\Phi}}$

$$\begin{array}{ccc} E & \xrightarrow{\tilde{\mathbf{\Phi}}} & U_{r,N} \\ \downarrow & & \downarrow \pi \\ X & \xrightarrow{\mathbf{\Phi}} & G_{r,N} \end{array}$$

which commutes with the mapping $\mathbf{\Phi}$ and which is injective on each fibre. We define $\tilde{\mathbf{\Phi}}(x, v), x \in X, v \in E_x$, by setting

$$\tilde{\mathbf{\Phi}}(x, v) = (\langle v, \xi_1(x)\rangle, \ldots, \langle v, \xi_N(x)\rangle),$$

where $\langle\, ,\rangle$ denotes the bilinear pairing between E and E^*. Thus $\tilde{\mathbf{\Phi}}|_{E_x}$ is a $\mathbf{C}$-linear mapping into $\mathbf{C}^N$, and we claim that (a) $\tilde{\mathbf{\Phi}}|_{E_x}$ is injective and (b) $\tilde{\mathbf{\Phi}}(E_x) = \pi^{-1}(\mathbf{\Phi}(x))$, where π is the projection in the universal bundle. Let f be a frame for E near $x \in X$ and let f^* be a dual frame for E^*; i.e., if $f = (e_1, \ldots, e_r)$ and $f^* = (e_1^*, \ldots, e_r^*)$, then $\langle e_\rho, e_\sigma^*\rangle = \delta_{\rho\sigma}$. Then we see that the mapping $\tilde{\mathbf{\Phi}}$ can be represented at x by the matrix product

$$\tilde{\mathbf{\Phi}}(x, v) = {}^t v(f) \cdot M(f^*), \tag{4.2}$$

where $M(f^*)$ is defined by (4.1) and is of maximal rank. Thus $\tilde{\mathbf{\Phi}}$ is injective on fibres. But (4.2) shows that the image of $\tilde{\mathbf{\Phi}}(E_x)$ is contained in the subspace of $\mathbf{C}^N$ spanned by the rows of $M(f^*)$, which implies that $\tilde{\mathbf{\Phi}}(E_x) = \pi^{-1}(\mathbf{\Phi}(x))$.

Q.E.D.

It follows from Proposition 4.2 and Theorem 3.6(a) that $c(E) = \mathbf{\Phi}^*(c(U_{r,N}))$. In particular, one can show easily from this that line bundles have integral Chern classes. Let $\tilde{H}^q(X, \mathbf{Z})$ denote the image of $H^q(X, \mathbf{Z})$ in $H^q(X, \mathbf{R})$ under the natural homomorphism induced by the inclusion of the constant sheaves $\mathbf{Z} \subset \mathbf{R}$ [this means that $\tilde{H}^q(X, \mathbf{Z})$ is integral cohomology modulo torision].

Proposition 4.3: Let $E \to X$ be a complex line bundle, Then $c_1(E) \in \tilde{H}^2(X, \mathbf{Z})$.

Proof: Since $c_1(E) = \Phi^*(c_1(U_{1,N}))$, where Φ is the mapping in Proposition 4.2, we see that it suffices to show that

$$c_1(U_{1,N} \in H^2(\mathbf{P}_{N-1}, \mathbf{Z})$$

[$H^2(\mathbf{P}_{N-1}, \mathbf{Z})$ has no torsion; see the discussion below]. In Sec. 2 we have computed the curvature for the canonical connection $D(h)$ associated with the natural metric h on the universal bundle $U_{1,N}$, and thus, by (2.10), we see that

$$c_1(U_{1,N}, D(h)) = \frac{1}{2\pi i} \frac{|f|^2 \sum d\xi_j \wedge d\bar{\xi}_j - \sum \bar{\xi}_j \xi_k d\xi_j \wedge d\bar{\xi}_k}{|f|^4}, \tag{4.3}$$

where $f = (\xi_1, \ldots, \xi_N)$ is a frame for $U_{1,N}$. Now, it is well known that

$$H^q(\mathbf{P}_n(\mathbf{C}), \mathbf{Z}) \cong \mathbf{Z}, \quad q \text{ even}, \quad q \le 2n$$

$$H^q(\mathbf{P}_n(\mathbf{C}), \mathbf{Z}) \cong 0, \quad q \text{ odd}, \quad \text{or } q > 2n,$$

which can be shown easily using singular cohomology (see Greenberg [1]). In fact there is a cell decomposition

$$\mathbf{P}_0 \subset \mathbf{P}_1 \subset \cdots \subset \mathbf{P}_{N-1}$$

where $\mathbf{P}_{j-1} \subset \mathbf{P}_j$ is a linear hyperplane, and $\mathbf{P}_j - \mathbf{P}_{j-1} \cong \mathbf{C}^j$. The submanifold $\mathbf{P}_j \subset \mathbf{P}_{N-1}$ is a generator for $H_{2j}(\mathbf{P}_N, \mathbf{Z})$, and there are no torsion elements. A closed differential form φ of degree $2j$ will be a representative of an integral cohomology class in $H^{2j}(\mathbf{P}_{N-1}, \mathbf{Z})$ if and only if

$$\int_{\mathbf{P}_j} \varphi \in \mathbf{Z}.$$

Thus, to see that $c_1(U_{1,N}) \in H^2(\mathbf{P}_{N-1}, \mathbf{Z})$, it suffices to compute

$$\int_{\mathbf{P}_1} \alpha,$$

where α is defined by (4.3). We can take $\mathbf{P}_1 \subset \mathbf{P}_{N-1}$ to be defined by the subspace in homogeneous coordinates

$$\{(z_1, \ldots, z_N) : z_j = 0, j = 3, \ldots, N\}.$$

Consider the frame f for $U_{1,N} \to \mathbf{P}_{N-1}$, defined over $W = \{z : z_1 \neq 0\}$, given by

$$f([1, \xi_2, \ldots, \xi_N]) = (1, \xi_2, \ldots, \xi_N),$$

where $(\xi_2, \ldots, \xi_N)$ are coordinates for $\mathbf{P}_N$ in the open set W. Then $f|_{W \cap \mathbf{P}_1}$ is given by

$$f([1, \xi_2, 0, \ldots, 0)]) = (1, \xi_2, 0, .., 0),$$

and we can think of ξ_2 as coordinates in $W \cap \mathbf{P}_1$ for $\mathbf{P}_1$. Thus the differential form $\alpha|_{\mathbf{P}_1}$ is given by (letting $z = \xi_2$)

$$\alpha = \frac{1}{2\pi i} \frac{dz \wedge d\bar{z}}{(1 + |z|^2)^2},$$

and therefore we obtain

$$\begin{aligned}\int_{\mathbf{P}_1} \alpha = \int_{\mathbf{P}_1 \cap W} \alpha &= \frac{1}{2\pi i} \int_{\mathbf{C}} \frac{dz \wedge d\bar{z}}{(1+|z|^2)^2} \\ &= -\frac{1}{\pi} \int_{\mathbf{R}^2} \frac{dx \wedge dy}{(1+|z|^2)^2} \\ &= -2 \int_0^{\infty} \frac{r dr}{(1+r^2)^2} \\ &= -1.\end{aligned}$$

This shows that $c_1(U_{1,N}) \in H^2(\mathbf{P}_{N-1}, \mathbf{Z})$ and hence that $c_1(E) \in \tilde{H}^2(X, \mathbf{Z})$.
Q.E.D.

Remark: This approach generalizes to vector bundles. In fact, $G_{r,N}(\mathbf{C})$ has a cell decomposition similar to that given above for $G_{1,N}(\mathbf{C})$ and has non-vanishing cohomology only in even degrees and has no torsion. The generalization of the cycles $\{\mathbf{P}_j \subset \mathbf{P}_N\}$ generating the homology are called Schubert varieties. Moreover, one can show that the Chern classes of the universal bundle $U_{r,N}$, appropriately normalized, are integral cohomology classes, and thus a version of Proposition 4.3 is valid for vector bundles (see Chern [2]). In algebraic topology, one defines the Chern classes as the pullbacks under the classifying map of the Chern classes of the universal bundle, thus admitting torsion elements. However, the proof of Theorem 3.6 in that context is considerably different and perhaps not quite so simple.

So far we have encountered two different approaches to Chern class theory: the differential-geometric definition in Sec. 3 and the classifying space approach discussed in the above remark. A third approach is to define Chern classes only for line bundles, extend the definition to direct sums of line bundles by using the required behavior on direct sums, and show that any vector bundle can be decomposed as a direct sum of line bundles by modifying the base space appropriately (see Hirzebruch [1]). For a comparison of almost all definitions possible, see Appendix I in Borel and Hirzebruch [1]. We shall present a simple sheaf-theoretic definition of Chern class for a complex line bundle and show that it is compatible with the differential-geometric (and consequently classifying space) definition. We shall assume a knowledge of Čech cohomology as presented in Sec. 4 in Chap. II.

Consider, first, holomorphic line bundles over a complex manifold X. Let $\mathcal{O}$ be the structure sheaf of X and let $\mathcal{O}^*$ be the sheaf of nonvanishing holomorphic functions on X.

Lemma 4.4: There is a one-to-one correspondence between the equivalence classes of holomorphic line bundles on X and the elements of the cohomology group $H^1(X, \mathcal{O}^*)$.

Proof: We shall represent $H^1(X, \mathcal{O}^*)$ by means of Čech cohomology. Suppose that $E \to X$ is a holomorphic line bundle. There is an open covering $\{U_\alpha\} = \mathfrak{U}$ and holomorphic functions

$$g_{\alpha\beta} : U_\alpha \cap U_\beta \longrightarrow GL(1, \mathbf{C}) = \mathbf{C} - \{0\} \tag{4.4}$$

such that

$$g_{\alpha\beta} \cdot g_{\beta\gamma} \cdot g_{\gamma\alpha} = 1 \quad \text{on } U_\alpha \cap U_\beta \cap U_\gamma. \tag{4.5}$$

$$g_{\alpha\alpha} = 1 \quad \text{on } U_\alpha.$$

Namely, the $\{g_{\alpha\beta}\}$ are the *transition functions* of the line bundle with respect to a suitable covering (see Sec. 2 of Chap. I). But the data $\{g_{\alpha\beta}\}$ satisfying (4.5) define a cocycle $g \in Z^1(\mathfrak{U}, \mathcal{O}^*)$ and hence a cohomology class in the direct limit $H^1(X, \mathcal{O}^*)$. Moreover, any line bundle $E' \to X$ which is isomorphic to $E \to X$ will correspond to the same class in $H^1(X, \mathcal{O}^*)$. This is easy to see by combining (via the isomorphism) the two sets of transition functions to get a single set of transition functions on a suitable refinement of the given $\{U_\alpha\}$ and $\{U'_\alpha\}$. Thus they will correspond to the same cohomology class. Conversely, given any cohomology class $\xi \in H^1(X, \mathcal{O}^*)$, it can be represented by a cocycle $g = \{g_{\alpha\beta}\}$ on some covering $\mathfrak{U} = \{U_\alpha\}$. By means of the functions $\{g_{\alpha\beta}\}$ one can construct a holomorphic line bundle having these transition functions. Namely, let

$$\tilde{E} = \cup\, U_\alpha \times \mathbf{C} \quad \text{(disjoint union)}$$

and identify

$$(x, z) \in U_\alpha \times \mathbf{C} \quad \text{with} \quad (y, w) \in U_\beta \times \mathbf{C}$$

if and only if

$$y = x \quad \text{and} \quad z = g_{\alpha\beta}(x) w.$$

This identification (or equivalence relation on $\tilde{E}$) gives rise to a holomorphic line bundle. Again, appealing to a common refinement argument, it is easy to check that one does obtain the desired one-to-one correspondence.

Q.E.D.

As we know from the differential-geometric definition, the Chern class of a line bundle depends only on its equivalence class, and this is most easily represented by a cocycle in $Z^1(\mathfrak{U}, \mathcal{O}^*)$ for a particular covering. Recall the exact sequence of sheaves in Example II.2.6,

$$0 \longrightarrow \mathbf{Z} \longrightarrow \mathcal{O} \xrightarrow{\exp} \mathcal{O}^* \longrightarrow 0,$$

and consider the induced cohomology sequence

$$\begin{array}{ccccccc} H^1(X, \mathcal{O}) & \longrightarrow & H^1(X, \mathcal{O}^*) & \xrightarrow{\delta} & H^2(X, \mathbf{Z}) & \longrightarrow & H^2(X, \mathcal{O}) \\ & & & \searrow & \downarrow & & \\ & & & & H^2(X, \mathbf{R}), & & \end{array}$$

where the vertical mapping is the natural homomorphism j induced by the inclusion of the constant sheaves $\mathbf{Z} \subset \mathbf{R}$ and δ is the Bockstein operator.

Note that the first Chern class, as defined in Sec. 3, gives a mapping (see the dashed arrow above)

$$c_1 : H^1(X, \mathcal{O}^*) \longrightarrow H^2(X, \mathbf{R}).$$

The following theorem tells us that we can compute Chern classes of line bundles by using the Bockstein operator δ.

Theorem 4.5: The diagram

$$\begin{array}{ccc} H^1(X, \mathcal{O}^*) & \xrightarrow{\delta} & H^2(X, \mathbf{Z}) \\ & \searrow^{c_1} & \downarrow j \\ & & H^2(X, \mathbf{R}) \end{array}$$

is commutative.

Proof: The basic element of the proof is to represent de Rham cohomology by Čech cohomology and then compute explicitly the Bockstein operator in this context. Suppose that $\mathfrak{U} = \{U_\alpha\}$ is a locally finite covering of X, and consider $\xi = \{\xi_{\alpha\beta\gamma}\} \in Z^2(\mathfrak{U}, \mathbf{R})$. We want to associate with ξ a closed 2-form φ on X. Since ξ also is an element of $Z^2(\mathfrak{U}, \mathcal{E})$ and $\mathcal{E}$ is fine, there exists a $\tau \in C^1(\mathfrak{U}, \mathcal{E})$ so that $\delta\tau = \xi$, for instance,

$$\tau_{\beta\gamma} = \sum_\alpha \varphi_\alpha \xi_{\alpha\beta\gamma},$$

where φ_α is a partition of unity subordinate to $\mathfrak{U}$. Exterior differentiation is well defined on cochains in $C^q(\mathfrak{U}, \mathcal{E}^p)$ and commutes with the coboundary operator, and so we obtain

$$\delta d\tau = d\delta\tau = d\xi = 0.$$

Then $d\tau \in Z^1(\mathfrak{U}, \mathcal{E}^1)$, but $\mathcal{E}^1$ is also fine, so do the same thing once more, writing

$$\mu_\beta = \sum_\alpha \varphi_\alpha d\tau_{\alpha\beta}.$$

Then $\mu \in C^0(\mathfrak{U}, \mathcal{E}^1)$ and $\delta\mu = d\tau$, and thus $d\mu \in C^0(\mathfrak{U}, \mathcal{E}^2)$. But

$$\delta d\mu = d\delta\mu = d^2\tau = 0,$$

and so $\varphi = -d\mu \in Z^0(\mathfrak{U}, \mathcal{E}^2) = \mathcal{E}^2(X)$ is a well-defined global differential form which is clearly d-closed. Thus to a cocycle $\xi \in Z^2(\mathfrak{U}, \mathbf{R})$ we have associated a closed differential form $\varphi(\xi)$. This induces a mapping at the cohomology level,

(4.6) $$\underset{\text{(Čech)}}{\check{H}^2(X, \mathbf{R})} \longrightarrow \underset{\text{(de Rham)}}{H^2(X, \mathbf{R})},$$

which one can show is well defined and is an isomorphism (cf. the proof of Theorem II. 3.13). Note that the mapping at the cocycle level depends on the choices made (τ and μ) but that the induced mapping on cohomology is independent of the choices made. This is thus an explicit representation for the isomorphism between de Rham cohomology and Cech cohomology. The choice of *sign* in this isomorphism was made so that the concept of "positivity" for Chern classes is compatible for the sheaf-theoretic and differential-geometric definitions of Chern classes (cf. Chap. VI; Kodaira [1]; Hirzebruch [1]; Borel-Hirzebruch [1; II-Appendix]).

Suppose now that $\mathfrak{U}$ is a covering of X, with the property that any intersection of elements of the covering is a cell (in particular is simply connected).† We want to use $\mathfrak{U}$ to describe the Bockstein operator δ:

$$\delta : H^1(X, \mathcal{O}^*) \longrightarrow H^2(X, \mathbf{Z}).$$

Suppose that $g = \{g_{\alpha\beta}\} \in Z^1(\mathfrak{U}, \mathcal{O}^*)$. Then $\sigma = \{\sigma_{\alpha\beta}\}$, defined by

$$\sigma_{\alpha\beta} = \frac{1}{2\pi i} \log g_{\alpha\beta} = \exp^{-1}(g_{\alpha\beta}),$$

defines an element of $C^1(\mathfrak{U}, \mathcal{O})$ (here we use the simply connectedness and any particular branch of the logarithm). Thus $\delta\sigma \in C^2(\mathfrak{U}, \mathcal{O})$, and since $\delta^2 = 0$, we see that $\delta\sigma \in Z^2(\mathfrak{U}, \mathcal{O})$. But

$$(\delta\sigma)_{\alpha\beta\gamma} = \frac{1}{2\pi i}(\log g_{\beta\gamma} - \log g_{\alpha\gamma} + \log g_{\alpha\beta}),$$

and this is integer-valued, since

$$g_{\alpha\beta} \cdot g_{\beta\gamma} = g_{\alpha\gamma};$$

i.e., $\{g_{\alpha\beta}\}$ is a cocycle in $Z^1(X, \mathcal{O}^*)$. Thus $\delta\sigma \in Z^2(\mathfrak{U}, \mathbf{Z})$ and is a representative for $\delta(g) \in H^2(X, \mathbf{Z})$.

Now let $g = \{g_{\alpha\beta}\}$ be the transition functions of a holomorphic line bundle $E \to X$ and let h be a Hermitian metric on E. Since $\{U_\alpha\}$ is a trivializing cover for E, we have frames f_α for E over U_α, and we set $h_\alpha = h(f_\alpha)$. Note that h_α is a positive C^∞ function defined in U_α. Thus

$$c_1(E, h) = \frac{i}{2\pi}\bar{\partial}(h_\alpha^{-1}\partial h_\alpha) \quad \text{in } U_\alpha,$$

which we rewrite as

$$c_1(E, h) = \frac{1}{2\pi i}\partial\bar{\partial}\log h_\alpha.$$

Note that the functions h_α satisfy

$$h_\alpha = |g_{\beta\alpha}|^2 h_\beta$$

on $U_\alpha \cap U_\beta$, which follows from the change of frame transformation (1.7) for the Hermitian metric h. We want to use the functions $\{h_\alpha\}$ in the transformation from Čech to de Rham representatives. As above, let

$$\delta\sigma \in Z^2(\mathfrak{U}, \mathbf{Z}), \quad \sigma_{\alpha\beta} = \frac{1}{2\pi i}\log g_{\alpha\beta},$$

be the Bockstein image of $\{g_{\alpha\beta}\}$ in $H^2(X, \mathbf{Z})$. We now want to associate to $\delta\sigma$ a closed 2-form via the construction giving (4.6), which will turn out to

†Such a covering always exists and will be a *Leray* covering for the constant sheaf; i.e., $H^q(|\sigma|, \mathbf{R}) = 0$ for any simplex σ of the covering $\mathfrak{U}$. If X is equipped with a Riemannian metric (considered as a real differentiable manifold), then every point $x \in X$ has a fundamental neighborhood system of *convex normal balis* (Helgason [1], p. 54), and the intersection of any finite number of such convex sets is again convex. Moreover, these convex sets are cells.

be the Chern form of E, concluding the proof of the theorem. Choose τ and μ in the construction of the mapping (4.6) by letting $\tau = \sigma$ and $\mu = \{\mu_\alpha\}$, where

$$\mu_\alpha = \frac{1}{2\pi i}\partial \log h_\alpha.$$

Then we see that this choice of $\mu = \{\mu_\alpha\}$ satisfies

$$\begin{aligned}(\delta\mu)_{\alpha\beta} = \mu_\beta - \mu_\alpha &= \frac{1}{2\pi i}\partial \log \frac{h_\beta}{h_\alpha}\\ &= \frac{1}{2\pi i}\partial \log g_{\alpha\beta}\bar{g}_{\alpha\beta}\\ &= \frac{1}{2\pi i}(\partial \log g_{\alpha\beta} + \partial \log \bar{g}_{\alpha\beta})\\ &= \frac{1}{2\pi i} d \log g_{\alpha\beta}\\ &= d\sigma_{\alpha\beta} = d\tau_{\alpha\beta}\end{aligned}$$

(here $\partial \log \bar{g}_{\beta\alpha} = 0$, since $g_{\beta\alpha}$ is holomorphic). Thus the closed 2-form associated with the cocycle $\delta\sigma$ is given by

$$\begin{aligned}\varphi = -d\mu &= d\left(\frac{i}{2\pi}\partial \log h_\alpha\right)\\ &= \frac{i}{2\pi}\bar{\partial}\partial \log h_\alpha = c_1(E, h).\end{aligned}$$

Q.E.D.

A modification of the above proof shows that Theorem 4.5 is also true in the C^∞ category. Namely, there is an exact sequence

$$0 \longrightarrow \mathbf{Z} \longrightarrow \mathcal{E} \longrightarrow \mathcal{E}^* \longrightarrow 0$$

on a differentiable manifold X, where $\mathcal{E}^*$ is the sheaf of nonvanishing C^∞ functions. The induced sequence in cohomology reads

$$\longrightarrow H^1(X, \mathcal{E}) \longrightarrow H^1(X, \mathcal{E}^*) \xrightarrow{\delta} H^2(X, \mathbf{Z}) \longrightarrow H^2(X, \mathcal{E}) \longrightarrow,$$

but $H^q(X, \mathcal{E}) = 0, q > 0$, since $\mathcal{E}$ is fine, and hence there is an isomorphism

$$H^1(X, \mathcal{E}^*) \xrightarrow[\cong]{\delta} H^2(X, \mathbf{Z}),$$

which asserts that all differentiable complex line bundles are determined by their Chern class in $H^2(X, Z)$ [but not necessarily by their real Chern class in $\tilde{H}^2(X, Z)$, as there may be some torsion lost]. For holomorphic line bundles, the situation is more complicated. Let X be a complex manifold and consider the corresponding sequence

(4.7)
$$\begin{array}{ccccccc} H^1(X, \mathcal{O}) & \longrightarrow & H^1(X, \mathcal{O}^*) & \xrightarrow{\delta} & H^2(X, \mathbf{Z}) & \longrightarrow & H^2(X, \mathcal{O}) \\ & & & \searrow^{c_1} & \downarrow j & & \\ & & & & \tilde{H}^2(X, \mathbf{Z}). & & \end{array}$$

Here we may have $H^1(X, \mathcal{O})$ or $H^2(X, \mathcal{O})$ nonvanishing, and line bundles would not be determined by their Chern class in $H^2(X, \mathbf{Z})$. We want to characterize the image of c_1 in the above diagram. Let $\tilde{H}^2_{1,1}(X, \mathbf{Z})$ be the cohomology classes in $\tilde{H}^2(X, \mathbf{Z})$ which admit a d-closed differential form of type (1, 1) as a representative, and let

$$H^2_{1,1}(X, \mathbf{Z}) = j^{-1}(\tilde{H}^2_{1,1}(X, \mathbf{Z})) \subset H^2(X, \mathbf{Z}).$$

Proposition 4.6: In (4.7),

$$c_1(H^1(X, \mathcal{O}^*)) = \tilde{H}^2_{1,1}(X, \mathbf{Z}).$$

Proof: It suffices to show that

$$\delta(H^1(X, \mathcal{O}^*)) = H^2_{1,1}(X, \mathbf{Z}) \quad \text{in (4.7)}.$$

To see this, it suffices to show that the image of $H^2_{1,1}(X, \mathbf{Z})$ in $H^2(X, \mathcal{O})$ is zero. Consider the following commutative diagram of sheaves (all natural inclusions),

$$\begin{array}{ccc} & \mathbf{C} & \\ \nearrow & & \searrow \\ \mathbf{Z} & \longrightarrow & \mathcal{O}, \end{array}$$

and the induced diagram on cohomology,

$$\begin{array}{ccc} & H^2(X, \mathbf{C}) & \\ \nearrow & & \searrow \\ H^2(X, \mathbf{Z}) & \longrightarrow & H^2(X, \mathcal{O}). \end{array}$$

Now $\tilde{H}^2_{1,1}(X, \mathbf{Z}) \subset H^2(X, \mathbf{C})$ and is the image of $H^2_{1,1}(X, \mathbf{Z})$ in the above diagram. Therefore it suffices to show that the image of $H^2_{1,1}(X, \mathbf{C})$ (defined as before) in $H^2(X, \mathcal{O})$ is zero. Consider the homomorphism of resolutions of sheaves

$$\begin{array}{ccccccccccc} 0 & \longrightarrow & \mathbf{C} & \longrightarrow & \mathcal{E}^0 & \xrightarrow{d} & \mathcal{E}^1 & \xrightarrow{d} & \mathcal{E}^2 & \xrightarrow{d} & \cdots \\ & & \downarrow{\scriptstyle i} & & \downarrow{\scriptstyle i} & & \downarrow{\scriptstyle \pi_{0,1}} & & \downarrow{\scriptstyle \pi_{0,2}} & & \\ 0 & \longrightarrow & \mathcal{O} & \longrightarrow & \mathcal{E}^{0,0} & \xrightarrow{\bar\partial} & \mathcal{E}^{0,1} & \xrightarrow{\bar\partial} & \mathcal{E}^{0,2} & \xrightarrow{\bar\partial} & \cdots \end{array}$$

where $\pi_{0,q} : \mathcal{E}^q \to \mathcal{E}^{0,q}$ is the projection on the submodule of forms of type $(0, q)$. Therefore the mapping

$$H^2(X, \mathbf{C}) \longrightarrow H^2(X, \mathcal{O})$$

is represented by mapping a d-closed differential form φ onto the $\bar\partial$-closed form $\pi_{0,2}\varphi$. It is then clear that the image of $H^2_{1,1}(X, \mathbf{C})$ in $H^2(X, \mathcal{O})$ is zero, since a class in $H^2_{1,1}(X, \mathbf{C})$ is represented by a d-closed form φ of type (1, 1), and thus $\pi_{0,2}\varphi = 0$.

Q.E.D.

Closely related to holomorphic line bundles is the concept of a divisor on a complex manifold X. Consider the exact sequence of multiplicative sheaves

$$0 \longrightarrow \mathcal{O}^* \longrightarrow \mathfrak{M}^* \longrightarrow \mathfrak{M}^*/\mathcal{O}^* \longrightarrow 0 \tag{4.8}$$

where $\mathcal{O}^*$ was defined above and $\mathfrak{M}^*$ is the sheaf of non-trivial meromorphic functions on X; i.e., the stalk $\mathfrak{M}_x^*$ is the group of non-zero elements of the quotient field of the integral domain $\mathcal{O}_x$ at a point $x \in X$ (see Gunning and Rossi [1], for the proofs of the algebraic structure of $\mathcal{O}_x$; i.e., $\mathcal{O}_x$ is a Noetherian local ring; moreover, $\mathcal{O}_x$ is an integral domain, with unique factorization). We let $\mathfrak{D} = \mathfrak{M}^*/\mathcal{O}^*$, and this is called the *sheaf of divisors* on X. A section of $\mathfrak{D}$ is called a *divisor*. If $D \in H^0(X, \mathfrak{D})$, then there is a covering $\mathfrak{A} = \{U_\alpha\}$ and meromorphic functions (sections of $\mathfrak{M}^*$) f_α defined in U_α such that

$$\frac{f_\beta}{f_\alpha} = g_{\alpha\beta} \in \mathcal{O}^*(U_\alpha \cap U_\beta). \tag{4.9}$$

Moreover,

$$g_{\alpha\beta} \cdot g_{\beta\gamma} \cdot g_{\gamma\alpha} = 1 \text{ on } U_\alpha \cap U_\beta \cap U_\gamma.$$

Thus a divisor gives rise to an equivalence class of line bundles represented by the cocycle $\{g_{\alpha\beta}\}$. This is seen more easily by looking at the exact sequence in cohomology induced by (4.8), namely

$$\begin{array}{ccccc} H^0(X, \mathfrak{M}^*) & \longrightarrow & H^0(X, \mathfrak{D}) & \longrightarrow & H^1(X, \mathcal{O}^*) \\ & & & & \downarrow{\scriptstyle \delta} \\ & & & & H^2(X, \mathbf{Z}), \end{array} \tag{4.10}$$

where we have added the vertical map coming from (4.7). From the sequence (4.8) we see that a divisor determines an equivalence class of holomorphic line bundles and that two different divisors give the same class if they "differ by" (multiplicatively) a global meromorphic function (this is called *linear equivalence* in algebraic geometry). Divisors occur in various ways, but very often as the divisor determined by a subvariety $V \subset X$ of codimension 1. Namely, such a sub-variety V can be defined by the following data: a covering $\{U_\alpha\}$ of X, holomorphic functions f_α in U_α, and $f_\beta/f_\alpha = g_{\alpha\beta}$ nonvanishing and holomorphic on $U_\alpha \cap U_\beta$. The subvariety V is then defined to be the zeros of the functions f_α in U_α. This then clearly gives rise to a divisor (see Gunning and Rossi [1] or Narasimhan [2] for a more detailed discussion of divisors and subvarieties). We shall need to use this concept later on only in the case of a nonsingular hypersurface $V \subset X$, which then gives rise to an equivalence class of holomorphic line bundles.

CHAPTER IV

ELLIPTIC OPERATOR THEORY

In this chapter we shall describe the general theory of elliptic differential operators on compact differentiable manifolds, leading up to a presentation of a general Hodge theory. In Sec. 1 we shall develop the relevant theory of the function spaces on which we shall do analysis, namely the Sobolev spaces of sections of vector bundles, with proofs of the fundamental Sobolev and Rellich lemmas. In Sec. 2 we shall discuss the basic structure of differential operators and their symbols, and in Sec. 3 this same structure is generalized to the context of pseudodifferential operators. Using the results in the first three sections, we shall present in Sec. 4 the fundamental theorems concerning homogeneous solutions of elliptic differential equations on a manifold. The pseudodifferential operators in Sec. 3 are used to construct a parametrix (pseudoinverse) for a given operator L. Using the parametrix we shall show that the kernel (null space) of L is finite dimensional and contains only C^∞ sections (regularity). In the case of self-adjoint operators, we shall obtain the decomposition theorem of Hodge, which asserts that the vector space of sections of a bundle is the (orthogonal) direct sum of the (finite dimensional) kernel and the range of the operator. In Sec. 5 we shall introduce elliptic complexes (a generalization of the basic model, the de Rham complex) and show that the Hodge decomposition in Sec. 4 carries over to this context, thus obtaining as a corollary Hodge's representation of de Rham cohomology by harmonic forms.

1. Sobolev Spaces

In this section we shall restrict ourselves to compact differentiable manifolds, for simplicity, although many of the topics that we shall discuss are certainly more general. Let X be a compact differentiable manifold with a strictly positive smooth measure μ.

We mean by this that $d\mu$ is a volume element (or density) which can be expressed in local coordinates $(x_1, \ldots, x_n)$ by

$$d\mu = \rho(x)dx = \rho(x)dx_1 \cdots dx_n$$

where the coefficients transform by

$$\rho(x)\,dx = \tilde{\rho}(y(x)) \left| \det \frac{\partial y(x)}{\partial x} \right| dx,$$

where $\tilde{\rho}(y)\,dy$ is the representation with respect to the coordinates $y = (y_1, \ldots, y_n)$, where $x \to y(x)$ and $\partial y/\partial x$ is the corresponding Jacobian matrix of the change of coordinates. Such measures always exist; take, for instance,

$$\rho(x) = |\det g_{ij}(x)|^{1/2},$$

where $ds^2 = \sum g_{ij}(x) dx_i \otimes dx_j$ is a Riemannian metric for X expressed in terms of the local coordinates $(x_1, \ldots, x_n)$.† If X is orientable, then the volume element $d\mu$ can be chosen to be a positive differential form of degree n (which can be taken as a definition of orientability).

Let E be a Hermitian (differentiable) vector bundle over X. Let $\mathcal{E}_k(X, E)$ be the kth order differentiable sections of E over $X, 0 \leq k \leq \infty$, where $\mathcal{E}_\infty(X, E) = \mathcal{E}(X, E)$. As usual, we shall denote the compactly supported sections‡ by $\mathcal{D}(X, E) \subset \mathcal{E}(X, E)$ and the compactly supported functions by $\mathcal{D}(X) \subset \mathcal{E}(X)$. Define an inner product (,) on $\mathcal{E}(X, E)$ by setting

$$(\xi, \eta) = \int_X \langle \xi(x), \eta(x) \rangle_E \, d\mu,$$

where $\langle\, , \rangle_E$ is the Hermitian metric on E. Let

$$\|\xi\|_0 = (\xi, \xi)^{1/2}$$

be the L^2-norm and let $W^0(X, E)$ be the completion of $\mathcal{E}(X, E)$. Let $\{U_\alpha, \varphi_\alpha\}$ be a finite trivializing cover, where, in the diagram

$$\begin{array}{ccc} E|_{U_\alpha} & \xrightarrow{\varphi_\alpha} & \tilde{U}_\alpha \times \mathbf{C}^m \\ \downarrow & & \downarrow \\ U_\alpha & \xrightarrow{\tilde{\varphi}_\alpha} & \tilde{U}_\alpha, \end{array}$$

φ_α is a bundle map isomorphism and $\tilde{\varphi}_\alpha \colon U_\alpha \to \tilde{U}_\alpha \subset \mathbf{R}^n$ are local coordinate systems for the manifold X. Then let

$$\varphi_\alpha^* \colon \mathcal{E}(U_\alpha, E) \longrightarrow [\mathcal{E}(\tilde{U}_\alpha)]^m$$

be the induced map. Let $\{\rho_\alpha\}$ be a partition of unity subordinate to $\{U_\alpha\}$, and define, for $\xi \in \mathcal{E}(X, E)$,

$$\|\xi\|_{s,E} = \sum_\alpha \|\varphi_\alpha^* \rho_\alpha \xi\|_{s,\mathbf{R}^n},$$

where $\| \quad \|_{s,\mathbf{R}^n}$ is the Sobolev norm for a compactly supported differentiable function

$$f \colon \mathbf{R}^n \longrightarrow \mathbf{C}^m,$$

†See any elementary text dealing with calculus on manifolds, e.g., Lang [1].

‡A section $\xi \in \mathcal{E}(X, E)$ has compact support on a (not necessarily compact) manifold X if $\{x \in X \colon \xi(x) \neq 0\}$ is relatively compact in X.

defined (for a scalar-valued function) by

$$\|f\|_{s,\mathbf{R}^n}^2 = \int |\hat{f}(y)|^2(1+|y|^2)^s\,dy, \tag{1.1}$$

where

$$\hat{f}(y) = (2\pi)^{-n}\int e^{-i\langle x,y\rangle} f(x)\,dx$$

is the Fourier transform in $\mathbf{R}^n$. We extend this to a vector-valued function by taking the s-norm of the Euclidean norm of the vector, for instance. Note that $\|\ \|_s$ is defined for all $s \in \mathbf{R}$, but we shall deal only with integral values in our applications. Intuitively, $\|\xi\|_s < \infty$, for s a positive integer, means that ξ has s derivatives in L^2. This follows from the fact that in $\mathbf{R}^n$, the norm $\|\ \|_{s,\mathbf{R}^n}$ is *equivalent* [on $\mathcal{D}(\mathbf{R}^n)$] to the norm

$$\left[\sum_{|\alpha|\le s}\int_{\mathbf{R}^n} |D^\alpha f|^2 dx\right]^{1/2}, \quad f \in \mathcal{D}(\mathbf{R}^n)$$

(see, e.g., Hörmander [1], Chap. 1). This follows essentially from the basic facts about Fourier transforms that

$$\widehat{D^\alpha f}(y) = y^\alpha \hat{f}(y),$$

where $y^\alpha = y_1^{\alpha_1}\cdots y_n^{\alpha_n}$, $D^\alpha = (-i)^{|\alpha|}\, D_1^{\alpha_1}\cdots D_n^{\alpha_n}$, $D_j = \partial/\partial x_j$, and $\|f\|_0 = \|\hat{f}\|_0$.

The norm $\|\ \|_s$ defined on E depends on the choice of partition of unity and the local trivialization. We let $W^s(X, E)$ be the completion of $\mathcal{E}(X, E)$ with respect to the norm $\|\ \|_s$. Then it is a fact, which we shall not verify here, that the *topology* on $W^s(X, E)$ is independent of the choices made; i.e., any two such norms are equivalent. Note that for $s = 0$ we have made two different choices of norms, one using the local trivializations and one using the Hermitian structure on E, and that these two L^2-norms are also equivalent.

We have a sequence of inclusions of the Hilbert spaces $W^s(X, E)$,

$$\cdots \supset W^s \supset W^{s+1} \supset \cdots \supset W^{s+j} \supset \cdots.$$

If we let H^* denote the antidual of a topological vector space over $\mathbf{C}$ (the conjugate-linear continuous functionals), then it can be shown that

$$(W^s)^* \cong W^{-s} \quad (s \ge 0).$$

In fact, we could have *defined* W^{-s} in this manner, using the definition involving the norms $\|\ \|_s$ for the nonnegative values of s. Locally this is easy to see, since we have for $f \in W^s(R^n)$, $g \in W^{-s}(\mathbf{R}^n)$ the duality (ignoring the conjugation problem by assuming that f and g are real-valued)

$$\langle f, g\rangle = \int f(x)\cdot g(x)dx = \int \hat{f}(\xi)\cdot\hat{g}(\xi)\,d\xi,$$

and this exists, since

$$|\langle f, g\rangle| \le \int |\hat{f}(\xi)|(1+|\xi|^2)^{s/2}|\hat{g}(\xi)|(1+|\xi|^2)^{-s/2}d\xi \le \|f\|_s\|g\|_{-s} < \infty.$$

The growth is the important thing here, and the patching process (being a

C^∞ process with compact supports) does not affect the growth conditions and hence the existence of the integrals. Thus the global result stated above is easily obtained. We have the following two important results concerning this sequence of Hilbert spaces.

Proposition 1.1 (Sobolev): Let $n = \dim_{\mathbf{R}} X$, and suppose that $s > [n/2] + k + 1$. Then

$$W^s(X, E) \subset \mathcal{E}_k(X, E).$$

Proposition 1.2 (Rellich): The natural inclusion

$$j\colon W^s(E) \subset W^t(E)$$

for $t < s$ is a completely continuous linear map.

Recall that *completely continuous* means that the image of a closed ball is relatively compact, i.e., j is a *compact* operator. In Proposition 1.2 the compactness of X is strongly used, whereas it is inessential for Proposition 1.1.

To give the reader some appreciation of these propositions, we shall give proofs of them in special cases to show what is involved. The general results for vector bundles are essentially formalism and the piecing together of these special cases.

Proposition 1.3 (Sobolev): Let f be a measurable L^2 function in $\mathbf{R}^n$ with $\|f\|_s < \infty$, for $s > [n/2] + k + 1$, a nonnegative integer. Then $f \in C^k(\mathbf{R}^n)$ (after a possible change on a set of measure zero).

Proof: Our assumption $\|f\|_s < \infty$ means that

$$\int_{\mathbf{R}^n} |\hat{f}(\xi)|^2 (1 + |\xi|^2)^s \, d\xi < \infty.$$

Let

$$\tilde{f}(x) = \int_{\mathbf{R}^n} e^{i\langle x, \xi\rangle} \hat{f}(\xi) \, d\xi$$

be the inverse Fourier transform, if it exists. We know that if the inverse Fourier transform exists, then $\tilde{f}(x)$ agrees with $f(x)$ almost everywhere, and we agree to say that $f \in C^0(\mathbf{R}^n)$ if this integral exists, making the appropriate change on a set of measure zero. Similarly, for some constant c,

$$D^\alpha f(x) = c \int e^{i\langle x, \xi\rangle} \xi^\alpha \hat{f}(\xi) \, d\xi$$

will be continuous derivatives of f if the integral converges. Therefore we need to show that for $|\alpha| \le k$, the integrals

$$\int e^{i\langle x, \xi\rangle} \xi^\alpha \hat{f}(\xi) \, d\xi$$

converge, and it will follow that $f \in C^k(\mathbf{R}^n)$. But, indeed, we have

$$\int |\hat{f}(\xi)||\xi|^{|\alpha|}\,d\xi = \int |\hat{f}(\xi)|(1+|\xi|^2)^{s/2}\frac{|\xi|^{|\alpha|}}{(1+|\xi|^2)^{s/2}}d\xi$$

$$\leq \|f\|_s\left(\int \frac{|\xi|^{2|\alpha|}}{(1+|\xi|^2)^s}d\xi\right)^{1/2}.$$

Now s has been chosen so that this last integral exists (which is easy to see by using polar coordinates), and so we have

$$\int |\hat{f}(\xi)||\xi|^{|\alpha|}\,d\xi < \infty,$$

and the proposition is proved.

Q.E.D.

Similarly, we can prove a simple version of Rellich's lemma.

Proposition 1.4 (Rellich): Suppose that $f_v \in W^s(\mathbf{R}^n)$ and that all f_v have compact support in $K \subset\subset \mathbf{R}^n$. Assume that $\|f_v\|_s \leq 1$. Then for any $t < s$ there exists a subsequence f_{vj} which converges in $\|\quad\|_t$.

Proof: We observe first that for $\xi, \boldsymbol{\eta} \in \mathbf{R}^n, s \in \mathbf{Z}^+$,

$$(1.2)\qquad (1+|\xi|^2)^{s/2} \leq 2^{s/2}(1+|\xi-\boldsymbol{\eta}|^2)^{s/2}(1+|\boldsymbol{\eta}|^2)^{s/2}.$$

To see this we write, using the Schwarz inequality,

$$1+|\zeta+\boldsymbol{\eta}|^2 \leq 1+(|\zeta|+|\boldsymbol{\eta}|)^2 \leq 1+2(|\zeta|^2+|\boldsymbol{\eta}|^2)$$

$$\leq 2(1+|\zeta|^2)(1+|\boldsymbol{\eta}|^2).$$

Now let $\xi = \zeta + \boldsymbol{\eta}$, and we obtain (1.2) easily.

Let $\varphi \in \mathcal{D}(\mathbf{R}^n)$ be chosen so that $\varphi \equiv 1$ near K. Then from a standard relation between the Fourier transform and convolution we have that

$$f_v = \varphi f_v$$

implies

$$(1.3)\qquad \hat{f}_v(\xi) = \int \hat{\varphi}(\xi-\boldsymbol{\eta})\hat{f}_v(\boldsymbol{\eta})\,d\boldsymbol{\eta}.$$

Therefore we obtain from (1.2) and (1.3) that

$$(1+|\xi|^2)^{s/2}|\hat{f}_v(\xi)|$$

$$\leq 2^{s/2}\int (1+|\xi-\boldsymbol{\eta}|^2)^{s/2}|\hat{\varphi}(\xi-\boldsymbol{\eta})|(1+|\boldsymbol{\eta}|^2)^{s/2}|\hat{f}_v(\boldsymbol{\eta})|\,d\boldsymbol{\eta}$$

$$\leq K_{s,\varphi}\|f_v\|_s \leq K_{s,\varphi},$$

where $K_{s,\varphi}$ is a constant depending on s and φ. Therefore $|\hat{f}_v(\xi)|$ is uniformly bounded on compact subsets of $\mathbf{R}^n$. Similarly, by differentiating (1.3) we obtain that all derivatives of $\hat{f}_v$ are uniformly bounded on compact subsets in the same manner. Therefore, there is, by Ascoli's theorem, a subsequence f_{vj} such that $\hat{f}_{vj}$ converges in the C^∞ topology to a C^∞ function on $\mathbf{R}^n$. Let us call $\{f_v\}$ this new sequence.

Let $\epsilon > 0$ be given. Suppose that $t < s$. Then there is a ball B_ϵ such that

$$\frac{1}{(1+|\xi|^2)^{s-t}} < \epsilon$$

for ξ outside the ball B_ϵ. Then consider

$$\|f_v - f_\mu\|_t^2 = \int_{\mathbf{R}^n} \frac{|(\hat{f}_v - \hat{f}_\mu)(\xi)|^2}{(1+|\xi|^2)^{s-t}}(1+|\xi|^2)^s\, d\xi$$

$$\leq \int_{B_\epsilon} |(\hat{f}_v - \hat{f}_\mu)(\xi)|^2(1+|\xi|^2)^t\, d\xi$$

$$+\epsilon \int_{\mathbf{R}^n - B_\epsilon} |(\hat{f}_v - \hat{f}_\mu)(\xi)|^2(1+|\xi|^2)^s\, d\xi$$

$$\leq \int_{B_\epsilon} |(\hat{f}_v - \hat{f}_\mu)(\xi)|^2(1+|\xi|^2)^t\, d\xi + 2\epsilon,$$

where we have used the fact that $\|f_v\|_s \leq 1$. Since we know that $\hat{f}_v$ converges on compact sets, we can choose v, μ large enough so that the first integral is $< \epsilon$, and thus f_v is a Cauchy sequence in the $\| \quad \|_t$ norm.

Q.E.D.

We now need to discuss briefly the concept of a *formal adjoint operator* in this setting.

Definition 1.5: Let

$$L\colon \mathcal{E}(X, E) \longrightarrow \mathcal{E}(X, F)$$

be a $\mathbf{C}$-linear map. Then a $\mathbf{C}$-linear map

$$S\colon \mathcal{E}(X, F) \longrightarrow \mathcal{E}(X, E)$$

is called an *adjoint* of L if

(1.4) $$(Lf, g) = (f, Sg)$$

for all $f \in \mathcal{E}(X, E), g \in \mathcal{E}(X, F)$.

It is an easy exercise, using the density of $\mathcal{E}(X, E)$ in $W^0(X, E)$, to see that an adjoint of an operator L is unique, if it exists. We denote this transpose by L^*. In later sections we shall discuss adjoints of various types of operators. This definition extends to Hilbert spaces over noncompact manifolds (e.g., $\mathbf{R}^n$) by using (1.4) as the defining relation for sections with compact support. This is then the *formal adjoint* in that context.

2. Differential Operators

Let E and F be differentiable $\mathbf{C}$-vector bundles over a differentiable manifold X.† Let

$$L\colon \mathcal{E}(X, E) \longrightarrow \mathcal{E}(X, F)$$

†The case of $\mathbf{R}$-vector bundles is exactly the same. For simplicity we restrict ourselves to the case of complex coefficients.

be a **C**-linear map. We say that L is a *differential operator* if for any choice of local coordinates and local trivializations there exists a linear partial differential operator $\tilde{L}$ such that the diagram (for such a trivialization)

$$\begin{array}{ccc} [\mathcal{E}(U)]^p & \xrightarrow{\tilde{L}} & [\mathcal{E}(U)]^q \\ \| \wr & & \| \wr \\ \mathcal{E}(U, U \times \mathbf{C}^p) & \longrightarrow & \mathcal{E}(U, U \times \mathbf{C}^q) \\ \cup & & \cup \\ \mathcal{E}(X, E)|_U & \xrightarrow{L} & \mathcal{E}(X, F)|_U \end{array}$$

commutes. That is, for $f = (f_1, \ldots, f_p) \in [\mathcal{E}(U)]^p$

$$\tilde{L}(f)_i = \sum_{\substack{j=1 \\ |\alpha| \le k}}^{p} a_\alpha^{ij} D^\alpha f_j, \quad i = 1, \ldots, q.$$

A differential operator is said to be of *order* k if there are no derivatives of order $\geq k+1$ appearing in a local representation. (For an intrinsic definition involving jet-bundles, see Palais [1], Chap. IV.) Let $\mathrm{Diff}_k(E, F)$ denote the vector space of all differential operators of order k mapping $\mathcal{E}(X, E)$ to $\mathcal{E}(X, F)$.

Suppose X is a compact differentiable manifold. We define $\mathrm{OP}_k(E, F)$ as the vector space of **C**-linear mappings

$$T: \mathcal{E}(X, E) \longrightarrow \mathcal{E}(X, F)$$

such that there is a continuous extension of T

$$\bar{T}_s: W^s(X, E) \longrightarrow W^{s-k}(X, F)$$

for all s. These are the *operators of order* k mapping E to F.

Proposition 2.1: Let $L \in \mathrm{OP}_k(E, F)$. Then L^* exists, and moreover $L^* \in \mathrm{OP}_k(F, E)$, and the extension

$$(\overline{L}^*)_s: W^s(X, F) \longrightarrow W^{s-k}(X, E)$$

is given by the adjoint map

$$(\overline{L}_{k-s})^*: W^s(X, F) \longrightarrow W^{s-k}(X, E).$$

This proposition is easy to prove since one has a candidate $(\overline{L}_{k-s})^*$ (for each s) which gives the desired adjoint when restricted to $\mathcal{E}(X, F)$ in a suitable manner. One uses the uniqueness of adjoints and Proposition 1.1.

Proposition 2.2: $\mathrm{Diff}_k(E, F) \subset \mathrm{OP}_k(E, F)$.

The proof of this proposition is not hard. Locally it involves, again, $\widehat{D^\alpha f}(\xi) = \xi^\alpha \hat{f}(\xi)$, and the definition of the s-norm.

We now want to define the *symbol* of a differential operator. The symbol will be used for the classification of differential operators into various types. First we have to define the set of all admissible symbols. Let $T^*(X)$ be the real cotangent bundle to a differentiable manifold X, let $T'(X)$ denote $T^*(X)$

with the zero cross section deleted (the bundle of nonzero cotangent vectors), and let $T'(X) \xrightarrow{\pi} X$ denote the projection mapping. Then π^*E and π^*F denote the pullbacks of E and F over $T'(X)$. We set, for any $k \in \mathbf{Z}$,

$$\begin{aligned}\mathrm{Smbl}_k(E, F) := \{\sigma \in \ &\mathrm{Hom}(\pi^*E, \pi^*F): \sigma(x, \rho v) \\ &= \rho^k \sigma(x, v), (x, v) \in T'(X), \rho > 0\}.\end{aligned}$$

We now define a linear map

$$\sigma_k: \mathrm{Diff}_k(E, F) \longrightarrow \mathrm{Smbl}_k(E, F), \tag{2.1}$$

where $\sigma_k(L)$ is called the *k-symbol* of the differential operator L. To define $\sigma_k(L)$, we first note that $\sigma_k(L)(x, v)$ is to be a linear mapping from E_x to F_x, where $(x, v) \in T'(X)$. Therefore let $(x, v) \in T'(X)$ and $e \in E_x$ be given. Find $g \in \mathcal{E}(X)$ and $f \in \mathcal{E}(X, E)$ such that $dg_x = v$, and $f(x) = e$. Then we define†

$$\sigma_k(L)(x, v)e = L\left(\frac{i^k}{k!}(g - g(x))^k f\right)(x) \in F_x.$$

This defines a linear mapping

$$\sigma_k(L)(x, v): E_x \longrightarrow F_x,$$

which then defines an element of $\mathrm{Smbl}_k(E, F)$, as is easily checked. It is also easy to see that the $\sigma_k(L)$, so defined, is independent of the choices made. We call $\sigma_k(L)$ the *k-symbol* of L.

Proposition 2.3: The symbol map σ_k gives rise to an exact sequence

$$0 \longrightarrow \mathrm{Diff}_{k-1}(E, F) \xrightarrow{j} \mathrm{Diff}_k(E, F) \xrightarrow{\sigma_k} \mathrm{Symbl}_k(E, F), \tag{2.2}$$

where j is the natural inclusion.

Proof: One must show that the k-symbol of a differential operator of order k has a certain form in local coordinates. Let L be a linear partial differential operator

$$L: [\mathcal{E}(U)]^p \longrightarrow [\mathcal{E}(U)]^q$$

where U is open in $\mathbf{R}^n$. Then it is easy to see that if

$$L = \sum_{|\nu| \le k} A_\nu D^\nu,$$

where $\{A_\nu\}$ are $q \times p$ matrices of C^∞ functions on U, then

$$\sigma_k(L)(x, v) = \sum_{|\nu| = k} A_\nu(x)\xi^\nu, \tag{2.3}$$

where $v = \xi_1 dx_1 + \cdots + \xi_n dx_n$. For each fixed (x, v), $\sigma_k(L)(x, v)$ is a linear mapping from $x \times \mathbf{C}^p \to x \times \mathbf{C}^q$, given by the usual multiplication of a vector in $\mathbf{C}^p$ by the matrix

$$\sum_{|\nu| = k} A_\nu(x)\xi^\nu.$$

†We include the factor i^k so that the symbol of a differential operator is compatible with the symbol of a pseudodifferential operator defined in Sec. 3 by means of the Fourier transform.

What one observes is that if $\sigma_k(L) = 0$, then the differential operator L has kth order terms equal to zero, and thus L is a differential operator of order $k-1$. Let us show that (2.3) is true. Choose $g \in \mathcal{E}(U)$ such that $v = dg = \sum \xi_j dx_j$; i.e., $D_j g(x) = \xi_j$. Let $e \in \mathbf{C}^p$. Then we have

$$\sigma_k(L)(x, v)e = \sum_{|\nu|\le k} A_\nu D^\nu \left(\frac{i^k}{k!}(g - g(x))^k e\right)(x).$$

Clearly, the evaluation at x of derivatives of order $\le k-1$ will give zero, since there will be a factor of $[g - g(x)]|_x = 0$ remaining. The only nonzero term is the one of the form (recalling that $D^\nu = (-i)^\nu D_1^{\nu_1} \cdots D_n^{\nu_n}$)

$$\sum_{|\nu|=k} A_\nu(x)\frac{k!}{k!}(D_1 g(x))^{\nu_1} \cdots (D_n g(x))^{\nu_n}$$

$$= \sum_{|\nu|=k} A_\nu(x)\xi_1^{\nu_1} \cdots \xi_n^{\nu_n} = \sum_{|\nu|=k} A_\nu(x)\xi^\nu,$$

which gives us (2.3). The mapping σ_k in (2.2) is well defined, and to see that the kernel is contained in $\mathrm{Diff}_{k-1}(E, F)$, it suffices to see that this is true for a local representation of the operator. This then follows from the local representation for the symbol given by (2.3).

Q.E.D.

We observe that the following property is true: If $L_1 \in \mathrm{Diff}_k(E, F)$ and $L_2 \in \mathrm{Diff}_m(F, G)$, then $L_2 L_1 = L_2 \circ L_1 \in \mathrm{Diff}_{k+m}(E, G)$, and, moreover,

(2.4) $$\sigma_{k+m}(L_2 L_1) = \sigma_m(L_2) \bullet \sigma_k(L_1),$$

where the right-hand product is the product of the linear mappings involved. The relation (2.4) is easily checked for local differential operators on trivial bundles (the chain rule for composition) and the general case is reduced to this one in a straightforward manner.

We now look at some examples.

Example 2.4: If $L: [\mathcal{E}(\mathbf{R}^n)]^p \to [\mathcal{E}(\mathbf{R}^n)]^q$ is an element of $\mathrm{Diff}_k(\mathbf{R}^n \times \mathbf{C}^p, \mathbf{R}^n \times \mathbf{C}^q)$, then

$$\sigma_k(L)(x, v) = \sum_{|\nu|=k} A_\nu(x)\xi^\nu,$$

where

$$L = \sum_{|\nu|\le k} A_\nu D^\nu, \quad v = \sum_{j=1}^n \xi_j dx_j,$$

the $\{A_\nu\}$ being $q \times p$ matrices of differentiable functions in $\mathbf{R}^n$ (cf. the proof of Proposition 2.3).

Example 2.5: Consider the de Rham complex

$$\mathcal{E}^0(X) \xrightarrow{d} \mathcal{E}^1(X) \xrightarrow{d} \cdots \xrightarrow{d} \mathcal{E}^n(X),$$

given by exterior differentiation of differential forms. Written somewhat differently, we have, for $T^* = T^*(X) \otimes \mathbf{C}$,

$$\mathcal{E}(X, \wedge^0 T^*) \xrightarrow{d} \mathcal{E}(X, \wedge^1 T^*) \xrightarrow{d} \cdots,$$

and we want to compute the associated 1-symbol mappings,

$$(2.5)\qquad \wedge^0 T_x^* \xrightarrow{\sigma_1(d)(x,v)} \wedge^1 T_x^* \xrightarrow{\sigma_1(d)(x,v)} \wedge^2 T_x^* \longrightarrow \cdots .$$

We claim that for $e \in \wedge^p T_x^*$, we have

$$\sigma_1(d)(x, v)e = iv \wedge e.$$

Moreover, the sequence of linear mappings in (2.5) is an exact sequence of vector spaces. These are easy computations and will be omitted.

Example 2.6: Consider the Dolbeault complex on a complex manifold X,

$$\mathcal{E}^{p,0}(X) \xrightarrow{\bar\partial} \mathcal{E}^{p,1}(X) \xrightarrow{\bar\partial} \cdots \xrightarrow{\bar\partial} \mathcal{E}^{p,n}(X) \longrightarrow 0.$$

Then this has an associated symbol sequence

$$\longrightarrow \wedge^{p,q-1} T_x^*(X) \xrightarrow{\sigma_1(\bar\partial)(x,v)} \wedge^{p,q} T_x^*(X) \xrightarrow{\sigma_1(\bar\partial)(x,v)} \wedge^{p,q+1} T_x^*(X) \longrightarrow,$$

where the vector bundles $\wedge^{p,q} T^*(X)$ are defined in Chap. I, Sec. 3. We have that $v \in T_x^*(X)$, considered as a real cotangent bundle. Consequently, $v = v^{1,0} + v^{0,1}$, given by the injection

$$\begin{aligned} 0 \longrightarrow T_x^*(X) \longrightarrow T_x^*(X) \otimes_{\mathbf{R}} \mathbf{C} &= T^*(X)^{1,0} \oplus T^*(X)^{0,1} \\ &= \wedge^{1,0} T^*(X) \oplus \wedge^{0,1} T^*(X). \end{aligned}$$

Then we claim that

$$\sigma_1(\bar\partial)(x, v)e = iv^{0,1} \wedge e,$$

and the above symbol sequence is exact. Once again we omit the simple computations.

Example 2.7: Let $E \longrightarrow X$ be a holomorphic vector bundle over a complex manifold X. Then consider the differentiable (p, q)-forms with coefficients in E, $\mathcal{E}^{p,q}(X, E)$, defined in (II.3.9), and we have the complex (II.3.10)

$$\longrightarrow \mathcal{E}^{p,q}(X, E) \xrightarrow{\bar\partial_E} \mathcal{E}^{p,q+1}(X, E) \longrightarrow,$$

which gives rise to the symbol sequence

$$\longrightarrow \wedge^{p,q} T_x^* \otimes E_x \xrightarrow{\sigma_1(\bar\partial_E)(x,v)} \wedge^{p,q+1} T_x^* \otimes E_x \longrightarrow .$$

We let $v = v^{1,0} + v^{0,1}$, as before, and we have for $f \otimes e \in \wedge^{p,q} T_x^* \otimes E$

$$\sigma_1(\bar\partial)(x, v) f \otimes e = (iv^{0,1} \wedge f) \otimes e,$$

and the symbol sequence is again exact.

We shall introduce the concept of *elliptic complex* in Sec. 5, which generalizes these four examples.

The last basic property of differential operators which we shall need is the existence of a formal adjoint.

Proposition 2.8: Let $L \in \mathrm{Diff}_k(E, F)$. Then L^* exists and $L^* \in \mathrm{Diff}_k(F, E)$.

Moreover, $\sigma_k(L^*) = \sigma_k(L)^*$, where $\sigma_k(L)^*$ is the adjoint of the linear map

$$\sigma_k(L)(x, v)\colon E_x \longrightarrow F_x.$$

Proof: Let $L \in \operatorname{Diff}_k(E, F)$, and suppose that μ is a strictly positive smooth measure on X and that h_E and h_E are Hermitian metrics on E and F. Then the inner product for any $\xi, \eta \in \mathcal{D}(X, E)$ is given by

$$(\xi, \eta) = \int_X \langle \xi, \eta \rangle_E \, d\mu,$$

and if ξ, η have compact support in a neighborhood where E admits a local frame f, we have

$$(\xi, \eta) = \int_{\mathbf{R}^n} {}^t\bar{\eta}(x) h_E(x) \xi(x) \rho(x)\, dx,$$

where $\rho(x)$ is a density,

$$\eta(x) = \eta(f)(x) = \begin{bmatrix} \eta^1(f)(x) \\ \cdot \\ \cdot \\ \cdot \\ \eta^r(f)(x) \end{bmatrix},$$

etc. Similarly, for $\sigma, \tau \in \mathcal{D}(X, F)$, we have

$$(\sigma, \tau) = \int_{\mathbf{R}^n} {}^t\bar{\tau}(x) h_F(x) \sigma(x) \rho(x)\, dx.$$

Suppose that $L\colon \mathcal{D}(X, E) \to \mathcal{D}(X, F)$ is a linear differential operator of order k, and assume that the sections have support in a trivializing neighborhood U which gives local coordinates for X near some point. Then we may write

$$(L\xi, \tau) = \int_{\mathbf{R}^n} {}^t\bar{\tau}(x) h_F(x) (M(x, D)\xi(x)) \rho(x)\, dx,$$

where

$$M(x, D) = \sum_{|\alpha| \le k} C_\alpha(x) D^\alpha$$

is an $s \times r$ matrix of partial differential operators; i.e., $C_\alpha(x)$ is an $s \times r$ matrix of C^∞ functions in R^n. Note that ξ and τ have compact support here. We can then write

$$(L\xi, \tau) = \int_{\mathbf{R}^n} \sum_{|\alpha| \le k} {}^t\bar{\tau}(x) \rho(x) h_F(x) C_\alpha(x) D^\alpha \xi(x)\, dx,$$

and we can integrate by parts, obtaining

$$\begin{aligned}(L\xi, \tau) &= \int_{\mathbf{R}^n} \sum_{|\alpha| \le k} (-1)^{|\alpha|} D^\alpha ({}^t\bar{\tau}(x) \rho(x) h_F(x) C_\alpha(x)) \xi(x)\, dx \\ &= \int_{\mathbf{R}^n} {}^t\overline{\Big(\sum_{|\alpha| \le k} \tilde{C}_\alpha(x) D^\alpha \tau(x)\Big)} h_E(x) \xi(x) \rho(x)\, dx,\end{aligned}$$

where $\tilde{C}_\alpha(x)$ are $r \times s$ matrices of smooth functions defined by the formula

$$(2.6) \qquad {}^t\overline{\Big(\sum_{|\alpha| \le k} \tilde{C}_\alpha D^\alpha \tau\Big)} = \sum_{|\alpha| \le k} (-1)^{|\alpha|} D^\alpha ({}^t\bar{\tau} \rho h_F C_\alpha) h_E^{-1} \rho^{-1},$$

and hence the $\tilde{C}_\alpha$ involve various derivatives of both metrics on E and F and of the density function $\rho(x)$ on X. This formula suffices to define a linear differential operator $L^*: \mathcal{D}(X, F) \to \mathcal{D}(X, E)$, which has automatically the property of being the adjoint of L. Moreover, we see that the symbol $\sigma_k(L^*)$ is given by the terms in (2.6) which only differentiate τ, since all other terms give lower-order terms in the expression $\sum_{|\alpha| \le k} \tilde{C}_\alpha(x)$. One checks that the symbol of L^* as defined above is the adjoint of the symbol of the operator L by representing $\sigma_k(L)$ with respect to these local frames and computing its adjoint as a linear mapping.

Q.E.D.

We have given a brief discussion of the basic elements of partial differential operators in a setting appropriate for our purposes. For more details on the subject, see Hörmander [1] for the basic theory of modern partial differential equations (principally in $\mathbf{R}^n$). Palais [1] has a formal presentation of partial differential operators in the context of manifolds and vector bundles, with a viewpoint similar to ours. In the next sections we shall generalize the concept of differential operators in order to find a class of operators which will serve as "inverses" for elliptic partial differential operators, to be studied in Sec. 4.

3. Pseudodifferential Operators

In this section we want to introduce an important generalization of differential operators called, appropriately enough, *pseudodifferential operators.* This type of operator developed from the study of the (singular) integral operators used in inverting differential operators (solving differential equations). On compact Riemannian manifolds a natural differential operator is the Laplacian operator, and our purpose here will be to give a sufficient amount of the recent theory of pseudodifferential operators in order to be able to "invert" such Laplacian operators, which will be introduced later in this chapter. This leads to the theory of harmonic differential forms introduced by Hodge in his study of algebraic geometry.

In defining differential operators on manifolds, we specified that they should locally look like the differential operators in Euclidean space with which we are all familiar. We shall proceed in the same manner with pseudo-differential operators, but we must spend more time developing the (relatively unknown) local theory. Once we have done this, we shall be able to obtain a general class of pseudodifferential operators mapping sections of vector bundles to sections of vector bundles on a differentiable manifold, in which class we can invert appropriate elliptic operators.

Recall that if U is an open set in $\mathbf{R}^n$ and if $p(x, \xi)$ is a polynomial in ξ of degree m, with coefficients being C^∞ functions in the variable $x \in U$, then we can obtain the most general linear partial differential operators in U by letting $P = p(x, D)$ be the differential operator obtained by replacing the vector $\xi = (\xi_1, \ldots, \xi_n)$ by $(-iD_1, \ldots, -iD_n)$, where we set $D_j = (\partial/\partial x_j)$

$(j = 1, \dots, n)$ and $D^\alpha = (-i)^{|\alpha|} D_1^{\alpha_1} \cdots D_n^{\alpha_n}$ replaces $\xi_1^{\alpha_1} \cdots \xi_n^{\alpha_n}$ in the polynomial $p(x, \xi)$. By using the Fourier transform we may write, for $u \in \mathcal{D}(U)$,

$$Pu(x) = p(x, D)u(x) = \int p(x, \xi)\hat{u}(\xi)e^{i\langle x,\xi\rangle}\, d\xi, \tag{3.1}$$

where $\langle x, \xi \rangle = \sum_{j=1}^n x_j \xi_j$ is the usual Euclidean inner product, and $\hat{u}(\xi) = (2\pi)^{-n} \int u(x)e^{-i\langle x,\xi\rangle}\, dx$ is the Fourier transform of u.

Thus (3.1) is an equivalent way (via Fourier transforms) to define the action of a differential operator $p(x, D)$ defined by a polynomial $p(x, \xi)$ on functions in the domain U. We use compact supports here so that there is no trouble with the integral existing near ∂U, and since $\mathcal{D}(U)$ is dense in most interesting spaces, it certainly suffices to know how the operator acts on such functions. Of course, $P(x, D)\colon \mathcal{D}(U) \to \mathcal{D}(U)$, since differential operators preserve supports.

To define the generalization of differential operators we are interested in, we can consider (3.1) as the *definition* of differential operator and generalize the nature of the function $p(x, \xi)$ which appears in the integrand.

To do this, we shall define classes of functions which possess, axiomatically, several important properties of the polynomials considered above.

Definition 3.1: Let U be an open set in $\mathbf{R}^n$ and let m be any integer.

(a) Let $\tilde{S}^m(U)$ be the class of C^∞ functions $p(x, \xi)$ defined on $U \times \mathbf{R}^n$, satisfying the following properties. For any compact set K in U, and for any multiindices α, β, there exists a constant $C_{\alpha,\beta,K}$, depending on α, β, K, and p so that

$$|D_x^\beta D_\xi^\alpha p(x, \xi)| \le C_{\alpha,\beta,K}(1 + |\xi|)^{m-|\alpha|}, \quad x \in K, \xi \in \mathbf{R}^n. \tag{3.2}$$

(b) Let $S^m(U)$ denote the set of $p \in \tilde{S}^m(U)$ such that

$$\text{The limit } \sigma_m(p)(x, \xi) = \lim_{\lambda\to\infty} \frac{p(x, \lambda\xi)}{\lambda^m} \text{ exists for } \xi \neq 0, \tag{3.3}$$

and, moreover,

$$p(x, \xi) - \psi(\xi)\sigma_m(p)(x, \xi) \in \tilde{S}^{m-1}(U),$$

where $\psi \in C^\infty(\mathbf{R}^n)$ is a cut-off function with $\psi(\xi) \equiv 0$ near $\xi = 0$ and $\psi(\xi) \equiv 1$ outside the unit ball.

(c) Let $\tilde{S}_0^m(U)$ denote the class of $p \in \tilde{S}^m(U)$ such that there is a compact set $K \subset U$, so that for any $\xi \in \mathbf{R}^n$, the function $p(x, \xi)$, considered as a function of $x \in U$, has compact support in K [i.e., $p(x, \xi)$ has uniform compact support in the x-variable]. Let $S_0^m(U) = S^m(U) \cap \tilde{S}_0^m(U)$.

We notice that if $p(x, \xi)$ is a polynomial of degree m (as before), then both properties (a) and (b) in Definition 3.1 above are satisfied. If the coefficients of p have compact support in U, then $p \in S_0^m(U)$. Property (a) expresses the growth in the ξ variable near ∞, whereas $\sigma_m(p)(x, \xi)$ is the mth order homogeneous part of the polynomial p, the lower-order terms having gone to zero in the limit. We shall also be interested in negative homogeneity, and the cut-off function in (b) is introduced to get rid of the singularity near $\xi = 0$, which occurs in this case.

A second example is given by an integral transform with a smooth kernel. Let $K(x, y)$ be a C^∞ function in $U \times U$ with compact support in the second variable. Then the operator

$$Lu(x) = \int_{\mathbf{R}^n} K(x, y)u(y)\, dy$$

can be written in the form (3.1) for an appropriate function $p(x, \xi)$; i.e., for $u \in \mathcal{D}(U)$,

$$Lu(x) = \int p(x, \xi)\hat{u}(\xi)e^{i\langle x,\xi\rangle}d\xi,$$

where $p \in \tilde{S}^m(U)$ for all m. To see this, we write, by the Fourier inversion formula,

$$\begin{aligned} Lu(x) &= \int K(x, y)\left[\int e^{i\langle y,\xi\rangle}\hat{u}(\xi)\, d\xi\right] dy \\ &= \int e^{i\langle x,\xi\rangle}\left[\int e^{i\langle y-x,\xi\rangle}K(x, y)dy\right]\hat{u}(\xi)d\xi \end{aligned}$$

and we let

$$p(x, \xi) = \int e^{i\langle y-x,\xi\rangle}K(x, y)\, dy,$$

which we rewrite as

$$p(x, \xi) = e^{-i\langle x,\xi\rangle}\int e^{i\langle y,\xi\rangle}K(x, y)\, dy.$$

Thus $p(x, \xi)$ is (except for the factor $e^{-i\langle x,\xi\rangle}$) for each fixed x the Fourier transform of a compactly supported function, and then it is easy to see (by integrating by parts) that $p(x, \xi)$, as a function of ξ, is rapidly decreasing at infinity; i.e.,

$$(1 + |\xi|)^N |p(x, \xi)| \longrightarrow 0$$

as $|\xi| \to \infty$ for all powers of N (this is the class $\mathcal{S}$ introduced by Schwartz [1]). It then follows immediately that $p \in \tilde{S}^m(U)$ for all m. Such an operator is often referred to as a *smoothing operator with C^∞ kernel.* The term smoothing operator refers to the fact that it is an operator of order $-\infty$, i.e., takes elements of any Sobolev space to C^∞ functions, which is a simple consequence of Theorem 3.4 below and Sobolev's lemma (Proposition 1.1).

Lemma 3.2: Suppose that $p \in S^m(U)$. Then $\sigma_m(p)(x, \xi)$ is a C^∞ function on $U \times (\mathbf{R}^n - \{0\})$ and is homogeneous of degree m in ξ.

Proof: It suffices (by the Arzela-Ascoli theorem) to show that for any compact subset of the form $K \times L$, where K is compact in U and L is compact in $\mathbf{R}^n - \{0\}$, we have the limit in (3.3) converging uniformly and that all derivatives in x and ξ of $p(x, \lambda\xi)/\lambda^m$ are uniformly bounded on $K \times L$ for $\lambda \in (1, \infty)$. But this follows immediately from the estimates in (3.2) since

$$D_x^\beta D_\xi^\alpha \left(\frac{p(x, \lambda\xi)}{\lambda^m}\right) = D_x^\beta D_\xi^\alpha p(x, \lambda\xi) \bullet \frac{\lambda^{|\alpha|}}{\lambda^m},$$

and hence, for all multiindices α, β,

$$\begin{aligned}\left|D_x^\beta D_\xi^\alpha \left(\frac{p(x, \lambda\xi)}{\lambda^m}\right)\right| &\leq C_{\alpha,\beta,K}(1 + \lambda|\xi|)^{m-|\alpha|} \bullet \frac{\lambda^{|\alpha|}}{\lambda^m} \\ &\leq C_{\alpha,\beta,K}(\lambda^{-1} + |\xi|)^{m-|\alpha|} \\ &\leq C_{\alpha,\beta,K} \sup_{\xi \in L}(1 + |\xi|)^{m-|\alpha|} < \infty.\end{aligned}$$

Therefore all derivatives are uniformly bounded, and in particular the limit in (3.3) is uniform. Showing homogeneity is even simpler. We write, for $\rho > 0$,

$$\begin{aligned}\sigma_m(x, \rho\xi) &= \lim_{\lambda\to\infty} \frac{p(x, \lambda\rho\xi)}{\lambda^{m}}\cdot \\ &= \lim_{\lambda\to\infty} \frac{p(x, \lambda\rho\xi)}{(\rho\lambda)^m} \bullet \rho^m \\ &= \lim_{\lambda\to\infty} \frac{p(x, \lambda'\xi)}{(\lambda')^m} \bullet \rho^m \quad (\lambda' = \rho\lambda) \\ &= \sigma_m(x, \xi) \bullet \rho^m.\end{aligned}$$

Q.E.D.

We now define the prototype (local form) of our pseudodifferential operator by using (3.1). Namely, we set, for any $p \in \tilde{S}^m(U)$ and $u \in \mathcal{D}(U)$,

$$L(p)u(x) = \int p(x, \xi)\hat{u}(\xi)e^{i\langle x,\xi\rangle}\, d\xi, \tag{3.4}$$

and we call $L(p)$ a *canonical pseudodifferential operator of order m*.

Lemma 3.3: $L(p)$ is a linear operator mapping $\mathcal{D}(U)$ into $\mathcal{E}(U)$.

Proof: Since $u \in \mathcal{D}(U)$, we have, for any multiindex α,

$$\xi^\alpha \hat{u}(\xi) = (2\pi)^{-n} \int D^\alpha u(x)e^{-i\langle x,\xi\rangle}\, dx,$$

and hence, since u has compact support, $|\xi^\alpha||\hat{u}(\xi)|$ is bounded for any α, which implies that for any large N,

$$|\hat{u}(\xi)| \leq C(1 + |\xi|)^{-N},$$

i.e., $\hat{u}(\xi)$ goes to zero at ∞ faster than any polynomial. Then we have the estimate for any derivatives of the integrand in (3.4),

$$\left|D_x^\beta p(x, \xi)\hat{u}(\xi)\right| \leq C(1 + |\xi|)^m(1 + |\xi|)^{-N},$$

which implies that the integral in (3.4) converges nicely enough to differentiate under the integral sign as much as we please, and hence $L(p)(u) \in \mathcal{E}(U)$.

It is clear from the same estimates that $L(p)$ is indeed a continuous linear mapping from $\mathcal{D}(U) \to \mathcal{E}(U)$.

Q.E.D.

Our next theorem tells us that the operators $L(p)$ behave very much like differential operators.

Theorem 3.4: Suppose that $p \in \tilde{S}_0^m(U)$. Then $L(p)$ is an operator of order m.

Remark: We introduce functions with compact support to simplify things somewhat. Our future interest is compact manifolds and the functions p which will arise will be of this form due to the use of a partition of unity.

Proof: We must show that if $u \in \mathcal{D}(U)$, then, for some $C > 0$,

$$\|L(p)u\|_s \leq C\|u\|_{s+m}, \tag{3.5}$$

where $\|\bullet\|_s = \|\bullet\|_{s,\mathbf{R}^n}$, as in (1.1), First we note that

$$\widehat{L(p)u}(\xi) = \int \hat{p}(\xi - \eta, \eta)\hat{u}(\eta)\, d\eta, \tag{3.6}$$

where $\hat{p}(\xi - \eta, \eta)$ denotes the Fourier transform of $p(x, \eta)$ in the first variable evaluated at the point $\xi - \eta$. Since p has compact support in the x-variable and because of the estimate (3.2), we have (as before) the estimate, for any large N,

$$|\hat{p}(\xi - \eta, \eta)| \leq C_N(1 + |\xi - \eta|^2)^{-N}(1 + |\eta|^2)^{m/2} \tag{3.7}$$

for $(\xi, \eta) \in \mathbf{R}^n \times \mathbf{R}^n$. We have to estimate

$$\|Lu\|_s^2 = \int |\widehat{L(p)u}(\xi)|^2(1 + |\xi|^2)^s d\xi$$

in terms of

$$\|u\|_{s+m}^2 = \int |\hat{u}(\xi)|^2(1 + |\xi|^2)^{m+s} d\xi.$$

We shall need Young's inequality, which asserts that if $f * g$ is the convolution of an $f \in L^1(\mathbf{R}^n)$ and $g \in L^p(\mathbf{R}^n)$, then

$$\|f * g\|_{L^p} \leq \|f\|_{L^1}\|g\|_{L^p}$$

(see Zygmund [1]).

Proceeding with the proof of (3.5) we obtain immediately from (3.6) and the estimate (3.7), letting C denote a sufficiently large constant in each estimate,

$$\begin{aligned} |\widehat{L(p)u}(\xi)| &\leq C \int (1 + |\xi - \eta|^2)^{-N}(1 + |\eta|^2)^{m/2}|\hat{u}(\eta)|\, d\eta \\ &\leq C \int \frac{(1 + |\xi - \eta|^2)^{-N}}{(1 + |\eta|^2)^{s/2}}(1 + |\eta|^2)^{(m+s)/2}|\hat{u}(\eta)|\, d\eta. \end{aligned}$$

Now, using (1.2), we easily obtain

$$|\widehat{L(p)u}(\xi)| \leq C(1+|\xi|^2)^{-s/2}\int(1+|\xi-\eta|^2)^{-N+s/2} \times (1+|\eta|^2)^{(m+s)/2}\,|\hat{u}(\eta)|\,d\eta.$$

Assume now that N is chosen large enough so that $f(\xi) = (1+|\xi|^2)^{-N+s/2}$ is integrable, and we see that

$$|\widehat{L(p)u}(\xi)|(1+|\xi|^2)^{s/2} \leq C\int(1+|\xi-\eta|^2)^{-N+s/2} \times (1+|\eta|^2)^{(m+s)/2}\,|\hat{u}(\eta)|\,d\eta.$$

By Young's inequality we obtain immediately

$$\begin{aligned}\|L(p)u\|_s &\leq C\|f\|_0 \bullet \|u\|_{s+m}\\ &\leq C\|u\|_{s+m}.\end{aligned}$$

Q.E.D.

We now want to define pseudodifferential operators in general. First we consider the case of operators on a differentiable manifold X mapping functions to functions.

Definition 3.6: Let L be a linear mapping $L\colon \mathcal{D}(X) \to \mathcal{E}(X)$. Then we say that L is a *pseudodifferential operator on* X if and only if for any coordinate chart $U \subset X$ and any open set $U' \subset\subset U$ there exists a $p \in S_0^m(U)$ (considering U as an open subset of $\mathbf{R}^n$) so that if $u \in \mathcal{D}(U')$, then [extending u by zero to be in $\mathcal{D}(X)$]

$$Lu = L(p)u;$$

i.e., by restricting to the coordinate patch U, there is a function $p \in S_0^m(U)$ so that the operator is a canonical pseudodifferential operator of the type introduced above.

More generally, if E and F are vector bundles over the differentiable manifold X, we make the natural definition.

Definition 3.7: Let L be a linear mapping $L\colon \mathcal{D}(X, E) \to \mathcal{E}(X, F)$. Then L is a *pseudodifferential operator* on X if and only if for any coordinate chart U with trivializations of E and F over U and for any open set $U' \subset\subset U$ there exists a $r \times p$ matrix (p^{ij}), $p^{ij} \in S_0^m(U)$, so that the induced map

$$L_U\colon \mathcal{D}(U')^p \longrightarrow \mathcal{E}(U)^r$$

with $u \in \mathcal{D}(U')^p \xrightarrow{L_U} Lu$, extending u by zero to be an element of $\mathcal{D}(X, E)$ [where $p = \text{rank } E, r = \text{rank } F$, and we identify $\mathcal{E}(U)^p$ with $\mathcal{E}(U, E)$ and $\mathcal{E}(U)^r$ with $\mathcal{E}(U, F)$], is a matrix of canonical pseudodifferential operators $L(p^{ij}), i = 1, \ldots, r, j = 1, \ldots, p$, defined by (3.4).

We see that this definition coincides (except for the restriction to the relatively compact subset U') with the definition of a differential operator given in Sec. 2, where all the corresponding functions p^{ij} are polynomials in $S^m(U)$.

Remark: The additional restriction in the definition of restricting the action of L_U to functions supported in $U' \subset\subset U$ is due to the fact that in general a pseudodifferential operator is not a *local operator*; i.e., it does not preserve supports in the sense that supp $Lu \subset$ supp u (which is easy to see for the case of a smoothing operator, for instance). In fact, differential operators can be characterized by the property of localness (a result of Peetre [1]). Thus the symbol of a pseudodifferential operator will depend on the choice of U' which can be considered as a choice of a cutoff function. The difference of two such local representations for pseudodifferential operators on $U' \subset\subset U$ and $U'' \subset\subset U$ will be an operator of order $-\infty$ acting on smooth functions supported in $U' \cap U''$.

Definition 3.8: The *local m-symbol* of a pseudodifferential operator $L\colon \mathcal{D}(X, E) \to \mathcal{E}(X, F)$ is, with respect to a coordinates chart U and trivializations of E and F over U, the matrix†

$$\sigma_m(L)_U(x, \xi) = [\sigma_m(p^{ij})(x, \xi)], \quad i = 1, \ldots, r, \quad j = 1, \ldots, p.$$

Note that in all these definitions the integer m may depend on the coordinate chart U. If X is not compact, then the integer m may be unbounded on X. We shall see that the smallest possible integer m in some sense will be the *order* of the pseudodifferential operator on X. But first we need to investigate the behavior of the local m-symbol under local diffeomorphisms in order to obtain a *global* m-symbol of a global operator L.

The basic principle is the same as for differential operators. If a differential operator is locally expressed as $L = \sum_{|\alpha|\le m} c_\alpha(x) D_x^\alpha$ and we make a change of coordinates $y = F(x)$, then we can express the same operator in terms of these new coordinates using the chain rule and obtain

$$\tilde{L} = \sum_{|\alpha|\le m} \tilde{c}_\alpha(y) D_y^\alpha$$

and

$$\tilde{L}(u(F(x)) = \sum_{|\alpha|\le m} \tilde{c}_\alpha(F(x)) D_y^\alpha u(F(x)).$$

Under this process, the order is the same, and, in particular, we still have a differential operator. Moreover, the mth order homogeneous part of the polynomial, $\sum_{|\alpha|=m} c_\alpha \xi^\alpha$, transforms by the Jacobian of the transformation $y = F(x)$ in a precise manner. We want to carry out this process for pseudodifferential operators, and this will allow us to generalize the symbol map given by Proposition 2.3 for differential operators. For simplicity we shall carry out this program here only for trivial line bundles over X, i.e., for

†$\sigma_m(L)_U$ will also depend on $U' \subset\subset U$, which we have suppressed here.

pseudodifferential operators mapping functions to functions, leaving the more general case of vector bundles to the reader.

The basic result we need can be stated as follows.

Theorem 3.9: Let U be open and relatively compact in $\mathbf{R}^n$ and let $p \in S_0^m(U)$. Suppose that F is a diffeomorphism of U onto itself [in coordinates $x = F(y), x, y \in \mathbf{R}^n$]. Suppose that $U' \subset\subset U$ and define the linear mapping

$$\tilde{L}: \mathcal{D}(U') \longrightarrow \mathcal{E}(U)$$

by setting

$$\tilde{L}v(y) = L(p)(F^{-1})^* v(F(y)).$$

Then there is a function $q \in S_0^m(U)$, so that $\tilde{L} = L(q)$, and, moreover,

$$\sigma_m(q)(y, \eta) = \sigma_m(p)\left(F(y), \left[{}^t\!\left(\frac{\partial F}{\partial y}\right)\right]^{-1} \eta\right).$$

Here $(F^{-1})^*: \mathcal{E}(U) \to \mathcal{E}(U)$ is given by $(F^{-1})^* v = v \circ F^{-1}$, and the basic content of the theorem is that pseudodifferential operators are invariant under local changes of coordinates and that the local symbols transform in a precise manner, depending on the Jacobian $(\partial F/\partial y)$ of the change of variables. Before we prove this theorem, we shall introduce a seemingly larger class of Fourier transform operators, which will arise naturally when we make a change of coordinates. Then we shall see that this class is no larger than the one we started with.

Let $p \in S_0^m(U)$ for U open in $\mathbf{R}^n$. Then we see easily that from (3.4) we obtain the representation

$$L(p)u(x) = (2\pi)^{-n} \iint e^{i\langle \xi, x-z\rangle} p(x, \xi) u(z)\, dz\, d\xi, \tag{3.8}$$

using the Fourier expression for $\hat{u}$. We want to generalize this representation somewhat by allowing the function p above to also depend on z. Suppose that we consider functions $q(x, \xi, z)$ defined and C^∞ on $U \times \mathbf{R}^n \times U$, with compact support in the x- and z-variables and satisfying the following two conditions (similar to those in Definition 3.1):

(a) $\left|D_\xi^\alpha D_x^\beta D_z^\gamma q(x, \xi, z)\right| \le C_{\alpha,\beta,\gamma}(1 + |\xi|)^{m-|\alpha|}$.

(3.9) (b) The limit $\displaystyle\lim_{\lambda\to\infty} \frac{q(x, \lambda\xi, x)}{\lambda^m} = \sigma_m(q)(x, \xi, x), \quad \xi \neq 0,$

exists and $\psi(\xi)\sigma_m(q)(x, \xi, x) - q(x, \xi, x) \in \tilde{S}^{m-1}(U)$.

Proposition 3.10: Let $q(x, \xi, z)$ satisfy the conditions in (3.9) and let the operator Q be defined by

$$Qu(x) = (2\pi)^{-n} \int e^{i\langle \xi, x-z\rangle} q(x, \xi, z) u(z)\, dz\, d\xi \tag{3.10}$$

for $u \in \mathcal{D}(U)$. Then there exists a $p \in S_0^m(U)$ such that $Q = L(p)$. Moreover,

$$\sigma_m(p)(x, \xi) = \lim_{\lambda\to\infty} \frac{q(x, \lambda\xi, x)}{\lambda^m}, \quad \xi \neq 0.$$

This proposition tells us that this "more general" type of operator is in fact one of our original class of operators, and we can compute its symbol.

Proof: Let $\hat{q}(x, \xi, \zeta)$ denote the Fourier transform of $q(x, \xi, z)$ with respect to the z-variable. Then we obtain, from (3.10),

$$Qu(x) = \iint e^{i\langle \xi, x\rangle} \hat{q}(x, \xi, \xi - \eta)\hat{u}(\eta)\, d\eta\, d\xi$$

$$= \int e^{i\langle \eta, x\rangle} \left\{ \int e^{i\langle \xi - \eta, x\rangle} \hat{q}(x, \xi, \xi - \eta)\, d\xi \right\} \hat{u}(\eta)\, d\eta.$$

Thus, if we set

$$p(x, \eta) = \int e^{i\langle \xi - \eta, x\rangle} \hat{q}(x, \xi, \xi - \eta)\, d\xi$$

(3.11)

$$= \int e^{i\langle \zeta, x\rangle} \hat{q}(x, \zeta + \eta, \zeta)\, d\zeta,$$

we have the operator Q represented in the form (3.4). First we have to check that $p(x, \eta) \in \tilde{S}^m(U)$, but this follows easily by differentiating under the integral sign in (3.11), noting that $\hat{q}(x, \zeta + \eta, y)$ decreases very fast at ∞ due to the compact support of $q(x, \xi, z)$ in the z-variable. We now use the mean-value theorem for the integrand:

$$\hat{q}(x, \zeta + \eta, \zeta) = \hat{q}(x, \eta, \zeta) + \sum_{|\alpha|=1} D_\eta^\alpha \hat{q}(x, \eta + \zeta_0, \zeta)\zeta^\alpha \tag{3.12}$$

for a suitable ζ_0 lying on the segment in $\mathbf{R}^n$ joining 0 to ζ. We have the estimate

$$|D_\eta^\alpha \hat{q}(x, \eta + \zeta_0, \zeta)| \le C_N(1 + |\eta + \zeta_0|)^{m-1}(1 + |\zeta|)^{-N}$$

for sufficiently large N, and since $|\zeta_0(\zeta)| \le |\zeta|$, we see that we obtain, with a different constant,

$$|D_\eta^\alpha \hat{q}(x, \eta + \zeta_0, \zeta)| \le \tilde{C}_N(1 + |\eta|)^{m-1}(1 + |\zeta|)^{-N+m-1}.$$

By inserting (3.12) in (3.11), choosing N sufficiently large, and integrating the resulting two terms we obtain

$$p(x, \eta) = q(x, \eta, x) + E(x, \eta),$$

where

$$|E(x, \eta)| \le C(1 + |\eta|)^{m-1}. \tag{3.13}$$

Therefore

$$\lim_{\lambda\to\infty} \frac{p(x, \lambda\eta)}{\lambda^m} = \lim_{\lambda\to\infty} \frac{q(x, \lambda\eta, x)}{\lambda^m} + \lim_{\lambda\to\infty} \frac{E(x, \lambda\eta)}{\lambda^m}, \qquad \eta \ne 0.$$

It follows that the limit on the left exists and that

$$\sigma_m(p)(x, \eta) = \lim_{\lambda\to\infty} \frac{q(x, \lambda\eta, x)}{\lambda^m}, \qquad \eta \ne 0,$$

since the last term above has limit zero because of the estimate (3.13).

The fact that $p(x, \xi) - \psi(\xi)\sigma_m(p)(x, \xi) \in \tilde{S}^{m-1}(U)$ follows easily from the hypothesis on $q(x, \eta, x)$ and the growth of $E(x, \eta)$.

Q.E.D.

We shall need one additional fact before we can proceed to our proof of Theorem 3.9. Namely, suppose that we rewrite (3.8) formally as

$$L(p)u(x) = (2\pi)^{-n} \int \left\{ \int e^{i\langle \xi, x-z\rangle} p(x, \xi)\, d\xi \right\} u(z)\, dz$$

and let

$$K(x, x - z) = \int e^{i\langle \xi, x-z\rangle} p(x, \xi)\, d\xi.$$

Then we have the following proposition.

Proposition 3.11: $K(x, w)$ is a C^∞ function of x and w provided that $w \neq 0$.

Proof: Suppose first that $m < -n$, then we have the estimate

$$|p(x, \xi)| \leq C(1 + |\xi|)^{-n-1}$$

from (3.2), and thus the integral

$$K(x, w) = \int e^{i\langle \xi, w\rangle} p(x, \xi)\, d\xi$$

converges. Integrating repeatedly by parts, and assuming that, for instance, $w_1 \neq 0$, we obtain

$$K(x, w) = (-1)^N \int \frac{e^{i\langle \xi, w\rangle}}{w_1^N} D_{\xi_1}^N p(x, \xi)\, d\xi$$

for any positive integer N. Hence

$$D_x^\alpha D_w^\beta K(x, w) = (-1)^N \int \frac{\xi^\beta e^{i\langle \xi, w\rangle}}{w_1^N} D_x^\alpha D_{\xi_1}^N p(x, \xi)\, d\xi.$$

Using the estimates (3.2) we see that the integral on the right converges for N sufficiently large. Thus $K(x, w)$ is C^∞ for $w \neq 0$, provided that $m < -n$. Suppose now that m is arbitrary; then we write, choosing $\rho > m + n$,

$$K(x, w) = \int e^{i\langle \xi, w\rangle} (1 + |\xi|^2)^{-\rho} (1 + |\xi|^2)^{\rho} p(x, \xi)\, d\xi$$

and we see that we have (letting $\Delta_w = \sum D_j^2$ be the usual Laplacian in the w variable)

$$K(x, w) = \int \left[(1 - \Delta_w)^\rho e^{i\langle \xi, w\rangle} \right] p(x, \xi)(1 + |\xi|^2)^{-\rho} d\xi$$

which is the same as

$$(1 - \Delta_w)^\rho \int e^{i\langle \xi, w\rangle} p(x, \xi)(1 + |\xi|^2)^{-\rho} d\xi$$

if the integral converges. But this then follows from the case considered above, since

$$p(x, \xi)(1 + |\xi|^2)^{-\rho} \in S_0^{-(n+1)}(U).$$

So, for $w \neq 0$, the above integral is C^∞, and thus $K(x, w)$ is C^∞ for $w \neq 0$.

Q.E.D.

We shall now use these propositions to continue our study of the behavior of a pseudodifferential operator under a change of coordinates.

Proof of Theorem 3.9: Now $p(x, \xi)$ has compact support in U (in the x-variable) by hypothesis. Let $\tilde{\psi}(x) \in \mathcal{D}(U)$ be chosen so that $\tilde{\psi} \equiv 1$ on $\operatorname{supp} p \cup U'$ and set $\psi(y) = \tilde{\psi}(F(y))$. We have, as in (3.8),

$$L(p)u(x) = (2\pi)^{-n} \iint e^{i\langle \xi, x-z\rangle} p(x, \xi) u(z)\, dz\, d\xi,$$

for $u \in \mathcal{D}(U') \subset \mathcal{D}(U)$. We write $z = F(w)$ and $v(w) = u(F(w))$ and obtain

$$L(p)u(F(y)) = (2\pi)^{-n} \iint e^{i\langle \xi, F(y)-F(w)\rangle} p(F(y), \xi) \left|\frac{\partial F}{\partial w}\right| v(w)\, dw\, d\xi,$$

where $|\partial F/\partial w|$ is the determinant of the Jacobian matrix $\partial F/\partial w$. By the mean-value theorem we see that

$$F(y) - F(w) = H(y, w) \bullet (y - w),$$

where $H(y, w)$ is a nonsingular matrix for w close to y and $H(w, w) = (\partial F/\partial w)(w)$. Let $\chi_1(y, w)$ be a smooth nonnegative function $\equiv 1$ near the diagonal $\mathbf{\Delta}$ in $U \times U$ and with support on a neighborhood of $\mathbf{\Delta}$ where $H(y, w)$ is invertible. Let $\chi_2 = 1 - \chi_1$. Thus we have

$$L(p)u(F(y)) = (2\pi)^{-n} \iint e^{i\langle \xi, H(y,w)\bullet(y-w)\rangle} p(F(y), \xi) \left|\frac{\partial F}{\partial w}\right| v(w)\, dw\, d\xi,$$

which we may rewrite, setting $\zeta = {}^t H(y, w)\xi$,

$$\begin{aligned} L(p)u(F(y)) &= (2\pi)^{-n} \Bigg[\iint e^{i\langle \zeta, y-w\rangle} p(F(y), [{}^t H(y, w)]^{-1}\zeta) \\ &\qquad \times \left|\frac{\partial F}{\partial w}\right| \psi(w) \frac{\chi_1(y, w)}{|H(y, w)|} v(w)\, dw\, d\zeta + Eu(F(y)) \Bigg] \\ &= (2\pi)^{-n} \left[\iint e^{i\langle \zeta, y-w\rangle} q_1(y, \zeta, w) v(w)\, dw\, d\zeta + Eu(F(y)) \right]. \end{aligned}$$

Here E is the term corresponding to χ_2 and

$$q_1(y, \zeta, w) = p(F(y), [{}^t H(y, w)]^{-1}\zeta) \left|\frac{\partial F}{\partial w}\right| \frac{\chi_1(y, w)}{|H(y, w)|} \psi(w),$$

while $\psi \in \mathcal{D}(U)$ as chosen above is identically 1 on a neighborhood of supp $v(y)$. Thus $q_1(y, \zeta, w)$ has compact support in the y and w variables, and it follows readily that conditions (a) and (b) of (3.9) are satisfied. Namely,

(a) will follow from the estimates for p by the chain rule, whereas for (b) we have

$$\sigma(q_1)(y,\xi,y) = \lim_{\lambda\to\infty} \frac{q_1(y,\lambda\xi,y)}{\lambda^m} = \lim_{\lambda\to\infty} \frac{p(F(y),[{}^t(\partial F/\partial y)]^{-1}\lambda\xi)}{\lambda^m}$$

$$= \sigma(p)\left(F(y),\left[{}^t\left(\frac{\partial F}{\partial y}\right)\right]^{-1}\zeta\right), \xi \neq 0;$$

moreover the desired growth of

$$\psi(\xi)\sigma(q_1)(y,\xi,y) - q_1(y,\xi,y)$$

follows easily from the hypothesized growth of

$$\psi(\xi)\sigma(p)(x,\xi) - p(x,\xi).$$

We still have to worry about the term E, which we claim is a smoothing operator of infinite order (see the example following Definition 3.1) and will give no contribution to the symbol. In fact, we have

$$Eu(x) = \int e^{i\langle\xi,x-z\rangle} p(x,\xi)\chi_2(x,z)u(z)\,dz\,d\xi$$

$$= \int\left\{\int e^{i\langle\xi,x-z\rangle} p(x,\xi)\chi_2(x,z)\,d\xi\right\}u(z)\,dz$$

$$= \int \chi_2(x,z)K(x,x-z)u(z)\,dz$$

$$= \int W(x,z)u(z)\,dz,$$

where

$$K(x,w) = \int e^{i\langle\xi,w\rangle} p(x,\xi)\,d\xi.$$

But we have seen earlier that $K(x,w)$ is a C^∞ function of x and w for $w \neq 0$.† Also, $\chi_2(x,z)$ vanishes identically near $x - z = 0$, so the product $\chi_2(x,z)K(x,x-z) = W(x,z)$ is a smooth function on $U \times \mathbf{R}^n$, and $Eu(x)$ is then a smoothing operator with C^∞ kernel, which we can write in terms of the new coordinates $y = F^{-1}(x)$, $w = F^{-1}(z)$,

$$Eu(F(y)) = \int W(F(y),F(w))u(F(w))\left|\frac{\partial F}{\partial w}\right|dw$$

$$= \int W_1(y,w)F^*u(w)\,dw,$$

where W_1 is a C^∞ function on $U \times U$, which we rewrite as

$$= \int W_1(y,w)\psi(w)F^*u(w)\,dw,$$

†Note that $K(x,w)$ has compact support in the first variable, since $p(x,\xi) = \varphi(x)p(x,\xi)$ for an appropriate $\varphi \in \mathcal{D}(U)$.

where we note that $\psi \equiv 1$ on supp F^*u. Then we have

$$Eu(F(y)) = \int W_2(y, w)v(w)\, dw,$$

where $v(w) = F^*u(w)$, as before, and

$$W_2(y, w) = W(F(y), F(w)) \left| \frac{\partial F}{\partial w} \right| \psi(w),$$

which is a smoothing operator of order $-\infty$ with C^∞ kernel with compact support in both variables, as discussed following Definition 3.1, Thus, by Proposition 3.11,

$$Eu(F(y)) = \int e^{i\langle w,\xi\rangle} q_2(y, \xi)\hat{v}(\xi)\, d\xi,$$

where

$$q_2(y, \xi) = \int e^{i\langle y-w,\xi\rangle} W_2(y, w)\, dw,$$

and $q_2 \in \tilde{S}_0^r(U)$ for all r. This implies easily that $\sigma_m(q_2)(y, \xi) \equiv 0$. Thus we can let $q = q_1 + q_2$, and we have

$$\tilde{L}v(y) = L(q),$$

and the symbols behave correctly [here we let $q_1(y, \xi, w)$ be replaced by $q_1(y, \xi)$, as given by Proposition 3.10].

Q.E.D.

We are now in a position to define the global symbol of a pseudodifferential operator on a differentiable manifold X. Again we treat the case of functions first, and we begin with the following definition.

Definition 3.12: Let X be a differentiable manifold and let $L\colon \mathcal{D}(X) \longrightarrow \mathcal{E}(X)$ be a pseudodifferential operator. Then L is said to be a *pseudodifferential operator of order m on X* if, for any choice of local coordinates chart $U \subset X$, the corresponding canonical pseudodifferential operator L_U is of order m; i.e., $L_U = L(p)$, where $p \in S^m(U)$. The class of all pseudodifferential operators on X of order m is denoted by $\mathrm{PDiff}_m(X)$.

Proposition 3.13: Suppose that X is a compact differentiable manifold. If $L \in \mathrm{PDiff}_m(X)$, then $L \in \mathrm{OP}_m(X)$.

Proof: This is immediate from Theorem 3.4 and the definition of Sobolev norms on a compact manifold, using a finite covering of X by coordinate charts.

Q.E.D.

This proposition tells us that the two definitions of "order" of a pseudodifferential operator are compatible. We remark that if $p \in S^m(U)$, for some $U \subset \mathbf{R}^n$, then $p \in S^{m+k}(U)$ for any positive k; moreover, in this case, $\sigma_{m+k}(p)$

$= 0, k > 0$. Thus we have the natural inclusion $\mathrm{PDiff}_m(X) \subset \mathrm{PDiff}_{m+k}(X)$, $k \geq 0$. Denote $\mathrm{Smbl}_m(X \times \mathbf{C}, X \times \mathbf{C})$ by $\mathrm{Smbl}_m(X)$ for simplicity.

Proposition 3.14: There exists a canonical linear map

$$\sigma_m\colon \mathrm{PDiff}_m(X) \longrightarrow \mathrm{Smbl}_m(X),$$

which is defined locally in a coordinate chart $U \subset X$ by

$$\sigma_m(L_U)(x, \xi) = \sigma_m(p)(x, \xi),$$

where $L_U = L(p)$ and where $(x, \xi) \in U \times (\mathbf{R}^n - \{0\})$ is a point in $T'(U)$ expressed in the local coordinates of U.

Proof: We merely need to verify that the local representation of $\sigma_m(L)$ defined above transforms correctly so that it is indeed globally a homomorphism of $T'(X) \times \mathbf{C}$ into $T'(X) \times \mathbf{C}$, which is homogeneous in the cotangent vector variable of order m (see the definition in Sec. 2). But this follows easily from the transformation formula for $\sigma_m(p)$ given in Theorem 3.9, under a local change of variables. The linearity of σ_m is not difficult to verify. Q.E.D.

This procedure generalizes to pseudodifferential operators mapping sections of vector bundles to sections of vector bundles, and we shall leave the formal details to the reader. We shall denote by $\mathrm{PDiff}_m(E, F)$ the space of pseudodifferential operators of order m mapping $\mathcal{D}(X, E)$ into $\mathcal{E}(X, F)$. Moreover, there is an analogue to Proposition 3.14, whose proof we omit.

Proposition 3.15: Let E and F be vector bundles over a differentiable manifold X. There exists a canonical linear map

$$\sigma_m\colon \mathrm{PDiff}_m(E, F) \longrightarrow \mathrm{Smbl}_m(E, F),$$

which is defined locally in a coordinate chart $U \subset X$ by

$$\sigma_m(L_U)(x, \xi) = [\sigma_m(p^{ij})(x, \xi)],$$

where $L_U = [L(p^{ij})]$ is a matrix of canonical pseudodifferential operators, and where $(x, \xi) \in U \times (\mathbf{R}^n - \{0\})$ is a point in $T'(U)$ expressed in the local coordinates of U.

One of the fundamental results in the theory of pseudodifferential operators on manifolds is contained in the following theorem.

Theorem 3.16: Let E and F be vector bundles over a differentiable manifold X. Then the following sequence is exact,

$$0 \longrightarrow K_m(E, F) \xrightarrow{j} \mathrm{PDiff}_m(E, F) \xrightarrow{\sigma_m} \mathrm{Smbl}_m(E, F) \longrightarrow 0, \tag{3.15}$$

where σ_m is the canonical symbol map given by Proposition 3.15, $K_m(E, F)$ is the kernel of σ_m, and j is the natural injection. Moreover, $K_m(E, F) \subset \mathrm{OP}_{m-1}(E, F)$ if X is compact.

Proof: We need to show that σ_m is surjective and that $\sigma_m(L) = 0$ implies that L is an operator of order $m - 1$. Doing the latter first, we note that $\sigma_m(L) = 0$ for some $L \in \text{PDiff}_m(E, F)$ means that in a local trivializing coordinate chart, L has the representation $L_U = [L(p^{ij})]$, $p^{ij} \in S^m(U)$. Since $\sigma_m(L)|_U = \sigma_m(L_U) = [\sigma_m(p^{ij})] = 0$, by hypothesis, it follows that $p^{ij} \in \tilde{S}^{m-1}(U)$, by hypothesis on the class S^m. Hence L_U is an operator of order $m - 1$, and thus L will be an operator of order $m - 1$. To prove that σ_m is surjective, we proceed as follows. Let $\{U_\mu\}$ be a locally finite cover of X by coordinate charts U_μ over which E and F are both trivializable. Let $\{\phi_\mu\}$ be a partition of unity subordinate to the cover $\{U_\mu\}$ and let $\{\psi_\mu\}$ be a family of functions $\psi_\mu \in \mathcal{D}(U_\mu)$, where $\psi_\mu \equiv 1$ on supp ϕ_μ. We then let χ be a C^∞ function on $\mathbf{R}^n$ with $\chi \equiv 0$ near $0 \in \mathbf{R}^n$ and $\chi \equiv 1$ outside the unit ball. Let $s \in \text{Smbl}_m(E, F)$ be given, and write $s = \sum_\mu \phi_\mu s = \sum_\mu s_\mu$; supp $s_\mu \subset$ supp $\phi_\mu \subset U_\mu$. Then with respect to a trivialization of E and F over U, we see that $s_\mu = [s_\mu^{ij}]$, a matrix of homogeneous functions $s_\mu^{ij}\colon U_\mu \times \mathbf{R}^n - \{0\} \longrightarrow \mathbf{C}$, and $s_\mu^{ij}(x, \rho\xi) = \rho^m s_\mu^{ij}(x, \xi)$, for $\rho > 0$. We let $p_\mu^{ij}(x, \xi) = \chi(\xi) s_\mu^{ij}(x, \xi)$. It follows from the homogeneity that $p_\mu^{ij} \in S_0^m(U)$ and that $\sigma_m(p_\mu^{ij}) = s_\mu^{ij}$. We now let

$$L_\mu\colon \mathcal{D}(U_\mu)^p \longrightarrow \mathcal{E}(U_\mu)^r$$

be defined by $L_\mu u = [L(p_\mu^{ij})]u$, with the usual matrix action of the matrix of operators on the vector u. If $u \in \mathcal{D}(X, E)$, then we let $u = \sum \phi_\mu u = \sum u_\mu$, considering each u_μ as a vector in $\mathcal{D}(U_\mu)^p$ by the trivializations. We then define

$$Lu = \sum_\mu \psi_\mu (L_\mu u_\mu),$$

and it is clear that

$$L\colon \mathcal{D}(X, E) \longrightarrow \mathcal{E}(X, F)$$

is an element of $\text{PDiff}_m(E, F)$, since locally it is represented by a matrix of canonical pseudodifferential operators of order m. Note that it is necessary to multiply by ψ_μ in order to sum, since $L_\mu u_\mu$ is C^∞, where we consider $L_\mu u_\mu$ as an element of $\mathcal{E}(U_\mu, F)$ in U_μ, but that it does not necessarily extend in a C^∞ manner to a C^∞ section of F over X. Thus we have constructed from s a pseudodifferential operator L in a noncanonical manner; it remains to show that $\sigma_m(L) = s$. But this is simple, since $(\psi_\mu L_\mu) \in \text{PDiff}_m(E, F)$ and

$$\begin{aligned}\sigma_m(\psi_\mu L_\mu)_U(x, \xi) &= \sigma_m(\psi_\mu(x) p_\mu^{ij}(x, \xi)) = \psi_\mu(x) \lim_{\lambda \to \infty} p_\mu^{ij}(x, \lambda\xi)/\lambda^m \\ &= \psi_\mu(x) s_\mu^{ij}(x, \xi) = s_\mu^{ij}(x, \xi),\end{aligned}$$

since $\psi_\mu \equiv 1$ on supp s_μ. It follows that

$$\sigma_m(\psi_\mu L_\mu) = s_\mu,$$

and by linearity of the symbol map

$$\sigma_m\Big(\sum_\mu \psi_\mu L_\mu\Big) = \sum_\mu s_\mu = s.$$

Q.E.D.

We need to show now that the direct sum $\mathrm{PDiff}(E, F) = \sum_m \mathrm{PDiff}_m(E, F)$ forms an algebra under composition, which is closed under transposition. We formulate this in the following manner.

Theorem 3.17: Let E, F, and G be vector bundles over a compact differentiable manifold X. Then

(a) If $Q \in \mathrm{PDiff}_r(E, F)$ and $P \in \mathrm{PDiff}_s(F, G)$, then the composition as operators $P \circ Q \in \mathrm{PDiff}_{r+s}(E, G)$, and, moreover,

$$\sigma_{r+s}(P \circ Q) = \sigma_s(P) \bullet \sigma_r(Q),$$

where the latter product is the composition product of the linear vector bundle maps

$$\pi^* E \xrightarrow{\sigma_r(Q)} \pi^* F \xrightarrow{\sigma_s(P)} \pi^* G.$$

(b) If $P \in \mathrm{PDiff}_m(E, F)$, then P^*, the adjoint of P, exists, where $P^* \in \mathrm{PDiff}_m(F, E)$, and, moreover,

$$\sigma_m(P^*) = \sigma_m(P)^*,$$

where $\sigma_m(P)^*$ denotes the adjoint of the linear map

$$\pi^* E \xrightarrow{\sigma_m(P)} \pi^* F.$$

Proof: To prove these facts it will suffice to consider local representations by canonical pseudodifferential operators, since this is how the action of the operator on functions is defined. First we consider the scalar case; i.e., E, F, and G are trivial line bundles, and we have the operators acting on C^∞ functions on X.

We begin by proving the existence of an adjoint in $\mathrm{PDiff}_m(X)$ and note that by Proposition 3.13 and Proposition 2.1, $P^* \in \mathrm{OP}_m(X)$ exists. Let U be a coordinate chart (considered as an open subset of $\mathbf{R}^n$) for X, and for any open set $U' \subset\subset U$, let $u, v \in \mathcal{D}(U')$. If $p \in S_0^m(U)$ such that $P_U = L(p)$, then by (3.4)

$$\begin{aligned}(u, P^* v) = (P_U u, v) &= \iint p(x, \xi) e^{i\langle x, \xi\rangle} \hat{u}(\xi) \overline{v(x)}\, d\xi\, dx \\ &= \iiint p(x, \xi) e^{i\langle x, \xi\rangle} (2\pi)^{-n} e^{-i\langle y, \xi\rangle} u(y) \overline{v(x)}\, dy\, d\xi\, dx \\ &= \int u(y) (2\pi)^{-n} \overline{\iint \overline{p(x, \xi)} e^{i\langle y - x, \xi\rangle} v(x)\, dx\, d\xi}\, dy.\end{aligned}$$

Let $r(y, \xi, x) = \overline{p(x, \xi)}$, and we have

$$(u, P^* v) = \int u(y) (2\pi)^{-n} \overline{\iint r(y, \xi, x) e^{i\langle y - x, \xi\rangle} v(x)\, dx\, d\xi}\, dy.$$

By Proposition 3.10, there exists $q \in S_0^m(U)$ such that

$$L(q) v(y) = (2\pi)^{-n} \iint r(y, \xi, x) e^{i\langle y - x, \xi\rangle} v(x)\, dx\, d\xi.$$

Therefore,

$$(u, P^*v) = \int u(y)\overline{L(q)v(y)}dy = (u, L(q)v)$$

and we have $P_U^* = L(q)$. Hence $P^* \in \text{PDiff}_m(X)$. Moreover,

$$\sigma_m(q)(x, \xi) = \lim_{\lambda\to\infty} \frac{r(x, \lambda\xi, x)}{\lambda^m} = \lim_{\lambda\to\infty} \frac{\overline{p(x, \lambda\xi)}}{\lambda^m} = \overline{\sigma_m(p)(x, \xi)},$$

and conjugation is the adjoint for trivial line bundles.

The composition formula now follows by a simple reduction to the adjoint problem. Note that $(Q^*)^* = Q$. For U, U' as above, let $P_U = L(p)$, $Q_U = L(q)$, and $Q_U^* = L(q')$ be representations in U. Then for $u \in \mathcal{D}(U')$, the proof of the adjoint property shows that

$$L(q)u(z) = (2\pi)^{-n} \iint \overline{q'(y, \xi)}e^{i\langle z-y,\xi\rangle}u(y)\,dy\,d\xi$$

$$= \int e^{i\langle z,\xi\rangle}((2\pi)^{-n} \int \overline{q'(y, \xi)}e^{-i\langle y,\xi\rangle}u(y)\,dy)\,d\xi.$$

Thus,

$$L(p) \circ L(q)u(x) = \int p(x, \xi)\widehat{L(q)u}(\xi)e^{i\langle x,\xi\rangle}d\xi$$

$$= (2\pi)^{-n} \iint p(x, \xi)\overline{q'(y, \xi)}e^{i\langle x-y,\xi\rangle}u(y)\,dy\,d\xi.$$

Let $s(x, \xi, y) = p(x, \xi)\overline{q'(y, \xi)}$. Then Proposition 3.10 shows that there exists a $t \in S_0^{r+s}(U)$ such that $P \circ Q|_U = L(t)$. Therefore $P \circ Q \in \text{PDiff}_{r+s}(X)$. Furthermore,

$$\sigma_{r+s}(t)(x, \xi) = \lim_{\lambda\to\infty} \frac{s(x, \lambda\xi, x)}{\lambda^{r+s}} = \lim_{\lambda\to\infty} \frac{p(x, \lambda s)}{\lambda^s} \cdot \lim_{\lambda\to\infty} \frac{\overline{q'(x, \lambda s)}}{\lambda^r}$$

$$= \sigma_s(p)(x, \xi)\overline{\sigma_r(q')(x, \xi)}$$

$$= \sigma_s(p)(x, \xi)\sigma_r(q)(x, \xi)$$

from the proof for the adjoint. Hence, $\sigma_{r+s}(P \circ Q) = \sigma_s(P) \cdot \sigma_r(Q)$ as desired.

The proofs for vector-valued functions (sections of bundles) are essentially the same as for scalar functions, with the added complication that we are dealing with matrices. Then the *order* of the terms in the integrals is crucial, since the matrix-valued entries will not, in general, commute. We shall omit any further details here.

Q.E.D.

For more detailed information about pseudodifferential operators on manifolds, consult the original papers of Seeley [1], Kohn and Nirenberg [1], and Hörmander [3, 4]. The expository article by Nirenberg [1] is an excellent reference.† Palais [1] has a development of the theory presented here along the lines of the Kohn and Nirenberg paper.

†Our presentation is simplified somewhat by the fact that we avoid the asymptotic expansion of a pseudodifferential operator (corresponding to the lower-order terms of a differential operator), since it is unnecessary for the applications to elliptic differential equations.

4. A Parametrix for Elliptic Differential Operators

In this section we want to restrict our attention to operators which generalize the classic Laplacian operator in Euclidean space and its inverse. These will be called elliptic operators. We start with a definition, using the same notation as in the preceding sections. Let E and F be vector bundles over a differentiable manifold X.

Definition 4.1: Let $s \in \mathrm{Smbl}_k(E, F)$. Then s is said to be *elliptic* if and only if for any $(x, \xi) \in T'(X)$, the linear map

$$s(x, \xi)\colon E_x \longrightarrow F_x$$

is an isomorphism.

Note that, in particular, both E and F must have the same fibre dimension. We shall be most interested in the case where $E = F$.

Definition 4.2: Let $L \in \mathrm{PDiff}_k(E, F)$. Then L is said to be *elliptic (of order k)* if and only if $\sigma_k(L)$ is an elliptic symbol.

Note that if L is an elliptic operator of order k, then L is also an operator of order $k + 1$, but clearly *not* an elliptic operator of order $k + 1$ since $\sigma_{k+1}(\mathrm{L}) = 0$. For convenience, we shall call any operator $L \in \mathrm{OP}_{-1}(E, F)$ a *smoothing operator*. We shall later see why this terminology is justified.

Definition 4.3: Let $L \in \mathrm{PDiff}(E, F)$. Then $\tilde{L} \in \mathrm{PDiff}(F, E)$ is called a *parametrix* (or *pseudoinverse*) for L if it has the following properties,

$$L \circ \tilde{L} - I_F \in \mathrm{OP}_{-1}(F)$$

$$\tilde{L} \circ L - I_E \in \mathrm{OP}_{-1}(E),$$

where I_F and I_E denote the identity operators on F and E, respectively.

The basic *existence* theorem for elliptic operators on a compact manifold X can be formulated as follows.

Theorem 4.4: Let k be any integer and let $L \in \mathrm{PDiff}_k(E, F)$ be elliptic. Then there exists a parametrix for L.

Proof: Let $s = \sigma_k(L)$. Then s^{-1} exists as a linear transformation, since s is invertible,

$$s^{-1}(x, \xi)\colon F_x \longrightarrow E_x,$$

and $s^{-1} \in \mathrm{Smbl}_{-k}(F, E)$. Let $\tilde{L}$ be any pseudodifferential operator in $\mathrm{PDiff}_{-k}(F, E)$ such that $\sigma_{-k}(\tilde{L}) = s^{-1}$, whose existence is guaranteed by Theorem 3.16. We have then that

$$\sigma_0(L \circ \tilde{L} - I_F) = \sigma_0(L \circ \tilde{L}) - \sigma_0(I_F),$$

and letting $\sigma_0(I_F) = 1_F$, the identity in $\text{Smbl}_0(F, F)$, we obtain

$$\sigma_0(L \circ \tilde{L} - I_F) = \sigma_k(L) \bullet \sigma_{-k}(\tilde{L}) - 1_F$$

$$= 1_F - 1_F = 0.$$

Thus, by Theorem 3.16, we see that

$$L \circ \tilde{L} - I_F \in \text{OP}_{-1}(F, F).$$

Similarly, $\tilde{L} \circ L - I_E$ is seen to be in $\text{OP}_{-1}(E, E)$.

Q.E.D.

This theorem tells us that modulo smoothing operators we have an inverse for a given elliptic operator. On compact manifolds, this turns out to be only a finite dimensional obstruction, as will be deduced later from the following proposition. First we need a definition. Let X be a compact differentiable manifold and suppose that $L \in \text{OP}_m(E, F)$. Then we say that L is *compact* (or *completely continuous*) if for every s the extension L_s: $W^s(E) \to W^{s-m}(F)$ is a compact operator as a mapping of Banach spaces.

Proposition 4.5: Let X be a compact manifold and let $S \in \text{OP}_{-1}(E, E)$. Then S is a compact operator of order 0.

Proof: We have for any s the following commutative diagram,

$$\begin{array}{ccc} W^s(E) & \xrightarrow{\tilde{S}_s} & W^s(E) \\ & {\scriptstyle S_s}\searrow \quad \nearrow{\scriptstyle j} & \\ & W^{s+1}(E), & \end{array}$$

where S_s is the extension of S to a mapping $W^s \to W^{s+1}$, given since $S \in \text{OP}_{-1}(E, E)$, and $\tilde{S}_s$ is the extension of S, as a mapping $W^s \to W^s$, given by the fact that $\text{OP}_{-1}(E, E) \subset \text{OP}_0(E, E)$. Since j is a compact operator (by Rellich's lemma, Proposition 1.2), then $\tilde{S}_s$ must also be compact.

Q.E.D.

In the remainder of this section we shall let E and F be fixed Hermitian vector bundles over a *compact* differentiable manifold X. Assume that X is equipped with a smooth positive measure μ (such as would be induced by a Riemannian metric, for example) and let $W^0(X, E) = W^0(E)$, $W^0(F)$ denote the Hilbert spaces equipped with L^2-inner products

$$(\xi, \eta)_E = \int_X \langle \xi(x), \eta(x) \rangle_E d\mu, \quad \xi, \eta \in \mathcal{E}(X, E)$$

$$(\sigma, \tau)_F = \int_X \langle \sigma(x), \tau(x) \rangle_F d\mu, \quad \sigma, \tau \in \mathcal{E}(X, F),$$

as in Sec. 1. We shall also consider the Sobolev spaces $W^s(E)$, $W^s(F)$, defined for all integral s, as before, and shall make use of these without further mention. If $L \in OP_m(E, F)$, denote by L_s: $W^s(E) \to W^{s-m}(F)$ the continuous extension of L as a continuous mapping of Banach spaces. We want to study the homogeneous and inhomogeneous solutions of the differential equation

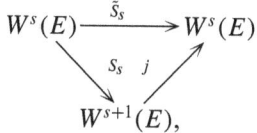

$L\xi = \sigma$, for $\xi \in \mathcal{E}(X, E), \sigma \in \mathcal{E}(X, F)$, where $L \in \text{Diff}_m(E, F)$, and L^* is the adjoint of L defined with respect to the inner products in $W^0(E)$ and $W^0(F)$; i.e.,

$$(L\xi, \tau)_F = (\xi, L^*\tau)_E,$$

as given in Proposition 2.8. If $L \in \text{Diff}_m(E, F)$, we set

$$\mathcal{H}_L = \{\xi \in \mathcal{E}(X, E)\colon L\xi = 0\},$$

and we let

$$\mathcal{H}_L^\perp = \{\eta \in W^0(E)\colon (\xi, \eta)_E = 0, \xi \in \mathcal{H}_L\}$$

denote the orthogonal complement in $W^0(E)$ of $\mathcal{H}_L$. It follows immediately that the space $\mathcal{H}_L^\perp$ is a closed subspace of the Hilbert space $W^0(E)$. As we shall see, under the assumption that L is elliptic $\mathcal{H}_L$ turns out to be finite dimensional [and hence a closed subspace of $W^0(E)$]. Before we get to this, we need to recall some standard facts from functional analysis, due to F. Riesz (see Rudin [1]).

Proposition 4.6: Let B be a Banach space and let S be a compact operator, $S\colon B \to B$. Then letting $T = I - S$, one has:

(a) Ker $T = T^{-1}(0)$ is finite dimensional.
(b) $T(B)$ is closed in B, and Coker $T = B/T(B)$ is finite dimensional.

In our applications the Banach spaces are the Sobolev spaces $W^s(E)$ which are in fact Hilbert spaces. Proposition 4.6(a) is then particularly easy in this case, and we shall sketch the proof for B, a Hilbert space. Namely, if the unit ball in a Hilbert space $\mathfrak{h}$ is compact, then it follows that there can be only a finite number of orthonormal vectors, since the distance between any two orthonormal vectors is uniformly bounded away from zero (by the distance $\sqrt{2}$). Thus $\mathfrak{h}$ must be finite dimensional. Proposition 4.6(a), for instance, then follows immediately from the fact that the unit ball in the Hilbert space $\mathfrak{h} = \text{Ker } T$ must be compact (essentially the definition of a compact operator). The proof that $T(B)$ is closed is more difficult and again uses the compactness of S. Since S^* is also compact, Ker T^* is finite dimensional, and the finite dimensionality of Coker T follows. More generally, the proof of Proposition 4.6 depends on the fundamental finiteness criterion in functional analysis which asserts that a locally compact topological vector space is necessarily finite dimensional. See Riesz and Nagy [1], Rudin [1], or any other standard reference on functional analysis for a discussion of this as well as a proof of the above proposition. (A good survey of this general topic can be found in Palais [1].)

An operator T on a Banach space is called a *Fredholm* operator if T has finite-dimensional kernel and cokernel. Then we immediately obtain the following from Theorem 4.4 and Propositions 4.5 and 4.6.

Theorem 4.7: Let $L \in \mathrm{PDiff}_m(E, F)$ be an elliptic pseudodifferential operator. Then there exists a parametrix P for L so that $L \circ P$ and $P \circ L$ have continuous extensions as Fredholm operators: $W^s(F) \to W^s(F)$ and $W^s(E) \to W^s(E)$, respectively, for each integer s.

We now have the important finiteness theorem for elliptic differential operators.

Theorem 4.8: Let $L \in \mathrm{Diff}_k(E, F)$ be elliptic. Then, letting $\mathcal{H}_{L_s} = \mathrm{Ker}\, L_s$: $W^s(E) \to W^{s-k}(F)$, one has

(a) $\mathcal{H}_{L_s} \subset \mathcal{E}(X, E)$ and hence $\mathcal{H}_{L_s} = \mathcal{H}_L$, all s.
(b) $\dim \mathcal{H}_{L_s} = \dim \mathcal{H}_L < \infty$ and $\dim W^{s-k}(F)/L_s(W^s(E)) < \infty$.

Proof: First we shall show that, for any s, $\dim \mathcal{H}_{L_s} < \infty$. Let P be a parametrix for L, and then by Theorem 4.7, it follows that

$$(P \circ L)_s\colon W^s(E) \longrightarrow W^s(F)$$

has finite dimensional kernel, and obviously $\mathrm{Ker} L_s \subset \mathrm{Ker}(P \circ L)_s$, since we have the following commutative diagram of Banach spaces:

$$\begin{array}{ccc} W^s(E) & \xrightarrow{(P \circ L)} & W^s(E) \\ & {\scriptstyle L_s} \searrow \quad \nearrow {\scriptstyle P_{s-k}} & \\ & W^{s+1}(E), & \end{array}$$

Hence $\mathcal{H}_{L_s}$ is finite dimensional for all s. By a similar argument, we see that L_s has a finite dimensional cokernel. Once we show that $\mathcal{H}_{L_s}$ contains only C^∞ sections of E, then it will follow that $\mathcal{H}_{L_s} = \mathcal{H}_L$ and that all dimensions are the same and, of course, finite.

To show that $\mathcal{H}_{L_s} \subset \mathcal{E}(X, E)$ is known as the *regularity* of the homogeneous solutions of an elliptic differential equation. We formulate this as a theorem stated somewhat more generally, which will then complete the proof of Theorem 4.8.

Theorem 4.9: Suppose that $L \in \mathrm{Diff}_m(E, F)$ is elliptic, and $\xi \in W^s(E)$ has the property that $L_s\xi = \sigma \in \mathcal{E}(X, F)$. Then $\xi \in \mathcal{E}(X, E)$.

Proof: If P is a parametrix for L, then $P \circ L - I = S \in OP_{-1}(E)$. Now $L\xi \in \mathcal{E}(X, F)$ implies that $(P \circ L)\xi \in \mathcal{E}(X, E)$, and hence

$$\xi = (P \circ L - S)\xi.$$

Since we assumed that $\xi \in W^s(E)$ and since $(P \circ L)\xi \in \mathcal{E}(X, E)$ and $S\xi \in W^{s+1}(E)$, it follows that $\xi \in W^{s+1}(E)$. Repeating this process, we see that $\xi \in W^{s+k}(E)$ for all $k > 0$. But by Sobolev's lemma (Proposition 1.1) it follows that $\xi \in \mathcal{E}_l(X, E)$, for all $l > 0$, and hence $\xi \in \mathcal{E}(X, E)(= \mathcal{E}_\infty(X, E))$.

Q.E.D.

We note that S is called a smoothing operator precisely because of the role it plays in the proof of the above lemma. It smooths out the weak solution $\xi \in W^s(E)$.

Remark: The above theorem did not need the compactness of X which is being assumed throughout this section for convenience. Regularity of the solution of a differential equation is clearly a local property, and the above proof can be modified to prove the above theorem for noncompact manifolds.

We have finiteness and regularity theorems for elliptic operators. The one remaining basic result is the existence theorem. First we note the following elementary but important fact, which follows immediately from the definition.

Proposition 4.10: Let $L \in \mathrm{Diff}_m(E, F)$. Then L is elliptic if and only if L^* is elliptic.

We can now formulate the following.

Theorem 4.11: Let $L \in \mathrm{Diff}_m(E, F)$ be elliptic, and suppose that $\tau \in \mathcal{H}_{L^*}^{\perp} \cap \mathcal{E}(X, F)$. Then there exists a unique $\xi \in \mathcal{E}(X, E)$ such that $L\xi = \tau$ and such that ξ is orthogonal to $\mathcal{H}_L$ in $W^0(E)$.

Proof: First we shall solve the equation $L\xi = \tau$, where $\xi \in W^0(E)$, and then it will follow from the regularity (Theorem 4.9) of the solution ξ that ξ is C^∞ since τ is C^∞, and we shall have our desired solution. This reduces the problem to functional analysis. Consider the following diagram of Banach spaces,

$$\begin{array}{ccc} W^m(E) & \xrightarrow{L_m} & W^0(F) \\ \updownarrow & & \updownarrow \\ W^{-m}(E) & \xleftarrow{L_m^*} & W^0(F), \end{array}$$

where we note that $(L_m)^* = (L^*)_0$, by the uniqueness of the adjoint, and denote same by L_m^*. The vertical arrows indicate the duality relation between the Banach spaces indicated. A well-known and elementary functional analysis result asserts that the closure of the range is perpendicular to the kernel of the transpose. Thus $L_m(W^m(E))$ is dense in $\mathcal{H}_{L_m^*}^{\perp}$. Moreover, since L_m has finite dimensional cokernel, it follows that L_m has closed range, and hence the equation $L_m\xi = \tau$ has a solution $\xi \in W^m(E)$. By orthogonally projecting ξ along the closed subspace $\mathrm{Ker}\, L_m$ ($= \mathcal{H}_L$ by Theorem 4.8), we obtain a unique solution.

Q.E.D.

Let $L \in \mathrm{Diff}_m(E) = \mathrm{Diff}_m(E, E)$. Then we say that L is *self-adjoint* if $L = L^*$. Using the above results we deduce easily the following fundamental decomposition theorem for self-adjoint elliptic operators.

Theorem 4.12: Let $L \in \mathrm{Diff}_m(E)$ be self-adjoint and elliptic. Then there exist linear mappings H_L and G_L

$$H_L\colon \mathcal{E}(X, E) \longrightarrow \mathcal{E}(X, E)$$

$$G_L\colon \mathcal{E}(X, E) \longrightarrow \mathcal{E}(X, E)$$

so that

(a) $H_L(\mathcal{E}(X, E)) = \mathcal{H}_L(E)$ and $\dim_{\mathbb{C}}\mathcal{H}_L(E) < \infty$.
(b) $L \circ G_L + H_L = G_L \circ L + H_L = I_E$, where $I_E =$ identity on $\mathcal{E}(X, E)$.
(c) H_L and $G_L \in OP_0(E)$, and, in particular, extend to bounded operators on $W^0(E)(= L^2(X, E))$.
(d) $\mathcal{E}(X, E) = \mathcal{H}_L(X, E) \oplus G_L \circ L(\mathcal{E}(X, E)) = \mathcal{H}_L(X, E) \oplus L \circ G_L(\mathcal{E}(X, E))$, and this decomposition is orthogonal with respect to the inner product in $W^0(E)$.

Proof: Let H_L be the orthogonal projection [in $W^0(E)$] onto the closed subspace $\mathcal{H}_L(E)$, which we know by Theorem 4.8 is finite dimensional. As we saw in the proof of Theorem 4.11, there is a bijective continuous mapping

$$L_m\colon W^m(E) \cap \mathcal{H}_L^{\perp} \longrightarrow W^0(E) \cap \mathcal{H}_L^{\perp}.$$

By the Banach open mapping theorem, L_m has a continuous linear inverse which we denote by G_0:

$$G_0\colon W^0(E) \cap \mathcal{H}_L^{\perp} \longrightarrow W^m(E) \cap \mathcal{H}_L^{\perp}.$$

We extend G_0 to all of $W^0(E)$ by letting $G_0(\xi) = 0$ if $\xi \in \mathcal{H}_L$, and noting that $W^m(E) \subset W^0(E)$, we see that

$$G_0\colon W^0(E) \longrightarrow W^0(E).$$

Moreover,

$$L_m \circ G_0 = I_E - H_L,$$

since $L_m \circ G_0 =$ identity on $\mathcal{H}_L^{\perp}$. Similarly,

$$G_0 \circ L_m = I_E - H_L$$

for the same reason. Since $G_0(\mathcal{E}(X, E)) \subset \mathcal{E}(X, E)$, by elliptic regularity (Theorem 4.9), we see that we can restrict the linear Banach space mappings above to $\mathcal{E}(X, E)$. Let $G_L = G_0|_{\mathcal{E}(X,E)}$, and it becomes clear that all of the conditions (a)–(d) are satisfied.

Q.E.D.

The above theorem was first proved by Hodge for the case where $E = \wedge^p T^*(X)$ and where $L = dd^* + d^*d$ is the Laplacian operator, defined with respect to a Riemannian metric on X (see Hodge [1] and de Rham [1]). Hodge called the homogeneous solutions of the equation $L\varphi = 0$ *harmonic p-forms*, since the operator L is a true generalization of the Laplacian in the plane. Following this pattern, we shall call the sections in $\mathcal{H}_L$, for L a self-adjoint elliptic operator, *L-harmonic sections*, and when there is no chance of confusion, simply *harmonic sections*. For convenience we shall refer to the operator G_L given by Theorem 4.12 as the *Green's operator* associated to L, also classical terminology.† The harmonic forms of Hodge and their generalizations will be used in our study of Kähler manifolds and algebraic geometry. We shall refine the above theorem in the next section dealing with elliptic complexes and at the same time give some examples of its usefulness.

Suppose that $E \to X$ is a differentiable vector bundle and $L: \mathcal{E}(X, E) \to \mathcal{E}(X, E)$ is an elliptic operator. Then the *index* of L is defined by

$$i(L) = \dim \operatorname{Ker} L - \dim \operatorname{Ker} L^*,$$

which is a well-defined integer (Theorem 4.8). The Atiyah-Singer index theorem asserts that $i(L)$ is a topological invariant, depending only on (a) the Chern classes of E and (b) a cohomology class in $H^*(X, \mathbf{C})$ defined by the top-order symbol of the differential operator L. Moreover, there is an explicit formula for $i(L)$ in terms of these invariants (see Atiyah and Singer [1, 2]). We shall see a special case of this in Sec. 5 when we discuss the Hirzebruch-Riemann-Roch theorem for compact complex manifolds.

We would like to give another application of the existence of the parametrix to prove a semicontinuity theorem for a family of elliptic operators. Suppose that $E \longrightarrow X$ is a differentiable vector bundle over a compact manifold X, and let $\{L_t\}$ be a *continuous family of elliptic operators*,

$$L_t: \mathcal{E}(X, E) \longrightarrow \mathcal{E}(X, E), \tag{4.1}$$

where t is a parameter varying over an open set $U \subset R^n$. By this we mean that for a fixed $t \in U$, L_t is an elliptic operator and that the coefficients of L_t in a local representation for the operator should be jointly continuous in $x \in X$ and $t \in U$.

Theorem 4.13: Let $\{L_t\}$ be a continuous family of elliptic differential operators of order m as in (4.1). Then dim Ker L_t is an upper semicontinuous function of the parameter t; moreover if $t_0 \in U$, then for $\epsilon > 0$ sufficiently small,

$$\dim \operatorname{Ker} L_t \leq \dim \operatorname{Ker} L_{t_0}$$

for $|t - t_0| < \epsilon$.

Proof: Suppose that $t_0 = 0$, let $B_1 = W^0(X, E)$ and $B_2 = W^{-m}(X, E)$, and let P be a parametrix for the operator $L = L_0$. Denoting the extensions of the operators L_t and P by the same symbols, we have

$$L_t: B_1 \longrightarrow B_2, \qquad t \in U$$

$$P: B_2 \longrightarrow B_1.$$

We shall continue the proof later, but first in this context we have the following lemma concerning the single operator $L = L_0$, whose proof uses

†Note that the Green's operator G_L is a *parametrix* for L, but such that $G_L \circ L - I = -H_L$ is a smoothing operator of infinite order which is orthogonal to G_L, a much stronger parametrix than that obtained from Theorem 4.4.

the existence of the parametrix P at $t = 0$. Let $\mathcal{H}_t = \text{Ker } L_t, t \in U$, and $\| \|_1, \| \|_2$ denote the norms in B_1 and B_2.

Lemma 4.14: There exists a constant $C > 0$ such that

$$\|u\|_1 \leq C\|L_0 u\|_2$$

if $u \in \mathcal{H}_0^\perp \subset B_1$ (orthogonal complement in the Hilbert space B_1).

Proof: Suppose the contrary. Then there exists a sequence $u_j \in \mathcal{H}_0^\perp$ such that

$$\|u_j\|_1 = 1$$

(4.2)
$$\|Lu_j\|_2 \leq \frac{1}{j}.$$

Consider

$$PLu_j = u_j + Tu_j,$$

where T is compact, Then

$$\begin{aligned}\|Tu_j\|_1 &\leq \|PLu_j\|_1 + \|u_j\|_1 \\ &\leq C\|Lu_j\|_2 + \|u_j\|_1 \\ &\leq C\left(\frac{1}{j}\right) + 1 \\ &\leq \tilde{C},\end{aligned}$$

where $C, \tilde{C}$ are constants which depend on the operator P (recall that P is a continuous operator from B_2 to B_1). Since $\|u_j\| = 1$, it follows that $\{Tu_j\}$ is a sequence of points in a compact subset of B_1, and as such, there is a convergent subsequence $y_{jn} = Tu_{jn} \to y_0 \in B_1$. Moreover, $y_0 \neq 0$, since $\lim_{n\to\infty} Lu_{jn} = 0$, by (4.2), and thus

$$0 = \lim_{n\to\infty} PLu_{jn} = \lim_{n\to\infty} u_{jn} + y_0,$$

which implies that $u_{jn} \to -y_0$ and

$$\|y_0\| = \lim_{n\to\infty} \|u_{jn}\| = 1.$$

However, $Ly_0 = -\lim_{n\to\infty} Lu_{jn} = 0$, as above, and this contradicts the fact that y_0 (which is the limit of u_{jn}) $\in \mathcal{H}_0^\perp$.

Q.E.D.

Proof of Theorem 4.13 continued: Let C be the constant in Lemma 4.14. We claim that for δ sufficiently small there exists a somewhat larger constant $\tilde{C}$ such that, for $u \in \mathcal{H}_0^\perp$,

(4.4)
$$\|u\|_1 \leq \tilde{C}\|L_t u\|_2,$$

provided that $|t| < \delta$, where $\tilde{C}$ is independent of t. To see this, we write

$$L_0 = L_t + L_0 - L_t,$$

and therefore (using the operator norm)

$$\|L_0\| \leq \|L_t\| + \|L_0 - L_t\|.$$

For any $\epsilon > 0$, there is a $\delta > 0$ so that

$$\|L_t - L_0\| < \epsilon,$$

for $|t| < \delta$, since the coefficients of L_t are continuous functions of the parameter t. Using Lemma 4.13, we have, for $u \in \mathcal{H}_0^\perp$,

$$\begin{aligned}\|u\|_1 &\le C\|L_0 u\|_2 \\ &\le C(\|L_t u\|_2 + \epsilon\|u\|_1),\end{aligned}$$

which gives

$$(1 - C\epsilon)\|u\|_1 \le C\|L_t u\|_2$$

for $|t| < \delta$. By choosing $\epsilon < C^{-1}$, we see that

$$\begin{aligned}\|u\|_1 &\le C(1 - C\epsilon)^{-1}\|L_t u\|_2 \\ &\le \tilde{C}\|L_t u\|_2,\end{aligned}$$

which gives (4.4). But $u \in \mathcal{H}_0^\perp$ by assumption, and it follows from the inequality (4.4) that $\mathcal{H}_0^\perp \cap \mathcal{H}_t = 0$ for $|t| < \delta$. Consequently, we obtain $\dim \mathcal{H}_t \le \dim \mathcal{H}_0$.

Q.E.D.

5. Elliptic Complexes

We now want to study a generalization of elliptic operators to be called elliptic complexes. The basic fact of generalization is that instead of considering a pair of vector bundles we now want to study a finite sequence of vector bundles connected by differential operators. Thus, let $E_0, E_1, \ldots, E_N$ be a sequence of differentiable vector bundles defined over a compact differentiable manifold X. Suppose that there is a sequence of differential operators, of some fixed order k, $L_0, L_1, \ldots, L_{N-1}$ mapping as in the following sequence†:

$$\mathcal{E}(E_0) \xrightarrow{L_0} \mathcal{E}(E_1) \xrightarrow{L_1} \mathcal{E}(E_2) \longrightarrow \cdots \xrightarrow{L_{N-1}} \mathcal{E}(E_N). \tag{5.1}$$

Associated with the sequence (5.1) is the associated symbol sequence (using the notation of Sec. 2)

$$0 \longrightarrow \pi^* E_0 \xrightarrow{\sigma(L_0)} \pi^* E_1 \xrightarrow{\sigma(L_1)} \pi^* E_2 \longrightarrow \cdots \xrightarrow{\sigma(L_{N-1})} \pi^* E_N \longrightarrow 0. \tag{5.2}$$

Here we denote by $\sigma(L_j)$ the k-symbol of the operator L_j. In most of our examples we shall have first-order operators.

Definition 5.1: The sequence of operators and vector bundles E (5.1) is called a *complex* if $L_i \circ L_{i-1} = 0$, $i = 1, \ldots, N-1$. Such a complex is called an *elliptic complex* if the associated symbol sequence (5.2) is exact.

†For simplicity we denote in this section $\mathcal{E}(X, E_j)$ by $\mathcal{E}(E_j)$, not to be confused with the sheaf of sections of E_j.

Suppose that E is a complex as defined above. Then we let

$$(5.3) \qquad H^q(E) = \frac{\operatorname{Ker}(L_q\colon \mathcal{E}(E_q) \longrightarrow \mathcal{E}(E_{q+1}))}{\operatorname{Im}(L_{q-1}\colon \mathcal{E}(E_{q-1}) \longrightarrow \mathcal{E}(E_q))} = \frac{Z^q(E)}{B^q(E)}$$

be the *cohomology groups* (vector spaces) of the complex $E, q = 0, \ldots, N$ [where $Z^q(E)$ and $B^q(E)$ denote the numerator and denominator, respectively]. For this definition to make sense, we make the convention that $L_{-1} = L_N = E_{-1} = E_{N+1} = 0$ (i.e., we make a trivial extension to a complex larger at both ends).

A single elliptic operator $L\colon \mathcal{E}(E_0) \to \mathcal{E}(E_1)$ is a simple example of an elliptic complex. Further examples are given in Sec. 2, namely, the de Rham complex (Example 2.5), the Dolbeault complex (Example 2.6), and the Dolbeault complex with vector bundle coefficients (Example 2.7). Elliptic complexes were introduced by Atiyah and Bott [1] and we refer the reader to this paper for further examples.

Let E denote an elliptic complex of the form (5.1). Then we can equip each vector bundle E_j in E with a Hermitian metric and the corresponding Sobolev space structures as in Sec. 1. In particular $W^0(E_j)$ will denote the L^2 space with inner product

$$(\xi, \eta)_{E_j} = \int_X \langle \xi(x), \eta(x) \rangle_{E_j} \, d\mu,$$

for an appropriate strictly positive smooth measure μ. Associated with each operator $L_j\colon \mathcal{E}(E_j) \to \mathcal{E}(E_{j+1})$, we have the adjoint operator $L_j^*\colon \mathcal{E}(E_{j+1}) \to \mathcal{E}(E_j)$, and we define the *Laplacian operators* of the elliptic complex E by

$$\mathbf{\Delta}_j = L_j^* L_j + L_{j-1} L_{j-1}^*\colon \mathcal{E}(E_j) \to \mathcal{E}(E_j), \quad j = 0, 1, \ldots, N.$$

It follows easily from the fact that the complex E is elliptic that the operators $\mathbf{\Delta}_j$ are well-defined elliptic operators of order $2k$. Moreover, each $\mathbf{\Delta}_j$ is self-adjoint. Namely,

$$\begin{aligned} \sigma(\mathbf{\Delta}_j) &= \sigma(L_j^*)\sigma(L_j) + \sigma(L_{j-1})\sigma(L_{j-1}^*) \\ &= [\sigma(L_j)^*\sigma(L_j) + \sigma(L_{j-1})\sigma(L_{j-1})^*], \end{aligned}$$

which is an isomorphism and, in fact, either positive or negative definite. The fact that $\sigma(\mathbf{\Delta}_j)$ is an isomorphism follows easily from the following linear algebra argument. If we have a diagram of finite dimensional Hilbert spaces and linear mappings,

$$\begin{array}{ccccc} U & \xrightarrow{A} & V & \xrightarrow{B} & W \\ \updownarrow & & \updownarrow & & \updownarrow \\ U & \xleftarrow{A^*} & V & \xleftarrow{B^*} & W, \end{array}$$

which is exact at V, where the vertical maps are the duality pairings in U, V, and W, then we see that $V = \operatorname{Im}(A) \oplus \operatorname{Im}(B^*)$. Moreover, AA^* is injective on $\operatorname{Im}(A)$ and vanishes on $\operatorname{Im}(B^*)$, while B^*B is injective on $\operatorname{Im}(B^*)$ and vanishes on $\operatorname{Im}(A)$. Thus $AA^* + B^*B$ is an isomorphism on V and in fact is positive definite. The self-adjointness of $\mathbf{\Delta}_j$ follows easily from the fact that $(L_j^*)^* = L_j$ and that the adjoint operation is linear.

Since each $\mathbf{\Delta}_j$ is self-adjoint and elliptic, we can, by Theorem 4.12, associate to each Laplacian operator a Green's operator $G_{\mathbf{\Delta}j}$, which we shall denote by G_j. Moreover, we let

$$\mathcal{H}(E_j) = \mathcal{H}_{\mathbf{\Delta}j}(E_j) = \text{Ker } \mathbf{\Delta}_j\colon \mathcal{E}(E_j) \longrightarrow \mathcal{E}(E_j)$$

be the $\mathbf{\Delta}_j$-harmonic sections, and let

$$H_j\colon \mathcal{E}(E_j) \longrightarrow \mathcal{E}(E_j)$$

be the orthogonal projection onto the closed subspace $\mathcal{H}(E_j)$.

To simplify the notation somewhat, we proceed as follows. Denote by

$$\mathcal{E}(E) = \bigoplus_{j=0}^{N} \mathcal{E}(E_j)$$

the graded vector space so obtained with the natural grading. We define operators L, L^*, $\mathbf{\Delta}$, G, H on $\mathcal{E}(E)$, by letting

$$L(\xi) = L(\xi_0 + \cdots + \xi_N) = L_0\xi_0 + \cdots + L_N\xi_N,$$

where $\xi = \xi_0 + \cdots + \xi_N$ is the decomposition of $\xi \in \mathcal{E}(E)$ into homogeneous components corresponding to the above grading. The other operators are defined similarly. We then have the formal relations still holding,

$$\mathbf{\Delta} = LL^* + L^*L$$

$$I = H + G\mathbf{\Delta} = H + \mathbf{\Delta}G,$$

which follow from the identities in each of the graded components, coming from Theorem 4.12. We note that these operators, so defined, respect the grading, that L is of degree $+1$, that L^* is of degree -1, and that $\mathbf{\Delta}$, G, and H are all of degree 0 (i.e., they increase or decrease the grading by that amount). This formalism corresponds to that of the d or $\bar{\partial}$ operator in the de Rham and Dolbeault complexes, these operators also being graded operators on graded vector spaces. Our purpose is to drop the somewhat useless subscripts when operating on a particular subspace $\mathcal{E}(E_j)$. We also extend the inner product on $\mathcal{E}(E_j)$ to $\mathcal{E}(E)$ in the usual Euclidean manner, i.e.,

$$(\xi, \eta)_E = \sum_{j=0}^{N} (\xi_j, \eta_j)_{E_j},$$

a consequence of which is that elements of different homogeneity are orthogonal in $\mathcal{E}(E)$. Let us denote by $\mathcal{H}(E) = \oplus\mathcal{H}(E_j)$ the total space of $\mathbf{\Delta}$-harmonic sections.

Using this notation we shall denote a given elliptic complex by the pair $(\mathcal{E}(E), L)$, and we shall say that the elliptic complex has an inner product if it has an inner product in the manner described above, induced by L^2-inner products on each component. Examples would then be $(\mathcal{E}^*(X), d)$ for X a differentiable manifold and $(\mathcal{E}^{p,*}(X), \bar{\partial})$ for p fixed and X a complex manifold (see Examples 2.5 and 2.6).

We now have the following fundamental theorem concerning elliptic complexes (due to Hodge for the case of the de Rham complex).

Theorem 5.2: Let $(\mathcal{E}(E), L)$ be an elliptic complex equipped with an inner product. Then

(a) There is an orthogonal decomposition
$$\mathcal{E}(E) = \mathcal{H}(E) \oplus LL^*G\mathcal{E}(E) \oplus L^*LG\mathcal{E}(E),$$

(b) The following commutation relations are valid:
 (1) $I = H + \mathbf{\Delta} G = H + G\mathbf{\Delta}$.
 (2) $HG = GH = H\mathbf{\Delta} = \mathbf{\Delta} H = 0$.
 (3) $L\mathbf{\Delta} = \mathbf{\Delta} L, L^*\mathbf{\Delta} = \mathbf{\Delta} L^*$.
 (4) $LG = GL, L^*G = GL^*$.

(c) $\dim_C \mathcal{H}(E) < \infty$, and there is a canonical isomorphism
$$\mathcal{H}(E_j) \cong H^j(E).$$

Proof: From Theorem 4.12 we obtain immediately the orthogonal decomposition
$$\mathcal{E}(E) = \mathcal{H}(E) \oplus (LL^* + L^*L)G\mathcal{E}(E).$$
If we show that the two subspaces of $\mathcal{E}(E)$,
$$LL^*G\mathcal{E}(E) \quad \text{and} \quad L^*LG\mathcal{E}(E),$$
are orthogonal, then we shall have part (a). But this is quite simple. Suppose that $\xi, \eta \in \mathcal{E}(E)$. Then consider the inner product (dropping the subscript E on the inner product symbol)
$$(LL^*G\xi, L^*LG\eta) = (L^2L^*G\xi, LG\eta),$$
and the latter inner product vanishes since $L^2 = 0$.

Part (b), (1) and (2), follow from the corresponding statements in Theorem 4.12 and its proof. Part (b), (3) follows immediately from the definition of L and $\mathbf{\Delta}$. In part (b), (4), we shall show that $LG = GL$, leaving the other commutation relation to the reader. First we have a simple proposition of independent interest, whose proof we shall give later.

Proposition 5.3: Let $\xi \in \mathcal{E}(E)$. Then $\mathbf{\Delta}\xi = 0$ if and only if $L\xi = L^*\xi = 0$; moreover, $LH = HL = L^*H = HL^* = 0$.

Using this proposition and the construction of G, we observe that both L and G vanish on $\mathcal{H}(E)$. Therefore it suffices to show that $LG = GL$ on $\mathcal{H}(E)^\perp$, and it follows immediately from the decomposition in Theorem 4.12 that any smooth $\xi \in \mathcal{H}(E)^\perp$ is of the form $\xi = \mathbf{\Delta}\varphi$ for some φ in $\mathcal{E}(E)$. Therefore we must show that $LG\mathbf{\Delta}\varphi = GL\mathbf{\Delta}\varphi$ for all $\varphi \in \mathcal{E}(E)$. To do this, we write, using $I = H + G\mathbf{\Delta}$,
$$\begin{aligned} L\varphi &= H(L\varphi) + G\mathbf{\Delta} L\varphi \\ &= HL\varphi + GL\mathbf{\Delta}\varphi, \end{aligned}$$

since $L\mathbf{\Delta} = \mathbf{\Delta}L$. We also have

$$\varphi = H\varphi + G\mathbf{\Delta}\varphi,$$

and applying L to this, we obtain

$$L\varphi = LH\varphi + LG\mathbf{\Delta}\varphi.$$

Setting the two expressions above for $L\varphi$ equal to each other, we obtain

$$GL\mathbf{\Delta}\varphi - LG\mathbf{\Delta}\varphi = LH\varphi - HL\varphi,$$

and by Proposition 5.3 we see that the right-hand side is zero.

In part (c), it is clear that the finiteness assertion is again a part of Theorem 4.12. To prove the desired isomorphism, we recall that $H^q(E) = Z^q(E)/B^q(E)$, as defined in (5.3), and let

$$\mathbf{\Phi}\colon Z^q(E) \longrightarrow \mathcal{H}(E_q)$$

be defined by $\mathbf{\Phi}(\xi) = H(\xi)$. It then follows from Proposition 5.3 that $\mathbf{\Phi}$ is a surjective linear mapping. We must then show that Ker $\mathbf{\Phi} = B^q(E)$. Suppose that $\xi \in Z^q(E)$ and $H(\xi) = 0$. Then we obtain, by the decomposition in part (a),

$$\xi = H\xi + LL^*G\xi + L^*LG\xi.$$

Since $H\xi = 0$ and since $LG = GL$, we obtain $\xi = LL^*G\xi$, and hence $\xi \in B^q(E)$.

Q.E.D.

Proof of Proposition 5.3: It is trivial that if $L\xi = L^*\xi = 0$ for $\xi \in \mathcal{E}(E)$, then $\mathbf{\Delta}\xi = 0$. Therefore we consider the converse, and suppose that $\mathbf{\Delta}\xi = 0$ for some $\xi \in \mathcal{E}(E)$. We then have

$$\begin{aligned}(\mathbf{\Delta}\xi, \xi) &= (LL^*\xi + L^*L\xi, \xi)\\ &= (LL^*\xi, \xi) + (L^*L\xi, \xi)\\ &= (L^*\xi, L^*\xi) + (L\xi, L\xi)\\ &= \|L^*\xi\|^2 + \|L\xi\|^2 = 0.\end{aligned}$$

It now follows that $L^*\xi = L\xi = 0$, and, consequently, $LH = L^*H = 0$. To show that $HL = 0$, it suffices to show that $(HL\xi, \eta) = 0$ for all $\xi, \eta \in \mathcal{E}(E)$. But H is an orthogonal projection in Hilbert space, and as such it is self-adjoint. Therefore we have, for any $\xi, \eta \in \mathcal{E}(E)$,

$$(HL\xi, \eta) = (L\xi, H\eta) = (\xi, L^*H\eta) = 0,$$

and hence $HL = 0$. That $HL^* = 0$ is proved in a similar manner.

Q.E.D.

Remark: We could easily have defined an elliptic complex to have differential operators of various orders, and Theorem 5.2 would still be valid, in a slightly modified form (see Atiyah and Bott [1]). We avoid this complication, as we do not need the more general result later on in our applications.

We now want to indicate some applications of the above theorem.

Example 5.4: Let $(\mathcal{E}^*(X), d)$ be the de Rham complex on a compact differentiable manifold X. As we saw in our proof of de Rham's theorem (Theorem II.3.15)

$$H^r(X, \mathbf{C}) \cong {}_\Delta H^r(X, \mathbf{C}) \cong H^r(\mathcal{E}^*(X))$$

(using complex coefficients). The first group is abstract sheaf cohomology, which is defined for any topological space; the second is singular cohomology; and the first isomorphism holds when we assume that X has the structure of a topological manifold (for example). When X has a differentiable structure, as we are assuming, then differential forms are defined and the de Rham group on the right makes sense. Thus we can use differential forms to represent singular cohomology. For convenience, we shall let $H^r(X, \mathbf{C})$ denote the de Rham group when we are working on a differentiable manifold, which will almost always be the case, making the isomorphisms above an identification. One further step in this direction of more specialized information about the homological topology of a manifold comes about when we assume that X is compact and that there is a Riemannian metric on X. This induces an inner product on $\wedge^p T^*(X)$ for each p, and hence $(\mathcal{E}^*(X), d)$ becomes an elliptic complex with an inner product. We denote the associated Laplacian by $\mathbf{\Delta} = \mathbf{\Delta}_d = dd^* + d^*d$. Let

$$\mathcal{H}^r(X) = \mathcal{H}_{\mathbf{\Delta}}(\wedge^r T^*(X))$$

be the vector space of $\mathbf{\Delta}$-*harmonic r-forms* on X. We shall call them simply *harmonic forms*, a metric and hence a Laplacian being understood. We thus obtain by Theorem 5.2.(c) that

$$H^r(X, \mathbf{C}) \cong \mathcal{H}^r(X).$$

This means that for each cohomology class $c \in H^r(X, \mathbf{C})$ there exists a unique harmonic form φ representing this class c, which is, by Proposition 5.3, d-closed. If we change the metric, we change the representation, but, nevertheless, for a given metric we have a distinguished r-form to represent a given class. It will turn out that this representative has more specialized information about the original manifold than an arbitrary representative might, in particular when the metric is chosen carefully (to be Kähler, for example, as we shall see in the next chapter). Thus we have continued the chain of representations of the sheaf cohomology on X with coefficients in $\mathbf{C}$, but for the first time we have a specific vector space representation; there are no equivalence classes to deal with, as in the previous representations. A consequence of Theorem 5.2 is that

$$\dim_{\mathbf{C}} H^q(X, \mathbf{C}) = \dim_{\mathbf{C}} \mathcal{H}^q(X) = b_q < \infty.$$

This finiteness is not obvious from the other representations, and, in fact, the harmonic theory we are developing here is one of the basic ways of obtaining finiteness theorems in general.† The numbers $b_q, q = 0, 1, \ldots, \dim_R(X)$,

†Of course, we could represent the de Rham groups by singular cohomology and prove that a compact topological manifold has a finite cell decomposition. This is the point of view of algebraic topology.

are the celebrated *Betti numbers* of the compact manifold X. By our results above the Betti numbers are *topological invariants* of X; i.e., (a) they depend only on the topological structure of X, and (b) they are invariant under homeomorphisms.

In the study of manifolds these numbers play an important role in their classification, and this is no less so if the manifold happens to be complex, as we shall see. We define

$$\chi(X) = \sum_{q=0}^{\dim X} (-1)^q b_q,$$

the *Euler characteristic* of X, also a topological invariant.

Example 5.5: Let X be a compact complex manifold of complex dimension n, and consider the elliptic complex

$$\cdots \xrightarrow{\bar\partial} \mathcal{E}^{p,q}(X) \xrightarrow{\bar\partial} \mathcal{E}^{p,q+1}(X) \xrightarrow{\bar\partial} \mathcal{E}^{p,q+2}(X) \xrightarrow{\bar\partial} \cdots,$$

for a fixed p, $0 \le p \le n$. As we saw in our previous study of this example (Example 2.6), this is elliptic, and in Chap. II (Theorem 3.17) we saw that

$$H^q(X, \mathbf{\Omega}^p) \cong H^q(\wedge^p, {}^*T^*(X), \bar\partial)$$

(Dolbeault's theorem), where $\mathbf{\Omega}^p$ is the sheaf of germs of holomorphic p-forms. We want to represent these cohomology groups by means of harmonic forms. Let $\wedge^{p,q}T^*(X)$ be equipped with a Hermitian metric, $0 \le p, q, \le n$ [induced by a Hermitian metric on $T(X)$, for example]. Then the complex above becomes an elliptic complex with an inner product (parametrized by the integer p). Denote the Laplacian by

$$\overline{\square} = \bar\partial\bar\partial^* + \bar\partial^*\bar\partial,$$

and let

$$\mathcal{H}^{p,q}(X) = \mathcal{H}_{\overline{\square}}(\wedge^{p,q}T^*(X))$$

be the $\overline{\square}$-harmonic (p, q)-forms, which we shall call simply *harmonic* (p, q)-forms when there is no confusion about which Laplacian is meant in a given context.

Similar to the de Rham situation, we have the following canonical isomorphism (using Theorem 5.2 along with Dolbeault's theorem):

$$H^q(X, \mathbf{\Omega}^p) \cong \mathcal{H}^{p,q}(X).$$

We define, for $0 \le p, q \le n$,

$$h^{p,q} = \dim_{\mathbb{C}} H^q(X, \mathbf{\Omega}^p) = \dim_{\mathbb{C}} \mathcal{H}^{p,q}(X),$$

which are called the *Hodge numbers* of the compact complex manifold X. Note that these numbers are invariants of the complex structure of X and do not depend on the choice of metric. The finite dimensionality again comes from Theorem 5.2.† The following theorem shows us how the Hodge numbers and the Betti numbers are related, in general (on Kähler manifolds, more will be true).

Theorem 5.6: Let X be a compact complex manifold. Then

$$\chi(X) = \sum (-1)^r b_r(X) = \sum (-1)^{p+q} h^{p,q}(X).$$

The proof of this theorem is a simple consequence of the fact that there is a spectral sequence (Fröhlicher [1])

$$E_1^{p,q} \cong H^q(X, \mathbf{\Omega}^p) \Longrightarrow H^r(X, \mathbf{C})$$

relating the Dolbeault and de Rham groups, and we omit the details as we do not need this result in later chapters. For Kähler manifolds this results from the Hodge theory developed in Chapter V.

Example 5.7: Let E be a holomorphic vector bundle over a compact complex manifold X and let $(\mathcal{E}^{p,*}(X, E), \bar{\partial})$ be the elliptic complex of (p, q)-forms with coefficients in E. By the generalization of Dolbeault's theorem given in Theorem II.3.20 the cohomology groups $H^q(X, \mathbf{\Omega}^p(E))$ represent the cohomology of the above complex, where $\mathbf{\Omega}^p(E) \cong \mathcal{O}(\wedge^p T^*(X) \otimes E)$ is the sheaf of germs of E-valued holomorphic p-forms. The bundles in the complex are of the form $\wedge^{p,q} T^*(X) \otimes E$, and equipping them with a Hermitian metric [induced from a Hermitian metric on $T(X)$ and E, for instance], we can then define a Laplacian

$$\overline{\square} = \bar{\partial}\bar{\partial}^* + \partial^*\bar{\partial} \colon \mathcal{E}^{p,q}(X, E) \longrightarrow \mathcal{E}^{p,q}(X, E),$$

as before. Letting $\mathcal{H}^{p,q}(X, E) = \mathcal{H}_{\overline{\square}}(\wedge^{p,q} T^*(X) \otimes E)$ be the $\overline{\square}$-harmonic E-valued (p, q)-forms in $\mathcal{E}^{p,q}(X, E)$ we have, by Theorem 5.2, the isomorphism (and harmonic representation)

$$H^q(X, \mathbf{\Omega}^p(E)) \cong \mathcal{H}^{p,q}(X, E),$$

a generalization of the previous example to vector bundle coefficients. We let

$$h^{p,q}(E) = h^{p,q}(X, E) = \dim_{\mathbf{C}} \mathcal{H}^{p,q}(X, E),$$

where we drop the notational dependence on X unless there are different manifolds involved. As before, it follows from Theorem 5.2 that $h^{p,q}(E) < \infty$, and we can define the *Euler characteristic of the holomorphic vector bundle* E to be

$$\chi(E) = \chi(X, E) = \sum_{q=0}^{n} (-1)^q h^{0,q}(E).$$

As before, the generalized Hodge numbers $h^{p,q}(E)$ depend only on the complex structures of X and E, since the dimensions are independent of the particular metric used. However, it is a remarkable fact that the Euler characteristic of

†A general theorem of Cartan and Serre asserts that the cohomology groups of any coherent analytic sheaf on a compact complex manifold are finite dimensional. This and the next example are special cases of this more general result, which is proved by different methods, involving Čech cohomology (cf. Gunning and Rossi [1]).

a holomophic vector bundle can be expressed in terms of topological invariants of the vector bundle E (its Chern classes) and of the complex manifold X itself (the Chern classes of the tangent bundle to X). This is the celebrated Riemann-Roch theorem of Hirzebruch, which we formulate below.

Let E be a complex (differentiable) vector bundle over X, where $r =$ rank E and X is a differentiable manifold of real dimension m. Let

$$c(E) = 1 + c_1(E) + \cdots + c_r(E)$$

be the total Chern class of E, which is an element of the cohomology ring $H^*(X, \mathbf{C})$, as we saw in Chap. III. Recall that the multiplication in this ring is induced by the exterior product of differential forms, using the de Rham groups as a representation of cohomology.† We introduce a formal factorization

$$c(E) = \prod_{i=1}^{r}(1 + x_i),$$

where the $x_i \in H^*(X, \mathbf{C})$. Then any formal power series in $x_1, \ldots, x_r$ which is symmetric in $x_1, \ldots, x_r$ is also a power series in $c_1(E), \ldots, c_r(E)$. This follows from the fact that the $c_j(E)$ are the elementary symmetric functions of the $(x_1, \ldots, x_r)$ (analogous to the case of the coefficients of a polynomial). Therefore we define

$$\mathcal{T}(E) = \prod_{i=1}^{r} \frac{x_i}{1 - e^{-x_i}}$$

$$ch(E) = \sum_{i=1}^{r} e^{x_i},$$

which are formal power series, symmetric in $x_1, \ldots, x_r$, and hence define a (more complicated-looking) formal power series in the Chern classes of E. We call $\mathcal{T}(E)$ the *Todd class* of E and $ch(E)$ is called the *Chern character* of E. Of course, there are only a finite number of terms in the expansion of the above formal power series since $H^q(X, \mathbf{C}) = 0$ for $q > \dim_{\mathbf{R}} X$.

We now recall that X is assumed to be compact, and then we let, for $c \in H^*(X, \mathbf{C})$,

$$c[X] = \int_X \varphi_m,$$

where φ_m is a closed differential form of degree m representing the homogeneous component in c of degree m; i.e., from the viewpoint of algebraic topology we evaluate the cohomology class on the fundamental cycle. By Stokes' theorem the above definition is a sensible one. We are now in a position to state the following theorem due to Hirzebruch for projective algebraic manifolds.

†Of course, the characteristic class theory is valid in a more general topological category, and the cohomology ring has the cup product of algebraic topology for multiplication, but on a differentiable manifold, the two theories are isomorphic.

Theorem 5.8 (Riemann-Roch-Hirzebruch): Let X be a compact complex manifold, and let E be a holomorphic vector bundle over X. Then

$$\chi(E) = \{ch(E) \cdot \mathcal{T}(T(X))\}[X].$$

Note that the left-hand side of the equality depends a priori on the complex structure of X and E, whereas the right-hand side is a priori a complex number (we could have made it a rational number had we worked with integral coefficients for our cohomology). Therefore two immediate consequences of the above formula is that these dependences are superfluous; i.e., the left-hand side depends only on the underlying topological structure, and the right-hand side is an integer.

This theorem is a special case of the Atiyah-Singer index theorem, discussed in Sec. 4, and was formulated and proved for projective algebraic manifolds by Hirzebruch in a famous monograph (Hirzebruch [1]) in 1956. The special case of a Kähler surface had been proved earlier by Kodaira. For $n = 1$ and E a line bundle, the above theorem is essentially the classic theorem of Riemann-Roch for Riemann surfaces (in the form proved by Serre [1]). This case is discussed thoroughly by Gunning [1]. For applications of the Riemann-Roch Theorem in this form to the study of compact complex surfaces (complex dimension 2), see Kodaira [5].

CHAPTER V

COMPACT COMPLEX MANIFOLDS

In this chapter we shall apply the differential equations and differential geometry of the previous two chapters to the study of compact complex manifolds. In Sec. 1 we shall present a discussion of the exterior algebra on a Hermitian vector space, introducing the fundamental 2-form and the Hodge $*$-operator associated with the Hermitian metric. In Sec. 2 we shall discuss and prove the principal results concerning harmonic forms on compact manifolds (real or complex), in particular, Hodge's harmonic representation for the de Rham groups, and special cases of Poincaré and Serre duality. In Sec. 3 we present the finite-dimensional representation theory for the Lie algebra $\mathfrak{sl}(2, \mathbf{C})$, from which we derive the Lefschetz decomposition theorem for a Hermitian exterior algebra. In Sec. 4 we shall introduce the concept of a Kähler metric and give various examples of Kähler manifolds (manifolds equipped with a Kähler metric). In terms of a Hermitian metric we define the Laplacian operators associated with the operators d, ∂, and $\bar{\partial}$ and show that when the metric is Kähler that the Laplacians are related in a simple way. We shall use this relationship in Sec. 5 to prove the Hodge decomposition theorem expressing the de Rham group as a direct sum of the Dolbeault groups (of the same total degree). In Sec. 6 we shall state and prove Hodge's generalization of the Riemann period relations for integrals of harmonic forms on a Kähler manifold. We shall then use the period relations and the Hodge decomposition to formulate the period mapping of Griffiths. In particular, we shall prove the Kodaira-Spencer upper semicontinuity theorem for the Hodge numbers on complex-analytic families of compact manifolds.

1. Hermitian Exterior Algebra on a Hermitian Vector Space

Let V be a real finite-dimensional vector space of dimension d which is equipped with an inner product $\langle\, ,\rangle$, a *Euclidean vector space*, and suppose that $\wedge V$ denotes the exterior algebra of V. Then for each degree p, the vector space $\wedge^p V$ has an inner product induced from the inner product of V. Namely, if $\{e_1, \ldots, e_d\}$ is an orthonormal basis for V, then $\{e_{i_1} \wedge \cdots \wedge e_{i_p}: 1 \leq i_1 < i_2 < \cdots < i_p \leq d\}$ is an orthonormal basis for $\wedge^p V$. An *orientation* on

V is a choice of ordering of a basis such as $\{e_1, \ldots, e_d\}$ up to an even permutation, which is equivalent to a choice of sign for a particular d-form, e.g., $e_1 \wedge \cdots \wedge e_d$.

We now define the *Hodge $*$-operator*. Choosing an orthonormal basis $\{e_1, \ldots, e_d\}$ for V as above, fix an orientation of V by specifying the d form $e_1 \wedge \cdots \wedge e_d$ which we will denote by vol (for volume element). The Hodge $*$-operator is a mapping

$$*: \wedge^p V \longrightarrow \wedge^{d-p} V$$

defined by setting

$$*(e_{i_1} \wedge \cdots \wedge e_{i_p}) = \pm e_{j_1} \wedge \cdots \wedge e_{j_{d-p}},$$

where $\{j_1, \ldots, j_{d-p}\}$ is the complement of $\{i_1, \ldots, i_p\}$ in $\{1, \ldots, d\}$, and we assign the plus sign if $\{i_1, \ldots, i_p, j_1, \ldots, j_{d-p}\}$ is an even permutation of $\{1, \ldots, d\}$, and the minus sign otherwise. In other words $*$ is defined so that

$$e_{i_1} \wedge \cdots \wedge e_{i_p} \wedge *(e_{i_1} \wedge \cdots \wedge e_{i_p}) = e_1 \wedge \cdots \wedge e_d = \text{vol}. \tag{1.1}$$

Extending $*$ by linearity to all of $\wedge^p V$ we find that if $\alpha, \beta \in \wedge^p V$, then

$$\alpha \wedge *\beta = \langle \alpha, \beta \rangle \text{ vol}, \tag{1.2}$$

where $\langle \alpha, \beta \rangle$ is the inner product induced on $\wedge^p V$ from V. Let us check that (1.2) is valid. Namely, if

$$\alpha = \sum_{|I|=p}{}' a_J e_J,$$

and

$$\beta = \sum_{|J|=p}{}' b_J e_J,$$

using multi-index notation, then

$$\alpha \wedge *\beta = \sum_{\substack{|I|=p \\ |J|=p}}{}' a_I b_J e_I \wedge *e_J.$$

We see that the wedge product in each term of the sum vanishes unless $I = \{i_1, \ldots, i_p\}$ coincides with $J = \{j_1, \ldots, j_p\}$, and then it follows immediately from (1.1) that

$$\begin{aligned} \alpha \wedge *\beta &= \sum_{|I|=p} a_I b_I \text{ vol} \\ &= \langle \alpha, \beta \rangle \text{ vol}. \end{aligned}$$

It is easily checked that the definition of the Hodge $*$-operator is independent of the choice of the orthonormal basis, and depends only on the inner product structure of V as well as a choice of orientation.†

†The classical references for the $*$-operator are Hodge [1], de Rham [1], and Weil [1].

We can extend (1.2) easily to complex-valued p-forms. Namely, if $\alpha, \beta \in \wedge^p V \otimes \mathbf{C}$, then $\bar{\beta}$ is well defined (cf. Sec. I.3). We write

$$\alpha = \sum_{|I|=p}{}' \alpha_I e_I, \quad \alpha_I \in \mathbf{C},$$

$$\beta = \sum_{|I|=p}{}' \beta_I e_I, \quad \beta_I \in \mathbf{C},$$

then we define an Hermitian inner product on $\wedge^p V \otimes \mathbf{C}$ by

$$\langle \alpha, \beta \rangle := \sum_{|I|=p}{}' \alpha_I \bar{\beta}_I.$$

If α, β are real, then we have the original inner product, so we use the same symbol $\langle\, ,\rangle$ for this complex extension. It follows then immediately that if $*$ is extended to $\wedge^* V \otimes \mathbf{C}$ by complex linearity, we obtain the relation

$$\alpha \wedge *\bar{\beta} = \langle \alpha, \beta \rangle \text{ vol.} \tag{1.3}$$

Let Π_r denote the projection onto homogeneous vectors of degree r,

$$\Pi_r \colon \wedge V \longrightarrow \wedge^r V,$$

and define the linear mapping $w \colon \wedge V \to \wedge V$ by setting

$$w = \Sigma(-1)^{dr+r} \Pi_r.$$

It is easy to see that $** = w$, and we remark that if d is even, then we have

$$w = \Sigma(-1)^r \Pi_r. \tag{1.4}$$

Let E be a complex vector space of complex dimension n. Let E' be the real dual space to the underlying real vector space of E, and let

$$F = E' \otimes_{\mathbf{R}} \mathbf{C}$$

be the complex vector space of complex-valued real-linear mappings of E to $\mathbf{C}$. Then F has complex dimension $2n$, and we let

$$\wedge F = \sum_{p=0}^{2n} \wedge^p F$$

be the $\mathbf{C}$-linear exterior algebra of F. We will refer to an $\omega \in \wedge^p F$ as a p-form or as a p-covector (on E). Now, as before, $\wedge F$ is equipped with a natural conjugation obtained by setting, if $\omega \in \wedge^p F$,

$$\overline{\omega}(v_1, \ldots, v_p) = \overline{\omega(v_1, \ldots, v_p)}, \quad v_j \in E.$$

We say that $\omega \in \wedge^p F$ is *real* if $\overline{\omega} = \omega$, and we will let $\wedge^p_{\mathbf{R}} F$ denote the real elements of $\wedge^p F$ (noting that $\wedge^p E' \cong \wedge^p_{\mathbf{R}} F$).

Let $\wedge^{1,0} F$ be the subspace of $\wedge^1 F$ consisting of complex-linear 1-forms on E, and let $\wedge^{0,1} F$ be the subspace of conjugate-linear 1-forms on E. Then we see that $\overline{\wedge^{1,0} F} = \wedge^{0,1} F$ and moreover

$$\wedge^1 F = \wedge^{1,0} F \oplus \wedge^{0,1} F,$$

and this induces (as in Sec. I.3) a bigrading on $\wedge F$,

$$\wedge F = \sum_{r=0}^{2n} \sum_{p+q=r} \wedge^{p,q} F,$$

and we see that if $\omega \in \wedge^{p,q} F$, then $\overline{\omega} \in \wedge^{q,p} F$.

Now we suppose than our complex vector space is equipped with a Hermitian inner product $\langle\, ,\rangle$. This inner product is a Hermitian symmetric sesquilinear† positive definite form, and can be represented in the following manner. If $\{z_1, \ldots, z_n\}$ is a basis for $\wedge^{1,0} F$, then $\{\bar{z}_1, \ldots, \bar{z}_n\}$ is a basis for $\wedge^{0,1} F$, and we can write, for $u, v \in E$,

$$\langle u, v\rangle = h(u, v),$$

where

$$h = \sum_{\mu,\nu} h_{\mu\nu} z_\mu \otimes \bar{z}_\nu,$$

and $(h_{\alpha\beta})$ is a positive definite Hermitian symmetric matrix. Now h is a complex-valued sesquilinear form acting on $E \times E$, and we can write

$$h = S + iA,$$

where S and A are real bilinear forms acting on E. One finds that S is a symmetric positive definite bilinear form, which represents the Euclidean inner product induced on the underlying real vector space of E by the Hermitian metric on E. Moreover one can calculate easily that

$$\begin{aligned} A &= \frac{1}{2i} \sum_{\mu,\nu} h_{\mu\nu} (z_\mu \otimes \bar{z}_\nu - \bar{z}_\nu \otimes z_\mu) \\ &= -i \sum_{\mu,\nu} h_{\mu\nu} z_\mu \wedge \bar{z}_\nu. \end{aligned}$$

Let us define

$$\mathbf{\Omega} = \frac{i}{2} \sum_{\mu,\nu} h_{\mu\nu} z_\mu \wedge \bar{z}_\nu, \tag{1.5}$$

the *fundamental 2-form* associated to the hermitian metric h. One sees immediately that

$$\mathbf{\Omega} = -\tfrac{1}{2} A = -\tfrac{1}{2} \operatorname{Im} h,$$

and thus

$$h = S - 2i\mathbf{\Omega}. \tag{1.6}$$

Moreover $\mathbf{\Omega} \in \wedge_{\mathbf{R}}^{1,1} F$, i.e., $\mathbf{\Omega}$ is a real 2-form of type (1, 1). We can always choose a basis $\{z_\mu\}$ of $\wedge^{1,0} F$ so that h has the form

$$h = \sum_{\mu} z_\mu \otimes \bar{z}_\mu. \tag{1.7}$$

†We recall that a mapping $f\colon E \times E \to \mathbf{C}$ is *sesquilinear* if f is real bilinear, and moreover, $f(\lambda u, v) = \lambda f(u, v)$, and $f(u, \lambda v) = \bar{\lambda} f(u, v)$, $\lambda \in \mathbf{C}$.

It then follows that, if we let

$$x_\mu = \frac{z_\mu + \bar{z}_\mu}{2}, \quad y_\mu = \frac{z_\mu - \bar{z}_\mu}{2i}$$

be the real and imaginary parts of $\{z_\mu\}$, then

$$h = \sum_\mu (x_\mu \otimes x_\mu + y_\mu \otimes y_\mu) - 2i \sum_\mu (x_\mu \wedge y_\mu), \tag{1.8}$$

and thus from (1.5), with respect to this basis,

$$\begin{aligned} S &= \sum x_\mu \otimes x_\mu + y_\mu \otimes y_\mu \\ \mathbf{\Omega} &= \sum x_\mu \wedge y_\mu = \frac{i}{2} \sum z_\mu \wedge \bar{z}_\mu. \end{aligned} \tag{1.9}$$

It follows from this that

$$\mathbf{\Omega}^n = n! x_1 \wedge y_1 \wedge \cdots \wedge x_n \wedge y_n. \tag{1.10}$$

Thus the fundamental 2-form associated to a Hermitian metric is a real form of type (1, 1) whose coefficient matrix is positive definite, and moreover, $\mathbf{\Omega}^n$ is a nonzero volume element of E'. Thus $\mathbf{\Omega}^n$ determines an orientation on E', and we see from (1.9) that $\{x_\mu, y_\mu\}$ is an orthonormal basis for E' in the induced Euclidean metric of E'. Thus we see that there is a naturally defined Hodge $*$-operator

$$*\colon \wedge^p E' \longrightarrow \wedge^{2n-p} E' \tag{1.11}$$

coming from the Hermitian structure of E. Namely, E' has the dual metric to the real underlying vector space of E, while E' is equipped with the orientation induced by the $2n$-form $\mathbf{\Omega}^n$ coming from the Hermitian structure of E. We define

$$\text{vol} = \frac{1}{n!} \mathbf{\Omega}^n, \tag{1.12}$$

which, with respect to the orthonormal basis used above, becomes

$$\text{vol} = x_1 \wedge y_1 \wedge \cdots \wedge x_n \wedge y_n.$$

Note that the definition (1.12) does not depend on the choice of the basis, and is an intrinsic definition of a volume element on E'.

We are now interested in defining various linear operators mapping $\wedge F \to \wedge F$ in terms of the above structure. Recall that we already defined w for an even dimensional vector space by (1.4), and this therefore defines

$$w\colon \wedge E' \longrightarrow \wedge E'$$

which we extend by complex-linearity to

$$w\colon \wedge F \longrightarrow \wedge F$$

where

$$\mathbf{\Pi}_r\colon \wedge F \longrightarrow \wedge^r F$$

is the natural projection. Similarly, since E has a Hermitian structure, as we saw above, there is a natural $*$-operator

$$*\colon \wedge^p E' \longrightarrow \wedge^{2n-p} E'$$

which we also extend as a complex-linear isomorphism to

$$*\colon \wedge^p F \longrightarrow \wedge^{2n-p} F.$$

Both w and $*$ are real operators. Now we let

$$\Pi_{p,q}\colon \wedge F \longrightarrow \wedge^{p,q} F$$

be the natural projection, and we define

$$J\colon \wedge F \longrightarrow \wedge F$$

by

$$J = \sum i^{p-q} \Pi_{p,q}.$$

Recall that the real operator J which represents the complex structure of the vector space F has the property that if $v \in \wedge^{1,0} F$, then $Jv = iv$, and if $v \in \wedge^{0,1} F$, then $Jv = -iv$. Thus we see immediately that J defined above is the natural multilinear extension of the complex structure operator J to the exterior algebra of F. We note also that $J^2 = w$ as linear operators.

We now define a linear mapping L in terms of $\mathbf{\Omega}$, the fundamental form associated to the Hermitian structure of E, namely, let

$$L\colon \wedge F \longrightarrow \wedge F$$

be defined by $L(v) = \mathbf{\Omega} \wedge v$. We see that

$$L\colon \wedge^p F \longrightarrow \wedge^{p+2} F$$

so it is homogeneous and of degree 2. Moreover,

$$L\colon \wedge^{p,q} F \longrightarrow \wedge^{p+1,q+1} F$$

and L is bihomogeneous of bidegree (1, 1), and it is apparent that L is a real operator since $\mathbf{\Omega}$ is a real 2-form. Recall from (1.3) that $\wedge^p F$ has a natural Hermitian inner product defined by

$$\langle \alpha, \beta \rangle \text{vol} = \alpha \wedge *\bar{\beta},$$

where $\text{vol} = (1/n!)\mathbf{\Omega}^n$ as before. With respect to this inner product L has a Hermitian adjoint

$$L*\colon \wedge^p F \longrightarrow \wedge^{p-2} F, \quad 2 \le p \le 2n,$$

and one finds that

$$L^* = w{*}L{*}. \tag{1.13}$$

To see that (1.13) holds we compute, for $\alpha \in \wedge^p F, \beta \in \wedge^{p+2} F$,

$$\begin{aligned}
\langle L\alpha, \beta \rangle \text{vol} &= \mathbf{\Omega} \wedge \alpha \wedge (*\bar{\beta}) \\
&= \alpha \wedge \mathbf{\Omega} \wedge (*\bar{\beta}) \\
&= \alpha \wedge L*\bar{\beta} \\
&= \alpha \wedge \overline{*w*L*\bar{\beta}} \\
&= \alpha \wedge *w*L*\bar{\beta} \\
&= \langle \alpha, w*L*\beta \rangle \text{vol} \\
&= \langle \alpha, L*\beta \rangle \text{vol}
\end{aligned}$$

using the fact that $*w* = id$, and $*$, L, and w are real operators. It follows from (1.13) that L^* is a real operator, homogeneous of degree -2. It will follow from the next proposition that L^* is bihomogeneous of degree $(-1, -1)$.

If M and N are two endomorphisms of a vector space, then we will denote by $[M, N] = MN - NM$ the commutator of the two endomorphisms. We now have a basic proposition giving fundamental relationships between the above operators.

Proposition 1.1: Let E be a Hermitian vector space of complex dimension n with fundamental form $\mathbf{\Omega}$ and associated operators w, J, L, and $L*$. Then

(a) $^*\Pi_{p,q} = \Pi^*_{n-q,n-p}$,

(b) $[L, w] = [L, J] = [L^*, w] = [L^*, J] = 0$,

(c) $[L^*, L] = \sum_{p=0}^{2n} (n-p)\Pi_p$.

To prove Proposition 1.1, it is necessary to introduce some notation which will allow us to effectively work with the convectors in $\wedge F$. Let $N = \{1, 2, \ldots, n\}$, and let us consider multi-indices $I = (\mu_1, \ldots, \mu_p)$, where $\mu_1, \ldots, \mu_p$ are distinct elements of N, and set $|I| = p$. Let $\{z_1, \ldots, z_n\}$ be a basis for $\wedge^{1,0} F$ such that the Hermitian metric h on E has the form $h = \sum_\mu z_\mu \otimes \bar{z}_\mu$ as in (1.7), with $\mathbf{\Omega}$ given by (1.9), and with $(1/n!)\mathbf{\Omega}^n =$ vol $= x_1 \wedge y_1 \wedge \cdots \wedge x_n \wedge y_n$ where $z_\mu = x_\mu + iy_\mu$, as in (1.10). The operator $*$ is now well-defined in terms of the orthonormal basis $\{x_1, y_1, \ldots, x_n, y_n\}$. If $I = (\mu_1, \ldots, \mu_p)$, then we let

$$\begin{aligned}
z_I &= z_{\mu_1} \wedge z_{\mu_2} \wedge \cdots \wedge z_{\mu_p} \\
x_I &= x_{\mu_1} \wedge x_{\mu_2} \wedge \cdots \wedge x_{\mu_p} \\
&\vdots
\end{aligned}$$

If M is a multiindex, we let

$$w_M = \prod_{\mu \in M} z_\mu \wedge \bar{z}_\mu = (-2i)^{|M|} \prod_{\mu \in M} x_\mu \wedge y_\mu.$$

In this last product it is clear that the ordering of the factors is irrelevant, since the terms commute with one another, and we shall use the same symbol M to denote the ordered p-tuple and its underlying set of elements, provided that this leads to no confusion. Any element of $\wedge F$ can be written in the form

$$\sum_{A,B,M}{}' c_{A,B,M} z_A \wedge \bar{z}_B \wedge w_M,$$

where $c_{A,B,M} \in \mathbf{C}$, and A, B, and M are (for a given term) mutually disjoint multiindices, and, as before, the prime on the summation sign indicates that the sum is taken over multiindices whose elements are strictly increasing sequences (what we shall call an increasing multiindex).

We have the following fundamental and elementary lemma which shows the interaction between the $*$-operator (defined in terms of the real structure) and the bigrading on $\wedge F$ (defined in terms of the complex structure).

Lemma 1.2: Suppose that A, B, and M are mutually disjoint increasing multiindices. Then

$$*(z_A \wedge \bar{z}_B \wedge w_M) = \gamma(a,b,m) z_A \wedge \bar{z}_B \wedge w_{M'}$$

for a nonvanishing constant $\gamma(a,b,m)$, where $a = |A|, b = |B|, m = |M|$, and $M' = N - (A \cup B \cup M)$. Moreover,

$$\gamma(a,b,m) = i^{a-b}(-1)^{p(p+1)/2+m}(-2i)^{p-n}$$

where $p = a + b + 2m$ is the total degree of $z_A \wedge \bar{z}_B \wedge w_M$.

Proof: Let $v = z_A \wedge \bar{z}_B \wedge w_M$. If $A = A_1 \cup A_2$ for some multiindex A, let

$$\epsilon_A^{A_1 A_2} = \begin{cases} 0 & \text{if } A_1 \cap A_2 \neq \varnothing \\ 1 & \text{if } A_1 A_2 \text{ is an even permutation of } A \\ -1 & \text{if } A_1 A_2 \text{ is an odd permutation of } A. \end{cases}$$

Using this notation it is easy to see that

$$z_A = \sum_{A=A_1\cup A_2}{}' \epsilon_A^{A_1 A_2} i^{a_2} x_{A_1} \wedge y_{A_2},$$

where the sum runs over all decompositions of A into increasing multiindices $A_1 \cup A_2$, and $a_1 = |A_1|$, etc. Thus we obtain

$$v = (-2i)^m \sum_{\substack{A=A_1\cup A_2 \\ B=B_1\cup B_2}}{}' \epsilon_A^{A_1A_2}\epsilon_B^{B_1B_2} i^{a_2-b_2} x_{A_1} \wedge y_{A_2} \wedge x_{B_1} \wedge y_{B_2} \wedge \prod_{\mu\in M} x_\mu \wedge y_\mu.$$

We now want to compute $*v$, having expressed v in terms of a real basis, and we shall do this term by term and then sum the result. To simplify the notation, consider the case where $B = \varnothing$. We obtain

$$*(z_A \wedge w_M) = (-2i)^m \sum_{A=A_1\cup A_2} \epsilon_A^{A_1A_2} i^{a_2} *\{x_{A_1} \wedge y_{A_2} \wedge \prod_{\mu\in M} x_\mu \wedge y_\mu\}. \tag{1.1}$$

It is clear that the result of $*$ acting on the bracketed expression is of the form

$$\pm x_{A_2} \wedge y_{A_1} \wedge \prod_{\mu\in M'} x_\mu \wedge y_\mu, \tag{1.2}$$

where $M' = N - (A \cup M)$. The only problem left is to determine the sign.

To do this it suffices (because of the commutativity of $\prod_{\mu\in M} x_\mu \wedge y_\mu)$ to consider the product (setting $a_2 = |A_2|$)

$$x_{A_1} \wedge y_{A_2} \wedge x_{A_2} \wedge y_{A_1} = (-1)^{{a_2}^2} x_{A_1} \wedge y_{A_1} \wedge x_{A_2} \wedge y_{A_2}.$$

Now, in general,

$$x_C \wedge y_C = (-1)^{|C|(|C|-1)/2} x_{\mu_1} \wedge y_{\mu_2} \wedge \cdots \wedge x_{\mu|C|} \wedge y_{\mu|C|},$$

and applying this to our problem above, we see immediately that the sign in (1.2) is of the form

$$(-1)^{{a_2}^2 + a_1(a_1-1)/2 + a_2(a_2-1)/2} = (-1)^r.$$

Putting this into (1.1), we obtain

$$(1.3)\quad *(z_A \wedge w_M) = (-2i)^m \sum_{A=A_1\cup A_2}{}' \epsilon_A^{A_1A_2} i^{a_2} (-1)^r x_{A_2} \wedge y_{A_1} \wedge \prod_{\mu\in M'} x_\mu \wedge y_\mu.$$

The idea now is to change variables in the summation. We write

$$\epsilon_A^{A_1A_2} = (-1)^{a_1a_2} \epsilon_A^{A_2A_1}$$

$$i^{a_2} = i^a (-1)^{a_1} i^{a_1},$$

and substituting in (1.3) we obtain

$$*(z_A \wedge w_M) = i^a (-2i)^m \sum_{A=A_1\cup A_2}{}' \epsilon_A^{A_2A_1} i^{a_1} \{(-1)^{r+a_1+a_1a_2}\}$$

$$\cdot x_{A_2} \wedge y_{A_1} \wedge \prod_{\mu\in M'} x_\mu \wedge y_\mu,$$

which is, modulo the bracketed term, of the right form to be const$(z_A \wedge w_M)$. A priori, the bracketed term depends on the decompositions $A = A_1 \cup A_2$; however, one can verify that in fact

$$(-1)^{r+a_1+a_1a_2} = (-1)^{a(a+1)/2} = (-1)^{p(p+1)/2+m},$$

and the bracketed constant pulls out in front the summation, and we obtain

$$*(z_A \wedge w_M) = i^a (-1)^{p(p+1)/2+m} (-2i)^{p-n} z_A \wedge w_{M'}.$$

The more general case is treated similarly.

Q.E.D.

Proof of Proposition 1.1: Part (a) follows immediately from Lemma 1.2. We note that (a) is equivalent to

(a′) $*|_{\wedge^{p,q} F} \colon \wedge^{p,q} F \longrightarrow \wedge^{n-q,n-p} F$ is an isomorphism.

Part (b) follows from the fact that L and Λ are homogeneous operators and are real.

We shall show part (c). Using the notation used in Lemma 1.2, we observe that

$$(1.4)\quad \begin{aligned} L(z_A \wedge \bar{z}_B \wedge w_M) &= \frac{i}{2}\Big(\sum_{\mu=1}^{n} z_\mu \wedge \bar{z}_\mu\Big) \wedge z_A \wedge \bar{z}_B \wedge w_M \\ &= \frac{i}{2} z_A \wedge \bar{z}_B \wedge \Big(\sum_{\mu\in M'} w_{M\cup\{\mu\}}\Big), \end{aligned}$$

where $M' = N - (A \cup B \cup M)$, as before. On the other hand, we see that, using Lemma 1.2 and the definition of Λ,

$$\Lambda(z_A \wedge \bar{z}_B \wedge w_M) = \frac{2}{i} z_A \wedge \bar{z}_B \wedge \Big(\sum_{\mu \in M} w_{M-|\mu|} \Big). \tag{1.5}$$

Using these formulas, one obtains easily, assuming that $z_A \wedge \bar{z}_B \wedge w_M$ has total degree p,

$$\Lambda L - L\Lambda = (n-p) z_A \wedge \bar{z}_B \wedge w_M,$$

and part (c) of Proposition 1.1 follows immediately.

Q.E.D.

2. Harmonic Theory on Compact Manifolds

In this section we want to give further applications of the theory of harmonic differential forms on compact (differentiable or complex) manifolds. As we have seen in Chap. IV, the Laplacian on a Riemannian manifold is defined by $dd^* + d^*d$, where $d*$ is the adjoint with respect to some inner product on the (elliptic) complex $\mathcal{E}^*(X)$ of complex-valued differential forms on X. We want to use the $*$-operator of Sec. 1 to define a *particular* inner product for the vector space of differential forms of a given degree, from which will follow a useful formula for the adjoint operator d^* (and related operators).

Suppose that X is a compact oriented Riemannian manifold of d dimensions. Then the orientation and Riemannian structure define the $*$-operator as in Sec. 1:

$$*\colon \wedge^p T_x^*(X) \xrightarrow{\cong} \wedge^{d-p} T_x^*(X)$$

at each point $x \in X$. Moreover, $*$ defines a smooth bundle map, since we can define it in the neighborhood of a point by choosing a smooth local (oriented) orthonormal frame. Hence $*$ induces an isomorphism of sections (assuming that we extend $*$ to $\wedge^p T^*(X) \otimes \mathbf{C}$ by complex linearity),

$$*\colon \mathcal{E}^p(X) \xrightarrow{\cong} \mathcal{E}^{d-p}(X),$$

where $d = \dim_{\mathbf{R}} X$.

Suppose that $\varphi \in \mathcal{E}^d(X)$. Then we can define, in a standard manner,

$$\int_X \varphi$$

by using a partition of unity $\{\varphi_\alpha\}$ subordinate to a finite covering of X by coordinate patches. Namely, let

$$f_\alpha\colon U_\alpha \underset{\text{open}}{\subset} \mathbf{R}^d \longrightarrow X$$

be the coordinate mappings, and set

$$\int_X \varphi = \sum_\alpha \int_{U_\alpha} f_\alpha^*(\varphi_\alpha \varphi) = \sum_\alpha \int_{\mathbf{R}^d} g_\alpha(x) dx_1 \wedge \cdots \wedge dx_d,$$

where the C^∞ function g_α has compact support in U_α. This is easily seen to be independent of the coordinate covering and partition of unity used.

If X is an oriented Riemannian manifold, then X carries a *volume element* dV, which is nothing but a d-form $\varphi \in \mathcal{E}^d(X)$, with the property that in any oriented system of local coordinates $U \subset X$

$$\varphi(x) = f(x)dx_1 \wedge \cdots \wedge dx_d,$$

where $f(x) > 0$ for all $x \in U$. By means of the $*$-operator it is easy to see that

$$\varphi = *(1), (1 \in \mathbf{C} \subset \mathcal{E}^0(X))$$

is indeed a volume element on X.

Remark: Denote in local coordinates the Riemannian metric on X by

$$ds^2 = g_{ij}dx^i \otimes dx^j,$$

using the summation convention, where g_{ij} is a symmetric positive definite matrix of functions. If we let g^{ij} be defined by

$$g^{ij}g_{jk} = \delta^i_k \quad \text{(Kronecker delta)},$$

and if we raise indices by setting

$$a^{i_1 \cdots i_p} = g^{i_1 j_1} \bullet g^{i_2 j_2} \bullet \ldots \bullet g^{i_p j_p} a_{i_1 \cdots i_p},$$

then we can express the $*$-operator given by the metric ds^2 explicitly in terms of these quantities (cf., deRham [1], pp. 119–122). Namely, we have, if

$$\alpha = \sum_{i_1 < \cdots < i_p} \alpha_{i_1 \cdots i_p} dx^{i_1} \wedge \cdots \wedge dx^{i_p},$$

then

$$(*\alpha) = \sum_{j_1 < \cdots < j_{d-p}} (*\alpha)_{j_1 \cdots j_{d-p}} dx^{j_1} \wedge \cdots \wedge dx^{j_{d-p}},$$

where

$$(*\alpha)_{j_1 \cdots j_{d-p}} = \pm\sqrt{\det(g_{ij})}\alpha^{i_1 \cdots i_p},$$

where $\{i_1, \ldots, i_p, j_1, \ldots, j_{d-p}\} = \{1, \ldots, d\}$, and we have the positive sign if the permutation is even and negative sign in the other case (just as in the case of an orthonormal basis). Thus in particular

$$*(1) = \sqrt{\det(g_{ij})}\, dx^1 \wedge \cdots \wedge dx^d$$

is the volume element in this case.

Define

$$(2.1) \qquad \begin{aligned} (\varphi, \psi) &= \int_X \varphi \wedge *\bar{\psi}, & \varphi, \psi &\in \mathcal{E}^p(X) \\ (\varphi, \psi) &= 0, & \varphi &\in \mathcal{E}^p(X), \psi \in \mathcal{E}^p(X), p \neq q \end{aligned}$$

and the integral is well defined since $\varphi \wedge *\bar{\psi}$ is a d-form on X. We can extend this definition to noncompact manifolds by considering only forms with compact support. We then have the following proposition.

Proposition 2.1: The form $(,)$ defined by (2.1) defines a positive definite, Hermitian symmetric, sesquilinear form on the complex vector space $\mathcal{E}*(X) = \oplus_{p=0}^{d} \mathcal{E}^p(X)$.

Proof: The Riemannian metric on X induces an Hermitian inner product $\langle , \rangle$ on $\wedge^p T*_x(X)$ for each $x \in X$, given by, for φ, ψ p-forms on X,

$$\varphi \wedge *\bar{\psi} = \langle \varphi, \psi \rangle \text{vol}$$

as we saw in (1.3). It is then clear that

$$(\varphi, \psi) = \int_X \varphi \wedge *\bar{\psi} = \int_X \langle \varphi, \psi \rangle \text{vol}$$

is a positive semidefinite, sesquilinear Hermitian form on $\mathcal{E}^p(X)$. To see that $(\, , \,)$ is positive definite, suppose that $\varphi \in \mathcal{E}^p(X)$ is not equal to zero at $x_0 \in X$, then near x_0, we can express φ in terms of a local oriented orthonormal frame for $T*(X) \otimes \mathbf{C}$, $\{e_1, \ldots, e_d\}$,

$$\varphi = \sum_{|I|=p}{}' \varphi_I e_I,$$

and

$$\varphi \wedge *\bar{\varphi} = \sum_{|I|=p}{}' |\varphi_I|^2 \text{ vol}$$

near x_0, and $\sum'_{|J|=p} |\varphi_I|^2 > 0$ near x_0. Then the contribution to the integral

$$(\varphi, \varphi) = \int_X \varphi \wedge *\bar{\varphi}$$

will be nonzero, and thus $(\varphi, \varphi) > 0$.

Q.E.D.

Thus the elliptic complex $(\mathcal{E}^*(X), d)$ is equipped with a canonical inner product depending only on the orientation and Riemannian metric of the base space X (in Sec. 5 of Chap. IV we had allowed arbitrary metrics on each of the vector bundles appearing in the complex). We would have arrived at the same inner product had we merely used the metric on $\wedge^p T^*(X)$ naturally induced by that of $T(X)$ and for our strictly positive measure $d\lambda$ used the volume element $*(1)$. However, the representation we have given here for the inner product on $\mathcal{E}^*(X)$ will prove to be very useful, as we shall see. For convenience, we shall call the inner product (2.1) on $\mathcal{E}^*(X)$ the *Hodge inner product on* $\mathcal{E}^*(X)$.

Suppose that X is a Hermitian complex manifold. Then we can define the Hodge inner product on $\mathcal{E}^*(X)$ with respect to the underlying Riemannian metric and a fixed orientation given by the complex structure (all complex manifolds are orientable).

Proposition 2.2: The direct sum decomposition $\mathcal{E}^r(X) = \sum_{p+q=r} \mathcal{E}^{p,q}(X)$ is an orthogonal direct sum decomposition with respect to the Hodge inner product.

Proof: Suppose that $\varphi \in \mathcal{E}^{p,q}(X)$ and that $\psi \in \mathcal{E}^{r,s}(X)$, where $p+q = r+s$. Then we see that $\varphi \wedge *\bar{\psi}$ is of type $(n-r+p, n-s+q)$, since $\bar{\psi}$ is of type (s, r) and $*\bar{\psi}$ is then of type $(n-r, n-s)$ by Proposition 1.1. Therefore $\varphi \wedge *\bar{\psi}$ is a $2n$-form if and only if $r = p$ and $s = q$. Otherwise, $\varphi \wedge *\bar{\psi}$ is identically zero. This proves the proposition.

Q.E.D.

Using the Hodge inner product, it will be very easy to compute the adjoints of various linear operators acting on $\mathcal{E}^*(X)$ (cf. the computation of $L*$ in Sec. 1). First we want to modify the $*$-operator in a manner which will be convenient for this purpose. On an oriented Riemannian manifold we define

$$\bar{*}\colon \mathcal{E}^*(X) \longrightarrow \mathcal{E}^*(X)$$

by setting $\bar{*}(\varphi) = *\bar{\varphi}$. Thus $\bar{*}$ is a conjugate-linear isomorphism of vector bundles,

$$\bar{*}\colon \wedge^p T^*(X)_c \longrightarrow \wedge^{m-p} T^*(X)_c,$$

where $m = \dim_{\mathbf{R}} X$. Suppose that X is now a Hermitian complex manifold and that $E \longrightarrow X$ is a Hermitian vector bundle. Let

$$\tau\colon E \longrightarrow E^*$$

be a conjugate-linear bundle isomorphism of E onto its dual bundle $E*$. The mapping τ depends on the Hermitian metric of E and is defined fibrewise in a standard manner. We then define

$$\bar{*}_E\colon \wedge^p T^*(X)_c \otimes E \longrightarrow \wedge^{2n-p} T^*(X)_c \otimes E^*$$

by setting

$$\bar{*}_E(\varphi \otimes e) = \bar{*}(\varphi) \otimes \tau(e)$$

for $\varphi \in \wedge^p T_x^*(X)_c$ and $e \in E_x$. Thus $\bar{*}_E$ is a conjugate-linear isomorphism of Hermitian vector bundles. We recall that we defined $\mathcal{E}^r(X, E)$ to be the sections of $\wedge^r T^*(X)_c \otimes E$ and that, moreover, there is a decomposition into bidegrees

$$\mathcal{E}^r(X, E) = \sum_{p+q=r} \mathcal{E}^{p,q}(X, E).$$

Thus we note that first the Hodge inner product on $\mathcal{E}^*(X)$ can be written as

$$(\varphi, \psi) = \int_X \varphi \wedge \bar{*}\psi,$$

and we extend this to a Hodge inner product on $\mathcal{E}^*(X, E)$ by setting

$$(\varphi, \psi) = \int_X \varphi \wedge \bar{*}_E \psi \tag{2.2}$$

if $\varphi, \psi \in \mathcal{E}^r(X, E)$. It is easy to see that $\varphi \wedge \bar{*}_E \psi$ does make sense and is a scalar $2n$-form which can be integrated over X (where $n = \dim_{\mathbf{C}} X$). In fact, if we let $\langle\,,\rangle$ represent the bilinear duality pairing between E and E^*, then we set, for $\varphi \in \wedge^p T_x^*(X)_c$, $e \in E_x$, $\psi \in \wedge^{2n-p} T_x^*(X)_c$, $f \in E_x^*$,

$$(\varphi \otimes e) \wedge (\psi \otimes f) = \varphi \wedge \psi \bullet \langle e, f\rangle \in \wedge^{2n} T_x^*(X)_c.$$

By using a basis for E and a dual basis for E^*, we can extend this exterior product to vector-bundle-valued differential forms, and it is easily checked that the resulting exterior product is independent of the choice of basis. Thus (2.4) defines what we shall call a Hodge inner product on $\mathcal{E}^*(X, E)$. Then it is easy to see that $\bar{*}_E$ preserves the bigrading on $\mathcal{E}^*(X, E)$, and that, in fact,

$$\bar{*}_E\colon \mathcal{E}^{p,q}(X, E) \xrightarrow{\sim} \mathcal{E}^{n-p,n-q}(X, E^*)$$

is a conjugate-linear isomorphism. It is then clear that Proposition 2.2 extends to this case.

We are now in a position to compute the adjoints of various operators with respect to the Hodge inner product. *Moreover, all adjoints in this and later sections of the book will be with respect to the Hodge inner product.*

Proposition 2.3: Let X be an oriented compact Riemannian manifold of real dimension m and let $\mathbf{\Delta} = dd^* + d^*d$, where the adjoint d^* is defined with respect to the Hodge inner product on $\mathcal{E}^*(X)$. Then

(a) $d^* = (-1)^{m+mp+1}\bar{*}d\bar{*} = (-1)^{m+mp+1}*d*$ on $\mathcal{E}^p(X)$.
(b) $*\mathbf{\Delta} = \mathbf{\Delta}*, \quad \bar{*}\mathbf{\Delta} = \mathbf{\Delta}\bar{*}$.

Proof: The basic fact we need is that $** = w$, as defined in Sec. 1. Suppose that $\varphi \in \mathcal{E}^{p-1}(X)$ and that $\psi \in \mathcal{E}^p(X)$. Then we consider

$$\begin{aligned}(d\varphi, \psi) &= \int_X d\varphi \wedge \bar{*}\psi \\ &= \int_X d(\varphi \wedge \bar{*}\psi) - (-1)^{p-1}\int_X \varphi \wedge d\bar{*}\psi,\end{aligned}$$

by the rule for differentiating a product of forms. Moreover, by Stokes' theorem, we see that the first term vanishes, and hence we obtain (noting that $** = \bar{*}\bar{*} = w$, since $*$ is real)

$$\begin{aligned}(d\varphi, \psi) &= (-1)^p \int_X \varphi \wedge \bar{*}(\bar{*}^{-1}d\bar{*})\psi \\ &= (-1)^p \int_X \varphi \wedge \bar{*}(\bar{*}wd\bar{*})\psi \\ &= (-1)^{m+mp+1}(\varphi, \bar{*}d\bar{*}\psi),\end{aligned}$$

and thus we have

$$d^* = (-1)^{m+mp+1}\bar{*}d\bar{*},$$

and since d is real, we also obtain

$$d^* = (-1)^{m+mp+1}*d*.$$

To prove (b), we compute, for $\varphi \in \mathcal{E}^p(X)$,

$$\begin{aligned}*\mathbf{\Delta}\varphi &= (-1)^{m+mp+1}(*d*d* + (-1)^m **d*d)\varphi \\ \mathbf{\Delta}*\varphi &= (-1)^{m+m(m-p)+1}(d*d** + (-1)^m *d*d*)\varphi,\end{aligned}$$

and so it suffices to show that (recall that $w = **$)

$$wd*d\varphi = d*dw\varphi.$$

But this is simple, since $w = \sum(-1)^{p+mp}\Pi_p$, and thus the right-hand side is $d*d(-1)^{p+mp}\varphi$, whereas the left-hand side has degree $m - p$, and so

$$wd*d\varphi = (-1)^{m-p+m(m-p)}d*d\varphi = (-1)^{p+mp}d*d\varphi.$$

Q.E.D.

We have a similar result for the Hermitian case. Note that $\bar{*}_{E*}$ is defined in the same way as $\bar{*}_E$ by using $\tau^{-1}: E^* \longrightarrow E$.

Proposition 2.4: Let X be a Hermitian complex manifold and let $E \longrightarrow X$ be a Hermitian holomorphic vector bundle. Then

(a) $\bar{\partial}: \mathcal{E}^{p,q}(X, E) \longrightarrow \mathcal{E}^{p,q+1}(X, E)$ has an adjoint $\bar{\partial}^*$ with respect to the Hodge inner product on $\mathcal{E}^{**}(X, E)$ given by

$$\bar{\partial}^* = -\bar{*}_{E^*}\bar{\partial}\bar{*}_E.$$

(b) If $\overline{\square} = \bar{\partial}\bar{\partial}^* + \bar{\partial}^*\bar{\partial}$ is the complex Laplacian acting on $\mathcal{E}^{**}(X, E)$, then

$$\overline{\square}\bar{*}_E = \bar{*}_E\overline{\square}.$$

Proof: In this case we also have $\bar{*}_E\bar{*}_{E^*} = w = \sum(-1)^p\Pi_p$, a simpler expression since the real dimension of X is even. The proof of (a) then follows as before, with minor modification. Suppose that $\varphi \in \mathcal{E}^{p,q-1}(X, E)$ and that $\psi \in \mathcal{E}^{p,q}(X, E)$. Then we have that $\varphi \wedge \bar{*}_E\psi$ is a scalar differential form of type $(n, n - 1)$, and hence $\bar{\partial}(\varphi \wedge \bar{*}_E\psi) = d(\varphi \wedge \bar{*}_E\psi)$. Moreover,

$$\bar{\partial}(\varphi \wedge \bar{*}_E\psi) = \bar{\partial}\varphi \wedge \bar{*}_E\psi + (-1)^{p+q-1}\varphi \wedge \bar{\partial}\bar{*}_E\psi.$$

Substituting into the inner product, we obtain, using Stokes' theorem as in the proof of Proposition 2.3,

$$\begin{aligned}(\bar{\partial}\varphi, \psi) &= (-1)^{p+q}\int_X \varphi \wedge \bar{\partial}\bar{*}_E\psi \\ &= (-1)^{p+q}\int \varphi \wedge \bar{*}_E(w\bar{*}_{E^*}\bar{\partial}\bar{*}_E\psi) \\ &= -\int \varphi \wedge \bar{*}_E(\bar{*}_{E^*}\bar{\partial}\bar{*}_E\psi) \\ &= (\varphi, -\bar{*}_{E^*}\bar{\partial}\bar{*}_E\psi),\end{aligned}$$

and hence (a) is proved. The proof of (b) is exactly the same as in Proposition 2.3 (Note that $\overline{\square}$ acting on $\mathcal{E}^{**}(X, E)$ and $\mathcal{E}^{**}(X, E^*)$ denotes two different operators).

Q.E.D.

Remark: We note that only $\bar{\partial}$ acts naturally on $\mathcal{E}^{p,q}(X, E)$ for a nontrivial holomorphic vector bundle E, whereas ∂ and hence d do not, since they do not annihilate the transition functions defining E. However, in the scalar case, we have $\partial\colon \mathcal{E}^{p,q}(X) \longrightarrow \mathcal{E}^{p+1,q}(X)$, and by the same calculation as above we obtain that $\partial^* = -\bar{*}\partial\bar{*}$ and that $\Box = \partial\partial^* + \partial^*\partial$ commutes with $\bar{*}$, exactly the same as the $\bar{\partial}$-operator case.

Using the above results we can derive two well-known duality theorems. We first remark that a finite dimensional complex vector space E is conjugate-linearly isomorphic to a complex vector space F if and only if F is complex-linearly isomorphic to E^*, the dual of E (and the bilinear pairing of E to F can be obtained from a Hermitian inner product on E).

Theorem 2.5 (Poincaré duality): Let X be a compact m-dimensional orientable differentiable manifold. Then there is a conjugate linear isomorphism

$$\sigma\colon H^r(X, \mathbf{C}) \longrightarrow H^{m-r}(X, \mathbf{C}),$$

and hence $H^{m-r}(X, \mathbf{C})$ is isomorphic to the dual of $H^r(X, \mathbf{C})$.

Proof: Introduce a Riemannian metric and an orientation on X and let $*$ be the associated $*$-operator. Then we have the commutative diagram

$$\begin{array}{ccc} \mathcal{E}^r(X) & \xrightarrow{\bar{*}} & \mathcal{E}^{m-r}(X) \\ \downarrow H_\Delta & & \downarrow H_\Delta \\ \mathcal{H}^r(X) & \xrightarrow{\bar{*}} & \mathcal{H}^{m-r}(X) \\ \| \wr & & \| \wr \\ H^r(X, \mathbf{C}) & \xrightarrow{\sigma} & H^{m-r}(X, \mathbf{C}), \end{array}$$

where H_Δ is the projection onto the harmonic forms given by Theorem IV.4.12, and the mapping $\bar{*}$ maps harmonic forms to harmonic forms since $\Delta\bar{*} = \bar{*}\Delta$, as we saw in Proposition 2.3. Moreover, the de Rham groups $H^r(X, \mathbf{C})$ are isomorphic to $\mathcal{H}^r(X)$ (Example IV.5.4), and σ is the induced conjugate linear isomorphism.

Q.E.D.

Remark: We could have restricted ourselves to real-valued differential forms and obtained the same result. Also, the more general Poincaré duality theorem of algebraic topology is true with coefficients in $\mathbf{Z}$ and is independent of any differentiable structure on X, but one needs a different type of proof for that (see, e.g., Greenberg [1]).

Corollary 2.6: Let X be as in Theorem 2.5. Then

$$b_r(X) = b_{m-r}(X), \quad r = 0, \ldots, m.$$

Our next result is more analytical in nature and depends very much on the complex structures involved, in contrast to the Poincaré duality above.

Theorem 2.7 (Serre duality): Let X be a compact complex manifold of complex dimension n and let $E \longrightarrow X$ be a holomorphic vector bundle over X. Then there is a conjugate linear isomorphism

$$\sigma: H^r(X, \mathbf{\Omega}^p(E)) \longrightarrow H^{n-r}(X, \mathbf{\Omega}^{n-p}(E^*)),$$

and hence these spaces are dual to one another.

Proof: By introducing Hermitian metrics on X and E, we can define the $\bar{*}_E$ operator. Then we obtain the following commutative diagram,

$$\begin{array}{ccc}
\mathcal{E}^{p,q}(X, E) & \xrightarrow{\bar{*}_E} & \mathcal{E}^{n-p,n-q}(X, E^*) \\
\downarrow H_{\bar{\square}} & & \downarrow H_{\bar{\square}} \\
\mathcal{H}^{p,q}(X, E) & \xrightarrow{\bar{*}_E} & \mathcal{H}^{n-p,n-q}(X, E^*) \\
\| \wr & & \| \wr \\
H^{p,q}(X, E) & \xrightarrow{\tau} & H^{n-p,n-q}(X, E^*) \\
\| \wr & & \| \wr \\
H^p(X, \mathbf{\Omega}^p(E)) & \xrightarrow{\sigma} & H^{n-q}(X, \mathbf{\Omega}^{n-p}(E^*)),
\end{array}$$

which proves the result immediately. Once again, $\bar{*}_E$ maps harmonic forms to harmonic forms by Proposition 2.4, and the $\{H^{p,q}(X, E)\}$ are the Dolbeault groups [the cohomology of the complex $(\mathcal{E}^{p,*}(X, E), \bar{\partial})$], which are isomorphic to $H^q(X, \mathbf{\Omega}^p(E))$, as we saw in Theorem II.3.20.

Q.E.D.

Remark: Serre proved this also in the case of noncompact manifolds, under certain closed range hypotheses on $\bar{\partial}$ and by using cohomology with compact supports, i.e., $H^q_*(X, \mathbf{\Omega}^p(E))$ is the topological dual of $H^{n-q}(X, \mathbf{\Omega}^{n-q}(E^*))$, where $H^q_*(\ )$ denotes cohomology with compact supports. In our case we have finite dimensional vector spaces (due to the harmonic theory), in which case Serre's hypothesis is fulfilled and the compact support is automatic. Serre's proof (in Serre [1]) used resolutions of $\mathbf{\Omega}^p(E)$ by both C^∞ forms and by distribution forms, and he was able to utilize the natural duality of these spaces to obtain his results. The proof above is due to Kodaira [1].

Corollary 2.8: Let X be a compact complex manifold of complex dimension n. Then

(a) $b_r(X) = b_{2n-r}(X), r = 0, \ldots, 2n.$

(b) $h^{p,q}(X) = h^{n-p,n-q}(X), p, q = 0, \ldots, n.$

3. Representations of $\mathfrak{sl}(2, \mathbf{C})$ on Hermitian Exterior Algebras

In this section we summarize the finite-dimensional complex representation theory for the Lie algebra $\mathfrak{sl}(2, \mathbf{C})$ of 2×2 complex matrices with trace zero, and then we will apply this theory to specific representations arising from Hermitian exterior algebras as in Sec. 1. This representation theory is available in various references (e.g., Serre [3], Varadarajan [1]), and we will survey the principal ideas needed for the applications we have in mind. We will use some elementary facts and terminology concerning Lie groups and Lie algebras as is found in any introduction to the subject (e.g., Chevalley [1], Helgason [1], Varadarajan [1]), such as the Lie algebra of a Lie group, and the associated exponential mapping, invariant measure on Lie groups, etc., although we will be using these concepts only for specific low-dimensional matrix groups and matrix algebras.

We recall that a *Lie algebra* is a vector space $\mathfrak{A}$ equipped with a Lie bracket product [,] which is anticommutative, and which satisfies the *Jacobi identity*

$$[X, [Y, Z]] + [Y, [Z, X]] + [Z, [X, Y]] = 0.$$

An algebra of matrices equipped with the commutator Lie bracket is the prototypical example of a Lie algebra. A *representation* of a Lie algebra $\mathfrak{A}$ on a complex vector space V is an algebra homomorphism

$$\pi \colon \mathfrak{A} \longrightarrow \operatorname{End}(V),$$

where $\operatorname{End}(V)$ is the Lie algebra of endomorphisms of V equipped with the commutator Lie bracket $[A, B] = AB - BA$. If $n = \dim V < \infty$, then we say that the representation has *dimension n*. If $\dim V = \infty$, then we say that π is an infinite-dimensional representation. A representation π is *irreducible* if there is no proper invariant subspace $V_0 \neq 0$ of V. Here V_0 is a proper invariant subspace if $0 \neq V_0 \neq V$, and

$$\pi(X)V_0 \subset V_0, \quad \text{for all } X \in \mathfrak{A}.$$

If π_1 and π_2 are representations on V_1 and V_2, respectively, then $\pi = \pi_1 \oplus \pi_2$ is a representation of $\mathfrak{A}$ on $V_1 \oplus V_2$ in a natural manner. Two representations π_1 and π_2 are *equivalent* if there is an isomorphism $S\colon V_1 \to V_2$ so that $\pi_1 = S^{-1}\pi_2 S$. A representation π is *completely reducible* if it is equivalent to a direct sum of irreducible representations. A representation of a Lie group (e.g., a matrix group) G on a finite-dimensional complex vector space V is a real-analytic homomorphism $\rho\colon G \to GL(V)$, where $GL(V)$ denotes the Lie group of nonsingular endomorphisms of the vector space V. In this case, one has the same notions of irreducibility, complete reducibility, etc. as discussed above for representations of Lie algebras.

The Lie algebra $\mathfrak{sl}(2, \mathbf{C})$ is, by definition, 2×2 complex matrices with trace zero. One finds that $\mathfrak{sl}(2, \mathbf{C})$ is the Lie algebra of the Lie group $SL(2, \mathbf{C})$, the group of 2×2 matrices with complex coefficients and determinant equal to 1. There is an exponential mapping

$$\exp\colon \mathfrak{sl}(2, \mathbf{C}) \longrightarrow SL(2, \mathbf{C}) \tag{3.1}$$

given by

$$\exp X = e^X = \sum_{n=0}^{\infty} X^n/n!,$$

which is norm convergent, and where, for $t \in C$,

$$
\begin{aligned}
e^{tX} &= I + tX + O(|t|^2),\\
e^{tX}e^{tY} &= I + t(X+Y) + O(|t|^2),\\
e^{tX}e^{tY}e^{-tX}e^{-tY} &= I + t^2([X,Y]) + O(|t|^3),
\end{aligned}
\tag{3.2}
$$

which indicates the basic relationship between the group law in $SL(2,\mathbf{C})$ and the Lie bracket in $\mathfrak{sl}(2,\mathbf{C})$.

Now consider the subgroup $SU(2)$ of $SL(2,\mathbf{C})$ consisting of unitary 2×2 matrices of determinant one. It follows readily from (3.2) that $\mathfrak{su}(2)$, the corresponding Lie algebra of $SU(2)$, consists of skew-Hermitian 2×2 matrices of trace zero, i.e., $X + X^* = 0, tr(X) = 0$, where $X^* = {}^t\overline{X}$ is the Hermitian adjoint. Thus we have the following diagram of groups and algebras, where i is the natural inclusion:

$$
\begin{array}{ccc}
\mathfrak{su}(2) & \xrightarrow{\;i\;} & \mathfrak{si}(2,\mathbf{C})\\
\big\downarrow{\scriptstyle\exp} & & \big\downarrow{\scriptstyle\exp}\\
SU(2) & \xrightarrow{\;i\;} & SL(2,\mathbf{C}).
\end{array}
\tag{3.3}
$$

For reference, we will write down explicit generators for these algebraic objects. First we note that $\mathfrak{sl}(2,\mathbf{C})$ has dimension 3 and a basis is given by

$$
X = \begin{bmatrix} 0 & 1\\ 0 & 0\end{bmatrix},\quad Y = \begin{bmatrix} 0 & 0\\ 1 & 0\end{bmatrix},\quad H = \begin{bmatrix} 1 & 0\\ 0 & -1\end{bmatrix}.
\tag{3.4}
$$

One checks that the commutation relations

$$
[X,Y] = H,\quad [H,X] = 2X,\quad [H,Y] = -2Y
\tag{3.5}
$$

hold. We see easily that $\mathfrak{su}(2)$ is a real form of $\mathfrak{sl}(2,\mathbf{C})$ (i.e., as vector spaces, $\mathfrak{sl}(2,\mathbf{C}) = \mathfrak{su}(2)\otimes_{\mathbf{R}}\mathbf{C}$), and has a basis (over $\mathbf{R}$) given by

$$
iH,\quad X - Y,\quad i(X+Y).
$$

We note that $i(X+Y)$ generates a one-parameter subgroup of $SU(2)$ given by

$$
\exp[it(X+Y)] = \begin{bmatrix} \cos t & i\sin t\\ i\sin t & \cos t\end{bmatrix},\qquad t\in\mathbf{R}.
$$

This can be checked by a direct computation or by noting that both 1-parameter subgroups have the same generator, namely

$$
i(X+Y) = i\begin{bmatrix} 0 & 1\\ 1 & 0\end{bmatrix} = \frac{d}{dt}\begin{bmatrix} \cos t & i\sin t\\ i\sin t & \cos t\end{bmatrix}\Bigg|_{t=0}
$$

Let

$$
w = \exp[\tfrac{1}{2}i\pi(X+Y)] = \begin{bmatrix} 0 & i\\ i & 0\end{bmatrix},
\tag{3.6}
$$

and we see that conjugation by w in $\mathfrak{sl}(2, \mathbf{C})$ gives rise to a reflection with respect to the above basis (the Weyl group reflection). Namely,

$$wHw^{-1} = -H, \quad wXw^{-1} = Y, \quad wYw^{-1} = X.$$

We return now to diagram (3.3). For each of the algebraic objects in (3.3) one considers representations on a complex vector space V, as we have done before:

$$\begin{array}{ccc} [\mathfrak{su}(2) \longrightarrow \mathrm{End}(V)]_{\mathbf{R}} & \xleftarrow{r_1} & [\mathfrak{sl}(2, \mathbf{C}) \longrightarrow \mathrm{End}(V)]_{\mathbf{C}} \\ {\scriptstyle d}\uparrow & & {\scriptstyle d}\uparrow \\ [SU(2) \longrightarrow GL(V)]_{\mathbf{R}} & \xleftarrow{r_2} & [SL(2, \mathbf{C}) \longrightarrow GL(V)]_{\mathbf{C}} \end{array} \tag{3.7}$$

Here $[\mathfrak{su}(2) \to \mathrm{End}(V)]_{\mathbf{R}}$ denotes $\mathbf{R}$-linear algebra homomorphisms, $[\mathfrak{sl}(2, \mathbf{C}) \to \mathrm{End}(V)]_{\mathbf{C}}$ denotes $\mathbf{C}$-linear algebra homomorphisms, $[SL(2) \to GL(V)]_{\mathbf{R}}$ denotes real-analytic group homomorphisms, and $[SL(2, \mathbf{C}) \to GL(V)]_{\mathbf{C}}$ denotes complex-analytic group homomorphisms. The mappings r_1 and r_2 are the natural restriction mappings, and d is the derivative mapping, recalling that the Lie algebra of a Lie group is the tangent space to the Lie group at the identity element, and noting that the derivative of a representation of a Lie group is indeed a representation of the associated Lie algebra.

We now have the following proposition.

Proposition 3.1: The mappings r_1, r_2 and d in (3.7) are all bijective, i.e., there is a one-to-one correspondence between representations of $SL(2, \mathbf{C})$, $\mathfrak{sl}(2, \mathbf{C})$, $SU(2)$ and $\mathfrak{su}(2)$.

Proof: First we see that r_1 is bijective since $\mathfrak{sl}(2, \mathbf{C})$ is the complexification of $\mathfrak{su}(2)$, and $\mathbf{R}$-linear homomorphisms defined on $\mathfrak{su}(2)$ extend naturally and uniquely as $\mathbf{C}$-linear homomorphisms on $\mathfrak{sl}(2, \mathbf{C})$. The mappings d are bijective since $SL(2, \mathbf{C})$ and $SU(2)$ are both connected and simply-connected $[SL(2, \mathbf{C}) \cong S^3 \times \mathbf{R}^3, SU(2) \cong S^3]$, thus insuring that the inverse of exp (the "logarithm") is well-defined on $SU(2)$ and $SL(2, \mathbf{C})$. The diagram is commutative, and we conclude that r_2 is bijective. In fact, if

$$\rho\colon \mathfrak{sl}(2, \mathbf{C}) \longrightarrow \mathrm{End}(V)$$

is given, and if $g = e^X \in SL(2, \mathbf{C})$ where $X \in \mathfrak{sl}(2, \mathbf{C})$, then the representation

$$\pi\colon SL(2, \mathbf{C}) \longrightarrow GL(V)$$

corresponding to the given ρ is of the form

$$\pi(e^X) = e^{\rho(X)}. \tag{3.8}$$

It is clear that $d\pi = \rho$.

Q.E.D.

Thus we have that representations of $\mathfrak{sl}(2, \mathbf{C})$ are in one-to-one correspondence with representations of $SU(2)$, a compact Lie group. We now have the following important theorem of H. Weyl concerning complete reducibility of representations of compact Lie groups (the "unitary trick"). We state the theorem in full generality, but will use it only for $G = SU(2)$.

Theorem 3.2: Let G be a compact Lie group, and let $\rho: G \to GL(V)$ be a representation on a finite-dimensional complex vector space. Then ρ is completely reducible.

Proof: Choose a basis for V so that $V \cong \mathbf{C}^n$. Let dg be the natural left invariant measure on the Lie group G which can be constructed from left invariant differential forms dual to the left invariant vector fields which comprise the Lie algebra of G (see Helgason [1], Chapter X, §1). Then

$$M(g) = \rho(g)\rho(g)^*$$

is a Hermitian positive definite matrix for each $g \in G$. Define

$$M = \int_G M(g)dg,$$

and it follows that M is Hermitian positive definite also. Then consider

$$\begin{aligned}
\rho(g)M\rho(g)^* &= \int_G \rho(g)\rho(\tau)\rho(\tau)^*\rho(g)^*d\tau \\
&= \int_G \rho(g\tau)\rho(g\tau)^*d\tau \\
&= \int_G \rho(\tau)\rho(\tau)^*d\tau = M,
\end{aligned}$$

using the invariance of $d\tau$ under the action of G on itself by left translation. Since M is positive definite, we can write

$$M = NN^*$$

where N is positive definite. Then we see that $\tilde{\rho} = N^{-1}\rho N$ is equivalent to ρ and moreover

$$\begin{aligned}
\tilde{\rho}(g)\tilde{\rho}(g)^* &= (N^{-1}\rho(g)N)(N^{-1}\rho(g)N)^* \\
&= N^{-1}\rho(g)NN^*\rho(g)^*(N^{-1})^* \\
&= N^{-1}M(N^{-1})^* \\
&= I,
\end{aligned}$$

and thus $\tilde{\rho}(g)$ is a unitary matrix for all $g \in G$. Now we check that $\tilde{\rho}$ is completely reducible. Suppose that V_0 is any subspace of V invariant under the action of $\tilde{\rho}$. Then let $V_0^\perp$ be the orthogonal complement to V_0 with respect to the usual Hermitian metric on $\mathbf{C}^n$. Then $\tilde{\rho}(V_0) \subset V_0$, and it follows immediately that $\tilde{\rho}(V_0^\perp) \subset V_0^\perp$, since $\tilde{\rho}(g)$, being unitary, preserves the inner product in $\mathbf{C}^n$ for each $g \in G$.

Q.E.D.

Corollary 3.3: Let ρ be a representation of $\mathfrak{sl}(2, \mathbf{C})$ on a finite-dimensional complex vector space, then ρ is completely reducible.

Proof: This follows immediately from Proposition 3.1, Theorem 3.2 and the fact that the bijections in (3.7) are natural and preserve irreducibility and direct sums.

Q.E.D.

Now we know that any representation ρ of $\mathfrak{sl}(2, \mathbf{C})$ on a finite-dimensional complex vector space V is the direct sum of irreducible representations. We now turn to an explicit description of these irreducible representations, which can be characterized, up to equivalence, by the dimension of the representation space, as we shall see. We start with a definition.

Definition 3.4: Let ρ be a representation of $\mathfrak{sl}(2, \mathbf{C})$ on a finite-dimensional complex vector space V. Let V^λ be the eigenvectors of $\rho(H)$ with eigenvalue λ, i.e., for $\lambda \in \mathbf{C}$,

$$V^\lambda = \{v \in V\colon \rho(H)v = \lambda v\}.$$

We say that $v \in V^\lambda$ has *weight* λ. A vector $v \in V$ is said to be *primitive of weight* λ if v is nonzero, $v \in V^\lambda$ and $\rho(X)v = 0$.

We now have some elementary lemmas which lead up to the basic canonical form for a representation of $\mathfrak{sl}(2, \mathbf{C})$. We assume a fixed finite-dimensional representation ρ on $\mathfrak{sl}(2, \mathbf{C})$ on a complex vector space V.

Lemma 3.5:

(a) The sum $\sum_{\lambda \in \mathbf{C}} V^\lambda$ is a direct sum,

(b) If v is of weight λ, then $\rho(X)v$ is of weight $\lambda + 2$ and $\rho(Y)v$ is of weight $\lambda - 2$.

Proof: (a) is simply the assertion that eigenvectors corresponding to different eigenvalues are linearly independent. For (b) we observe that

$$\begin{aligned}
\rho(H)\rho(X)v &= (\rho(H)\rho(X) - \rho(X)\rho(H))v + \rho(X)\rho(H)v \\
&= \rho([H, X])v + \lambda\rho(X)v \\
&= \rho(2X)v + \lambda\rho(X)v \\
&= (\lambda + 2)\rho(X)v.
\end{aligned}$$

Similarly, $\rho(H)\rho(Y)v = (\lambda - 2)\rho(Y)v$.

Q.E.D.

Lemma 3.6: Every representation ρ of $\mathfrak{sl}(2, \mathbf{C})$ on a finite-dimensional complex vector space has at least one primitive vector.

Proof: Let v_0 be an eigenvector of $\rho(H)$, and consider the sequence of eigenvectors of $\rho(H)$

$$v_0, \rho(X)v_0, \rho(X)^2v_0, \ldots, \rho(X)^n v_0, \ldots.$$

The nonzero terms in this sequence are linearly independent, since they are eigenvectors with differing eigenvalues (Lemma 3.5), so the sequence must terminate, and hence for some fixed k, $\rho(X)^k v_0 = 0$, $\rho(X)^{k-1} v_0 \neq 0$, and thus $v = \rho(X)^{k-1} v_0$ is a primitive vector.

Q.E.D.

We now have the basic description of an irreducible representation of $\mathfrak{sl}(2, \mathbf{C})$ on a finite-dimensional complex vector space.

Theorem 3.7: Let ρ be an irreducible representation of $\mathfrak{sl}(2, \mathbf{C})$ on a finite-dimensional complex vector space V. Let $v_0 \in V$ be a primitive vector of weight λ for the representation ρ. Then, letting $v_{-1} = 0$, and setting

$$v_n = (1/n!)\rho(Y^n)v_0, \qquad n = 0, 1, \ldots, m, \ldots,$$

one obtains, for $n \geq 0$,

(a) $\rho(H)v_n = (\lambda - 2n)v_n$,
(b) $\rho(Y)v_n = (n+1)v_{n+1}$,
(c) $\rho(X)v_n = (\lambda - n + 1)v_{n-1}$.

Moreover, $\lambda = m$, where $m + 1 = \dim_{\mathbf{C}} V$, and

$$\rho(Y^n)v_0 = 0, \qquad n > m.$$

Proof: (a) asserts that v_n is of weight $\lambda - 2n$, which follows immediately from Lemma 3.5. (b) is clear from the definition of v_n, while (c) follows by induction on n. Namely, for $n = 0$, we have $\rho(X)v_0 = 0$, since v_0 was primitive, and $v_{-1} = 0$. Suppose we know (c) for $n - 1$, then we compute

$$\begin{aligned} n\rho(X)v_n &= \rho(X)\rho(Y)v_{n-1} = \rho(Y)\rho(X)v_{n-1} + \rho([X, Y])v_{n-1} \\ &= (\lambda - n + 2)\rho(Y)v_{n-2} + \rho(H)v_{n-1} \\ &= (\lambda - n + 2)(n-1)v_{n-1} + (\lambda - 2n + 2)v_{n-1} \\ &= n(\lambda - n + 1)v_{n-1}, \end{aligned}$$

and we obtain (c) after dividing by n.

We now show that λ is necessarily an integer. Since V is finite-dimensional, there is an integer $m \geq 0$ such that

$$\begin{aligned} &v_0, \ldots, v_m \qquad \text{are nonzero} \\ &v_{m+1}, \ldots, v_{m+k}, \ldots = 0 \end{aligned}$$

recalling that the nonzero v_j's are eigenvectors of $\rho(H)$ with differing eigenvalues. Now apply (c) to v_{m+1}, obtaining

$$\begin{aligned} 0 = \rho(X)v_{m+1} &= (\lambda - (m+1) + 1)v_m \\ &= (\lambda - m)v_m, \end{aligned}$$

and since $v_m \neq 0$, it follows that $\lambda = m$.

Let V_m be the vector space spanned by $\{v_0, \ldots, v_m\}$. Then we claim that

V_m is invariant under the action of ρ on V. Suppose $v = \sum_{n=0}^{m} \alpha_n v_n$, $\alpha_j \in \mathbf{C}$, then

$$\rho(H)v = \sum_{n=0}^{m} \alpha_n (m - 2n) v_n$$

$$\rho(Y)v = \sum_{n=0}^{m} \alpha_n (n + 1) v_{n+1}$$

$$\rho(X)v = \sum_{n=0}^{m} \alpha_n (m - n + 1) v_{n-1},$$

so $\rho(\mathfrak{sl}(2, \mathbf{C}))V_m \subset V_m$. Thus V_m is a nonzero invariant subspace, and since ρ is assumed irreducible, it follows that $V = V_m$, and that $m + 1 = \dim V$.

Q.E.D.

Remark: We see that the basis $\{v_n\}$ in Theorem 3.7 gives a canonical form for the matrices representing the linear mappings $\rho(H)$, $\rho(Y)$ and $\rho(X)$ acting on V. Namely

$$\rho(H) = \begin{bmatrix} m & 0 & \cdots & 0 \\ 0 & m-2 & & \vdots \\ & \ddots & & \\ \vdots & & \ddots & 0 \\ 0 & \cdots & 0 & -m \end{bmatrix}$$

$$\rho(X) = \begin{bmatrix} 0 & m & & 0 & \cdots & 0 \\ & \ddots & & & & \\ 0 & & \ddots & m-1 & & \vdots \\ \vdots & & & \ddots & \ddots & 1 \\ & & & & \ddots & \\ 0 & & & \cdots & 0 & 0 \end{bmatrix}$$

$$\rho(Y) = \begin{bmatrix} 0 & 0 & & \cdots & & 0 \\ & \ddots & & & & \\ 1 & & \ddots & & & \\ & \ddots & & \ddots & & \vdots \\ 0 & & \ddots & & \ddots & \\ \vdots & & & m-1 & \ddots & 0 \\ 0 & \cdots & & 0 & m & 0 \end{bmatrix}$$

which for $m = 1$ gives the original 2×2 matrices in (3.4), showing that they are in the same canonical form.

Next we see that there is, up to equivalence, only one irreducible representation of dimension $m + 1$. Somewhat later we will describe an explicit example of an $(m + 1)$-dimensional irreducible representation, arising from symmetric tensor products.

Theorem 3.8: Let V be a complex vector space of dimension $m+1$, with $m \geq 0$, and let $\{v_0, \ldots, v_m\}$ be a basis for V. Then define a representation ρ of $\mathfrak{sl}(2, \mathbf{C})$ on V by setting

$$\begin{aligned} &\text{(a)} \quad \rho(H)v_n = (m-2n)v_n, \\ (3.9) \qquad &\text{(b)} \quad \rho(Y)v_n = (n+1)v_{n+1}, \\ &\text{(c)} \quad \rho(X)v_n = (m-n+1)v_{n-1}, \end{aligned}$$

where $n = 0, \ldots, m$, and $v_{-1} = v_{m+1} = 0$. This representation is irreducible, and any irreducible complex representation of dimension $m+1$ is equivalent to this one.

Proof: One checks readily that the mapping ρ: $\mathfrak{sl}(2, \mathbf{C}) \to \operatorname{End}(V)$ given by (3.9) is indeed a representation. Suppose now that V_0 is a nonzero subspace of V invariant under ρ. Then there is an eigenvector of $\rho(H)$ contained in V_0. The list of eigenvectors of $\rho(H)$ in (3.9a) is complete, so V_0 must contain one of the vectors v_k for some k. But then applying (3.9c) to v_k, we see that $v_0 \in V_0$. Then using (3.9b) we see that V_0 must contain $v_n, n = 0, \ldots, m$. Thus $V_0 = V$, and ρ is irreducible. It is clear from Theorem 3.7 that an arbitrary irreducible representation of dimension $m+1$ is equivalent to this one.

Q.E.D.

Corollary 3.9: Suppose ρ: $\mathfrak{sl}(2, \mathbf{C}) \to V$ is an irreducible representation of dimension $m+1, m \geq 0$. Let $\varphi \in V$ be an eigenvector of $\rho(H)$ of weight λ; then there exists a primitive vector of weight $\lambda + 2r$, for some integer $r \geq 0$, so that

$$\varphi = \rho(Y)^r \varphi_0,$$

and where

$$\varphi_0 = \frac{(m-r)!}{m!r!}\rho(X)^r \varphi.$$

Proof: Let $\{v_0, \ldots, v_m\}$ be a basis for V satisfying (3.9) for the given representation ρ. Then we see that for r fixed, $0 \leq r \leq m$, we have

$$\rho(X)v_r = (m-r+1)v_{r-1},$$

$$\rho(X)^2 v_r = (m-r+1)(m-r+2)v_{r-2},$$

etc., and thus

$$\rho(X)^r v_r = (m-r+1)\cdots(m)v_0 = \frac{m!}{(m-r)!}v_0.$$

Then applying the second "ladder operator," we see that

$$\rho(Y)\rho(X)^r v_r = \frac{m!}{(m-r)!}v_1,$$

$$\rho(Y)^2\rho(X)^r v_r = \frac{m!\,2}{(m-r)!}v_2,$$

etc., and thus

$$\rho(Y)^r \rho(X)^r v_r = \frac{m!r!}{(m-r)!} v_r,$$

and thus we obtain the useful identity

$$v_r = \frac{(m-r)!}{m!r!} \rho(Y)^r \rho(X)^r v_r. \tag{3.10}$$

Now suppose that φ is any eigenvector of $\rho(H)$. Then φ is a multiple of one of the eigenvectors $\{v_0, \ldots, v_m\}$ above, say, $\varphi = \alpha v_r$. Then it follows from (3.10) that

$$\varphi = \frac{(m-r)!}{m!r!} \rho(Y)^r \rho(X)^r \varphi,$$

and letting

$$\varphi_0 = \frac{(m-r)!}{m!r!} \rho(X)^r \varphi,$$

we see that φ_0 is primitive, and the corollary is proven.

Q.E.D.

We now introduce a specific representation of $SL(2, \mathbf{C})$ and its derived representation of $\mathfrak{sl}(2, \mathbf{C})$. Consider $\mathbf{C}^2$ as column vectors, and let

$$v_{1,0} = \begin{bmatrix} 1 \\ 0 \end{bmatrix}, \quad v_{1,1} \begin{bmatrix} 0 \\ 1 \end{bmatrix}$$

be standard basis vectors. Then $SL(2, \mathbf{C})$ acts on $\mathbf{C}^2$ by left matrix multiplication, and we call this representation π_1. Then if we consider $S^m(\mathbf{C}^2)$, the m-fold symmetric tensor product of $\mathbf{C}^2$ with itself, we define $\pi_m = S^m(\pi_1)$, where each matrix $\pi_m(g)$ is the multilinear extension of $\pi_1(g) = g$ to $S^m(\mathbf{C}^2)$, and we note that dim $S^m(\mathbf{C}^2) = m + 1$. The representation π_m induces a derived representation $\rho_m = d\pi_m$ of $\mathfrak{sl}(2, \mathbf{C})$ on $S^m(\mathbf{C}^2)$. We note that ρ_1 is simply matrix multiplication on the left by elements of $\mathfrak{sl}(2, \mathbf{C})$ [just as for the Lie group $SL(2, \mathbf{C})$, whereas $\rho_m(g) = d\pi_m(g)$ is the extension of the linear mapping $\rho_1(g)$ to $S^m(\mathbf{C}^2)$ as a *derivation*, which is easy to check.

Thus in particular we obtain the following results:

$$\rho_1(H)v_{1,0} = v_{1,0}, \quad \rho_1(X)v_{1,0} = 0, \quad \rho_1(Y)v_{1,0} = v_{1,1},$$

etc., and this representation satisfies the relations in (3.9) for $m = 1$. Now define

$$v_{m,k} = v_{1,0}^{m-k} v_{1,1}^k, \quad 0 \le k \le m,$$

$m+1$ elements of $S^m(\mathbf{C}^2)$. Then $\{v_{m,k}\}$ is a basis for $S^m(\mathbf{C}^2)$. Moreover, one can compute easily that

$$\begin{aligned}
\rho_m(H)v_{m,k} &= (m-2k)v_{m,k}, \quad 0 \le k \le m, \\
\rho_m(X)v_{m,0} &= 0, \\
\rho_m(Y)v_{m,m} &= 0, \\
\rho_m(X)v_{m,k} &= kv_{m,k-1}, \quad 1 \le k \le m, \\
\rho_m(Y)v_{m,k} &= (m-k)v_{m,k+1}, \quad 0 \le k \le m-1.
\end{aligned}$$

It is clear from these relations that any basis vector $v_{m,k}$ is expressible as powers of $\rho_m(X)$ and $\rho_m(Y)$ acting on $v_{m,0}$, which we see is a primitive vector of weight m. Thus $v_{m,0}$ generates $S^m(\mathbf{C}^2)$ by the action of ρ_m, and thus ρ_m is irreducible, and hence equivalent to the representation in Theorem 3.8. In fact, if we set

$$\varphi_k = \rho_m(Y)^k v_{m,0},$$

we see that

$$\varphi_k = \frac{m!}{(m-k)!} v_{1,0}^{m-k} v_{1,1}^k,$$

from which follows the irreducibility.

Now let us compute the action of w, the Weyl element in $SL(2, \mathbf{C})$, on φ_k. We see that

$$\begin{aligned}\pi_m(w)\varphi_k &= \frac{m!}{(m-k)!}\pi_m(w)(v_{1,0})^{m-k}(v_{1,1})^k \\ &= \frac{m!}{(m-k)!} S^m(\pi_1(w))(v_{1,0})^{m-k}(v_{1,1})^k.\end{aligned}$$

But

$$\pi_1(w)v_{1,0} = iv_{1,1}$$

$$\pi_1(w)v_{1,1} = iv_{1,0},$$

and hence

$$\pi_m(w)\varphi_k = i^m \frac{k!}{(m-k)!}\varphi_{m-k}.$$

Thus we obtain

$$\pi_m(w)\rho_m(Y)^k\varphi_0 = i^m \frac{k!}{(m-k)!}\rho_m(Y)^{m-k}\varphi_0. \tag{3.11}$$

Now we note that the identity (3.11) which involves both the representation of $SL(2, \mathbf{C})$ and $\mathfrak{sl}(2, \mathbf{C})$ was derived from this particular explicit representation, but we see from its form that it will be valid on any irreducible representation of $SL(2, \mathbf{C})$ and $\mathfrak{sl}(2, \mathbf{C})$ on a vector space of dimension $m+1$.

Now consider a specific representation of $\mathfrak{sl}(2, \mathbf{C})$ on the exterior algebra of forms on an Hermitian vector space E. We will use the notation and terminology of Sec. 1. Let E be a fixed Hermitian vector space of complex dimension n, and associate to E the algebra of forms $\wedge F$, and the operators L and L^*. We introduce the notation:

$$\Lambda := L^*$$

$$B := \sum_{p=0}^{2n}(n-p)\Pi_p.$$

We then define a representation

$$\alpha\colon \mathfrak{sl}(2, \mathbf{C}) \longrightarrow \operatorname{End}(\wedge F)$$

by setting

$$\alpha(X) = \Lambda, \quad \alpha(Y) = L, \quad \alpha(H) = B.$$

We see by Proposition 1.1 that α is indeed a representation of $SL(2, \mathbf{C})$, since the commutation relations $[B, L] = -2L$, $[B, \Lambda] = 2\Lambda$, and $[\Lambda, L] = B$ are easy to verify.

Definition 3.10: A p-form $\varphi \in \wedge^p F$ is said to be *primitive* if $\Lambda\varphi = 0$, i.e., if $\alpha(X)\varphi = 0$.

Remark: Recall that $B = \sum_{p=0}^{2n}(n - p)\Pi_p$, and thus any homogeneous form of degree p is an eigenvector of $\alpha(H)$ of weight $n-p$. Hence a primitive p-form is a primitive vector for the representation α of weight $n - p$.

If φ is a primitive p-form, then the action of α generates a subspace $F_\varphi \subset \wedge F$ of dimension $n - p + 1$ on which α acts irreducibly. Moreover, the action of α leaves the real forms $\wedge_{\mathbf{R}} F$ invariant since L, Λ, and B are real operators. The decomposition of $\wedge_{\mathbf{R}} F$ into irreducible components is called the *Lefschetz decomposition* of the exterior algebra, and this is compatible with the decomposition $F = \oplus \wedge^{p,q} F$, since L, Λ, and B are bihomogeneous operators. This is elaborated in the theorems which follow. By Proposition 3.1, we see that α induces a representation of $SL(2, \mathbf{C})$ on $\wedge F$, for which we will use the notation π_α. We can restrict π_α to $SU(2)$, and we observe that $\pi_\alpha|_{SU(2)}$ is unitary, which follows from the fact that $\alpha|_{\mathfrak{su}(2)}$ are skew-Hermitian operators, i.e.,

$$\alpha(iH) = iB, \quad \alpha(i(X + Y)) = i(\Lambda + L), \quad \alpha(X - Y) = \Lambda - L.$$

The following theorems are consequences of the representation theory of $\mathfrak{sl}(2, \mathbf{C})$ for the specific representation α on the Hermitian exterior algebra $\wedge F$. The first results can be proved directly without appealing to representation theory, as is done in Weil [1], but we prefer to use the representation theory as it gives more insight into the major results (cf. Chern [3] and Serre [3]. We can then give Hecht's elegant proof of the fundamental Kähler identities using the language developed here. Let $(x)^+ = \max(x, 0)$.

Theorem 3.11: Let E be an Hermitian vector space of complex dimension n.

(a) If $\varphi \in \wedge^p F$ is a primitive p-form, then $L^q\varphi = 0, q \geq (n - p + 1)^+$.
(b) There are no primitive forms of degree $p > n$.

Proof: Let φ be a primitive p-form, and let F_φ be the subspace of $\wedge F$ generated by the action of $\mathfrak{sl}(2, \mathbf{C})$ on φ by the representation α. Then $\rho(H)\varphi = m\varphi$, where $\dim F_\varphi = m + 1$. But $\rho(H)\varphi = (n - p)\varphi$, so $m = n - p$. Thus $\rho(Y)^q\varphi = L^q\varphi = 0$, for $q \geq (n - p + 1)^+$, by Theorem 3.7. Part (b) is a simple corollary of the fact that $\dim F_\varphi = n - p + 1$.

Q.E.D.

We will refer to the following theorem as the *Lefschetz decomposition theorem* for an Hermitian exterior algebra.

Theorem 3.12: Let E be an Hermitian vector space of complex dimension n, and let $\varphi \in \wedge^p F$ be a p-form, then

(a) One can write φ uniquely in the form

$$\varphi = \sum_{r \geq (p-n)^+} L^r \varphi_r, \tag{3.12}$$

where, for each $r \geq (p-n)^+, \varphi_r$ is a primitive $(p-2r)$-form. Moreover, each φ_r can be expressed in the form

$$\varphi_r = \sum_{r,s} a_{r,s} L^s \Lambda^{r+s} \varphi, \quad a_{r,s} \in \mathbf{Q}. \tag{3.13}$$

(b) If $L^m \varphi = 0$, then the primitive $(p-2r)$-forms φ_r appearing in the decomposition vanish if $r \geq (p-n+m)^+$, i.e.,

$$\varphi = \sum_{r=(p-n)^+}^{(p-n+m)^+} L^r \varphi_r,$$

(c) if $p \leq n$, and $L^{n-p}\varphi = 0$, then $\varphi = 0$.

Proof: The representation space $V = \wedge F$ of the representation α decomposes into a direct sum of irreducible subspaces $V = V_1 \oplus \cdots \oplus V_l$. Let φ be a p-form, then

$$\varphi = \psi^1 + \cdots + \psi^l,$$

$\psi^j \in V_j$. Then each ψ^j is an eigenvector of $\rho(H)$ of weight $n-p$, and hence by Corollary 3.9, we see that

$$\psi^j = L^{r_j} \chi_j,$$

where χ_j is a primitive $(p-2r_j)$-form, and

$$\chi_j = c_j \Lambda^{r_j} \psi^j, \quad c_j \in \mathbf{Q}. \tag{3.14}$$

Collecting the primitive forms of the same degree, we obtain a decomposition of φ of the form

$$\varphi = \sum_{r \geq (n-p)^+} L^r \varphi_r,$$

where each φ_r is primitive of degree $(p-2r)$.

To see that the decomposition is unique, we suppose that

$$0 = \varphi_0 + L\varphi_1 + \cdots + L^m \varphi_m, \tag{3.15}$$

where each φ_j is primitive $j = 0, \ldots, m \geq 1$. We note that it follows from Theorem 3.7 that

$$\Lambda^k L^k \varphi_k = c_k \varphi_k, \quad k = 1, \ldots, m \tag{3.16}$$

for a rational nonzero constant c_k, depending only on p, k and n. Applying Λ^m to (3.15) and using (3.16) we find that

$$0 = \Lambda^m \varphi_0 + \Lambda^{m-1}(\Lambda L)\varphi_1 + \cdots + \Lambda(\Lambda^{m-1}L^{m-1})\varphi_{m-1} + \Lambda^m L^m \varphi_m,$$

which implies immediately that $\varphi_m = 0$, contradicting our assumption that φ_m was primitive. Thus the decomposition (3.12) is unique.

To see that (3.13) holds we proceed in a similar manner. Let the p-form φ have the decomposition

$$\varphi = \varphi_0 + L\varphi_1 + \cdots + L^m \varphi_m,$$

where φ_j are primitive $(p - 2j)$-forms. Then

$$\begin{aligned}\Lambda^m \varphi &= \Lambda^m \varphi_0 + \Lambda^{m-1}(\Lambda L)\varphi_1 + \cdots + \Lambda^m L^m \varphi_m \\ &= 0 + \cdots + 0 + c_m \varphi_m,\end{aligned}$$

and so

$$\varphi_m = (1/c_m)\Lambda^m \varphi.$$

By induction from above, we get formulas of the type (3.13) for each $\varphi_j, j = 0, \ldots, m$.

Parts (b) and (c) follow simply from the uniqueness. Namely, for part (b), we see that

$$0 = L^m \varphi = \sum_{r \geq (p-n)^+} L^{m+r} \varphi_r.$$

Since φ_r is primitive, it follows from Theorem 3.11 that $L^q \varphi_r = 0$ if $q \geq (n - (p - 2r) + 1)^+$, which implies that $L^{r+m}\varphi_r = 0$ if $r < (p - n + m)$. Thus we have

$$0 = \sum_{r \geq (p-n+m)^+} L^{r+m} \varphi_r = \sum_{q \geq (p+2m-n)^+} L^q \varphi_{q-m}.$$

The total degree of each term is $2m + p$, and thus we have a primitive decomposition of the zero form of degree $p + 2m$, from which it follows that $\varphi_{q-m} = 0, q \geq (p + 2m - n)^+$, i.e., $\varphi_r = 0, r \geq (p - n + m)^+$, as desired. Finally, part (c) is a special case of part (b).

Q.E.D.

Corollary 3.13: Let φ be a p-form in $\wedge F$. Then a necessary and sufficient condition that φ be primitive is that both (a) $p \leq n$ and (b) $L^{n-p+1}\varphi = 0$.

This corollary is a simple consequence of the Lefschetz decomposition theorem (Theorem 3.12).

We now want to prove some fundamental results concerning the relationship between the operators $*$, L and Λ which are important in the theory of Kähler manifolds. The development we give here is due to Hecht [1] and differs from the more traditional viewpoint of Weil [1] in that a global representation of both $SL(2, \mathbf{C})$ and $\mathfrak{sl}(2, \mathbf{C})$ on the Hermitian exterior algebra is utilized, leading to some simple ordinary differential equations which simplifies some of the combinatorial arguments found in Weil [1].

Let E now be an Hermitian vector space with fundamental form

$$\boldsymbol{\Omega} = \sum_{\mu=1}^{n} x_\mu \wedge y_\mu,$$

given by (1.9) where $\{x_\mu, y_\mu\}$ is an orthonormal basis for E', as before. Now, if η is any p-form in $\wedge F$, we let

$$e(\eta)\varphi := \eta \wedge \varphi$$

be the operator acting on $\wedge F$ given by wedging with η. If η is a real 1-form, then we check easily that

$$e^*(\eta) = *e(\eta)*, \tag{3.17}$$

and if $\{e_1, \ldots, e_{2n}\}$ is a real oriented orthonormal basis for E', we see from Sec. 1 that

$$\begin{aligned} &e^*(e_{j_1})(e_{j_1} \wedge \cdots \wedge e_{j_k}) = e_{j_2} \wedge \cdots \wedge e_{j_k}, \\ &\text{if} \quad j_1 \notin \{j_2, \ldots, j_k\}, \quad \text{and 0 otherwise.} \end{aligned} \tag{3.18}$$

We note that

$$L = e(\boldsymbol{\Omega}) = \sum_{\mu=1}^{n} e(x_\mu)e(y_\mu),$$

$$\Lambda = e^*(\boldsymbol{\Omega}) = \sum_{\mu=1}^{n} e^*(y_\mu)e^*(x_\mu).$$

It is clear that

$$[L, e(\eta)] = 0, \quad \text{for any } \eta \in \wedge F, \tag{3.19}$$

since $\boldsymbol{\Omega}$ is a 2-form. On the other hand, we claim that

$$\begin{aligned} &\text{(a)} \quad [\Lambda, e(x_\mu)] = e^*(y_\mu), \\ &\text{(b)} \quad [\Lambda, e(y_\mu)] = -e^*(x_\mu), \end{aligned} \tag{3.20}$$

for $\mu = 1, \ldots, n$. We note that (3.20b) follows from (3.20a) by reversing the role of x_μ, y_μ in the definition of the operator L. To see that (3.20a) holds we consider

$$\begin{aligned} [\Lambda, e(x_j)] &= \sum_{\mu=1}^{n} e^*(y_\mu)e^*(x_\mu)e(x_j) - e(x_j)\sum_{\mu=1}^{n} e^*(y_\mu)e^*(x_\mu) \\ &= e^*(y_j)e^*(x_j)e(x_j) - e(x_j)e^*(y_j)e^*(x_j), \end{aligned} \tag{3.21}$$

since $e(x_j)$ commutes with $e^*(x_\mu)$ and $e^*(y_\mu)$ for $\mu \neq j$, which follows readily from (3.18). Now we consider the action of both $[\Lambda, e(x_j)]$ given by (3.21) and $e^*(y_j)$ on monomials, i.e., multiples of products of x_i's and y_i's. Then we see that if ψ is a given form, then

$$\psi = \psi_1 + x_j \wedge \psi_2 + y_j \wedge \psi_3 + x_j \wedge y_j \wedge \psi_4,$$

where ψ_1, ψ_2, ψ_3 and ψ_4 do not contain x_j or y_j or a wedge factor. It follows readily that

$$[\Lambda, e(x_j)]\psi = \psi_3 - x_j \wedge \psi_4$$

and also that

$$e^*(y_j)\psi = \psi_3 - x_j \wedge \psi_4,$$

so (3.20b) follows.

Now suppose that η is a (1, 0)-form. Then

$$\begin{aligned}[\Lambda, e(\eta)] &= -ie^*(\bar{\eta}), \\ [\Lambda, e(\bar{\eta})] &= ie^*(\eta).\end{aligned} \tag{3.22}$$

Moreover, if η is a real 1-form, then

$$[\Lambda, e(\eta)] = -Je^*(\eta)J^{-1}. \tag{3.23}$$

We see that (3.22) follows from (3.20), since it suffices to consider the special case of $\eta = x_j + iy_j$. To see that (3.23) is true, we simply note that any real 1-form can be written in the form $\eta = \varphi + \overline{\varphi}$, where φ is of type (1, 0), and then one checks that

$$\begin{aligned}-ie^*(\bar{\eta}) &= -Je^*(\bar{\eta})J^{-1}, \\ ie^*(\eta) &= -Je^*(\eta)J^{-1}.\end{aligned}$$

With these preparations made, we now want to prove two basic lemmas due to Hecht [1]. We introduce the following operator on $\wedge F$ induced by the action of $SL(2, \mathbf{C})$ on $\wedge F$ by the representation π_α. Let

$$\# = \pi_\alpha(w) = \exp(\tfrac{1}{2}i\pi\alpha(X+Y)) = \exp(\tfrac{1}{2}i\pi(\Lambda + L)).$$

The first lemma shows us that # is closely related to the $*$ operator.

Lemma 3.14: Let η be a real 1-form. Then

$$\#e(\eta)\#^{-1} = -iJe^*(\eta)J^{-1}. \tag{3.24}$$

Proof: We set, for $t \in \mathbf{C}$,

$$\begin{aligned}e_t(\eta) &= \exp(it\alpha(X+Y)) \bullet e(\eta) \bullet \exp(-it\alpha(X+Y)), \\ &= \exp(it[\Lambda + L]) \bullet e(\eta) \bullet \exp(-it(\Lambda + L)),\end{aligned}$$

and we note that $e_{\pi/2}(\eta) = \#e(\eta)\#^{-1}$. We will see that $e_t(\eta)$ satisfies a simple differential equation with initial condition $e_0(\eta) = e(\eta)$, which can be easily solved, and evaluating the solution at $t = \frac{1}{2}\pi$ will give the desired result. First we let

$$ad(X)Y = [X, Y]$$

for operators X and Y. Then one obtains

$$e_t(\eta) = \sum_{k=0}^{\infty}(1/k!)adk[it(\Lambda + L)]e(\eta). \tag{3.25}$$

This follows from the fact that if σ is any representation of $SL(2, \mathbf{C})$ on V, then (cf. (3.8))

$$\sigma(e^A) = e^{d\sigma(A)},$$

i.e., representations commute with the exponential mapping. In this case σ is conjugation by π_α, and $d\sigma$ is given by $ad(\alpha)$ (cf., Helgason [1] or Varadarajan [1]), and $ad\alpha(X+Y) = ad(\Lambda + L)$.

Now $ad^k(\Lambda + L)$ is a sum of monomials in $ad(\Lambda)$ and $ad(L)$. Since $\Lambda L = L\Lambda + B, ad(L)e(\eta) = 0$, and $ad(-B)e(\eta) = e(\eta)$ (since η is of degree 1), we see that $e_t(\eta)$ can be expressed in the form

$$e_t(\eta) = \sum_{t=0}^{\infty} a_k(t) ad^k(\Lambda)e(\eta),$$

where $a_k(t)$ are real-analytic functions in t. Now (3.23) implies that $ad^k(\Lambda)e(\eta) = 0$, for $k \geq 2$, since Λ commutes with J and $e^*(\eta)$. Thus

$$e_t(\eta) = a_0(t)e(\eta) + a_1(t)ad(\Lambda)e(\eta). \tag{3.26}$$

Let $f'(t)$ denote differentiation with respect to t. Then we see, by differentiating (3.25), that $e_t(\eta)$ satisfies the differential equation

$$\begin{aligned} &\text{(a)} \quad e_t'(\eta) = i(ad(\Lambda) + ad(L))e_t(\eta). \\ &\text{(b)} \quad e_0(\eta) = e(\eta). \end{aligned} \tag{3.27}$$

We can solve (3.27) by using (3.26). Namely, we have

$$e_t(\eta) = a_0'(t)e(\eta) + a_1'(t)ad(\Lambda)e(\eta) \tag{3.28}$$

must equal

$$i(ad(\Lambda + L))[a_0(t)e(\eta) + a_1(t)ad(\Lambda)e(\eta)]$$
$$= ia_0(t)ad(\Lambda)e(\eta) + ia_1(t)ad(L)ad(\Lambda)e(\eta),$$

using the fact that $ad^2(\Lambda)e(\eta) = 0$, and $ad(L)e(\eta) = 0$. But

$$\begin{aligned} ad(L)ad(\Lambda)e(\eta) &= ad([L, \Lambda])e(\eta) + ad(\Lambda)ad(L)e(\eta) \\ &= ad(-B)e(\eta) = e(\eta), \end{aligned}$$

and thus (3.28) must equal

$$ia_0(t)ad(\Lambda)e(\eta) + ia_1(t)e(\eta).$$

This will be satisfied if

$$a_0'(t) = ia_1(t),$$
$$a_1'(t) = ia_0(t).$$

Then letting $a_0(t) = \cos t, a_1(t) = i \sin t$, we find that

$$e_t(\eta) = \cos t \; e(\eta) + i \sin t \; ad(\Lambda)e(\eta) \tag{3.29}$$

is the unique solution to (3.27). Letting $t = \frac{1}{2}\pi$ in (3.29) yields

$$e_{\pi/2} = i[\Lambda, e(\eta)],$$

which by (3.23) gives (3.24) as desired.

Q.E.D.

The next lemma shows the precise relationship between $*$ and $\#$ acting on p-forms.

Lemma 3.15: Let $\varphi \in \wedge^p F$, then

$$*\varphi = i^{p^2-n} J^{-1} \# \varphi.$$

Proof: The $*$-operator satisfies

$$*1 = \mathrm{vol} = (1/n!) L^n(1), \tag{3.30}$$

$$*e(\eta) = (-1)^p e^*(\eta) *, \tag{3.31}$$

as an operator on $\wedge^p F$ for any real 1-form η. Relation (3.30) is clear. To see (3.31), let $\varphi \in \wedge^p F$, and write

$$*e(\eta)\varphi = *e(\eta)**^{-1}\varphi = (-1)^{2n-p} e^*(\eta) * \varphi.$$

Now $*$ is the only linear operator on $\wedge F$ satisfying both (3.30) and (3.31), as the forms obtained from 1 by repeated application of $e(\eta)$ span $\wedge F$. Now let

$$\tilde{*} = i^{p^2-n} J^{-1} \#$$

be an operator defined on $\wedge^p F$. We recall from (3.11) that

$$\#\alpha(Y)^k \varphi_0 = i^m \frac{k!}{(m-k)!} \alpha(Y)^{m-k} \varphi_0, \tag{3.32}$$

where φ_0 is primitive of weight m. But $\varphi_0 = 1$ is a primitive 0-form of weight n, so we have, using (3.32) for $k = 0$,

$$\#1 = (i^n/n!) L^n(1).$$

Thus

$$\tilde{*}1 = i^{-n}(i^n/n!) L^n(1) = \mathrm{vol}.$$

Similarly, if $\eta \in \wedge^1_{\mathbf{R}} F$, and $\varphi \in \wedge^p F$, we see that

$$\begin{aligned}
\tilde{*}e(\eta)\varphi &= i^{(p+1)^2-n} J^{-1} \# e(\eta)\varphi, \\
&= i^{p^2-n}(-1)^p i J^{-1} \# e(\eta) \#^{-1} \# \varphi, \\
&= i^{p^2-n}(-1)^p e^*(\eta) J^{-1} \# \varphi, \\
&= (-1)^p e^*(\eta) \tilde{*} \varphi,
\end{aligned}$$

thus verifying (3.31) for $\tilde{*}$. Thus $* = \tilde{*}$.

Q.E.D.

We now have an important relation between $*$ and L^r acting on primitive p-forms (cf., Weil [1]), the proof of which is due to Hecht [1].

Theorem 3.16: Let φ be a primitive p-form in $\wedge^p F$, then

$$*L^r\varphi = (-1)^{p(p+1)/2} \frac{r!}{(n-p-r)!} L^{n-p-r} J\varphi, \quad 0 \le r \le n-p.$$

Proof: Let F_φ be the subspace of $\wedge F$ generated by $\{L^r\varphi\}$, $0 \le r \le n-p$.

Then $\pi_\alpha|_{F_\varphi}$ is an irreducible representation of $SL(2, \mathbf{C})$, and we see by (3.32) that

$$\# L^r \varphi = i^{n-p} \frac{r!}{(n-p-r)!} L^{n-p-r} \varphi.$$

Hence, by Lemma 3.15,

$$\begin{aligned} * L^r \varphi &= i^{(p+2r)^2 - n} J^{-1} \# L^r \varphi \\ &= i^{p^2 - n} J^{-1} i^{n-p} \frac{r!}{(n-p-r)!} L^{n-p-r} \varphi \\ &= i^{p^2 - p} (J^{-1})^2 \frac{r!}{(n-p-r)!} L^{n-p-r} J \varphi \\ &= i^{p^2 - p} (-1)^p \frac{r!}{(n-p-r)!} L^{n-p-r} J \varphi \\ &= (-1)^{p(p+1)/2} \frac{r!}{(n-p-r)!} L^{n-p-r} J \varphi. \end{aligned}$$

Q.E.D.

4. Differential Operators on a Kähler Manifold

Let X be a Hermitian complex manifold with Hermitian metric h. Then there is associated to X and h a fundamental form $\mathbf{\Omega}$, which at each point $x \in X$ is the form of type (1, 1), which is the fundamental form associated as in (1.5) with the Hermitian bilinear form

$$h_x \colon T_x(X) \times T_x(X) \longrightarrow \mathbf{C},$$

given by the Hermitian metric.

Definition 4.1: A Hermitian metric h on X is called a *Kähler metric* if the fundamental form $\mathbf{\Omega}$ associated with h is closed; i.e., $d\mathbf{\Omega} = 0$.

Definition 4.2:

(a) A complex manifold X is said to be of *Kähler type* if it admits at least one Kähler metric.

(b) A complex manifold equipped with a Kähler metric is called a *Kähler manifold.*

We shall see later that not every complex manifold X admits a Kähler metric. On a complex manifold a Hermitian metric can be expressed in local coordinates by a Hermitian symmetric tensor

$$h = \sum h_{\mu\nu}(z) dz_\mu \otimes d\bar{z}_\nu,$$

where $h = \lfloor h_{\mu\nu} \rfloor$ is a positive definite Hermitian symmetric matrix (depending on z); i.e., $h = {}^t\bar{h}$ and ${}^t\bar{u} h u > 0$ for all vectors $u \in \mathbf{C}^n$. The associated fundamental form is then, in this notation,

$$\mathbf{\Omega} = \frac{i}{2}\sum h_{\mu\nu}(z)dz_\mu \wedge d\bar{z}_\nu.$$

In the notation of Chap. III,

$$h_{\mu\nu}(z) = h\left(\frac{\partial}{\partial z_\mu}, \frac{\partial}{\partial z_\nu}\right)(z).$$

Let us first give some examples of Kähler manifolds.

Example 4.3: Let $X = \mathbf{C}^n$ and let $h = \sum_{\mu=1}^n dz_\mu \otimes d\bar{z}_\mu$. Then

$$\mathbf{\Omega} = \frac{i}{2}\sum_{\mu=1}^{n} dz_\mu \wedge d\bar{z}_\mu = \sum_{\mu=1}^{n} dx_\mu \wedge dy_\mu,$$

where $z_\mu = x_\mu + iy_\mu$, $\mu = 1, \ldots, n$, is the usual notation for real and imaginary coordinates. Then, clearly, $d\mathbf{\Omega} = 0$, since $\mathbf{\Omega}$ has constant coefficients, and hence h is a Kähler metric on $\mathbf{C}^n$.

Example 4.4: Let $\omega_1, \ldots, \omega_{2n}$ be $2n$ vectors in $\mathbf{C}^n$ which are linearly independent over $\mathbf{R}$ and let $\mathbf{\Gamma}$ be the lattice consisting of all integral linear combinations of $\{\omega_1, \ldots, \omega_{2n}\}$. The lattice $\mathbf{\Gamma}$ acts in a natural way on $\mathbf{C}^n$ by translation, $z \to \gamma + z$, if $\gamma \in \mathbf{\Gamma}$. Let $X = \mathbf{C}^n/\mathbf{\Gamma}$ be the set of equivalence classes with respect to $\mathbf{\Gamma}$, where we say that z and w are equivalent with respect to $\mathbf{\Gamma}$ if $z = w + \gamma$ for some $\gamma \in \mathbf{\Gamma}$. By giving X the usual quotient topology, we see that X is in a natural manner a complex manifold† and that its universal covering space is $\mathbf{C}^n$. We call X a *complex torus*, and X is homeomorphic to $S^1 \times \cdots \times S^1$, with $2n$-factors. The Kähler metric h on $\mathbf{C}^n$, given above, is *invariant* under the action of $\mathbf{\Gamma}$ on $\mathbf{C}^n$; i.e., if $\gamma \in \mathbf{\Gamma}$ gives a mapping $\gamma\colon \mathbf{C}^n \to \mathbf{C}^n$, then $\gamma^*h = h$, where γ^* is the induced mapping on (covariant) tensors. Because of this invariance, we can find a Hermitian metric $\tilde{h}$ on X so that if $\pi\colon \mathbf{C}^n \to \mathbf{C}^n/\mathbf{\Gamma}$ is the holomorphic projection mapping, then $\pi^*(\tilde{h}) = h$. This is easy to see, and we omit any details here. Moreover, π is a local diffeomorphism, and hence in a neighborhood U of a point $z \in \mathbf{C}^n$, we have $\pi_U := \pi|_U$ is a biholomorphic mapping. Hence $(\pi_U^{-1})^*h|_U = \tilde{h}|_{\pi(U)}$, and similarly for the corresponding $\mathbf{\Omega}$ and $\tilde{\mathbf{\Omega}}$. Since d commutes with $(\pi_U^{-1})^*$, we have

$$d\tilde{\mathbf{\Omega}}|_{\pi(U)} = (\pi_U^{-1})^* d\mathbf{\Omega}|_U = 0.$$

Then $\tilde{h}$ defined on X is a Kähler metric, and all complex tori are then necessarily of Kähler type.

Example 4.5: One of the most important manifolds of Kähler type is $\mathbf{P}_n$. Let $(\xi_0, \ldots, \xi_n)$ be homogeneous coordinates for $\mathbf{P}_n$, and consider the differential form $\tilde{\mathbf{\Omega}}$,

$$\tilde{\mathbf{\Omega}} = \frac{i}{2}\,\frac{|\xi|^2 \sum\limits_{\mu=0}^{n} d\xi_\mu \wedge d\bar{\xi}_\mu - \sum\limits_{\mu,\nu=0}^{n} \bar{\xi}_\mu \xi_\nu d\xi_\mu \wedge d\bar{\xi}_\nu}{|\xi|^4}$$

†See Proposition 5.3 for a proof of this fact.

where we have let $|\xi|^2 = \xi_0^2 + \cdots + \xi_n^2$, as usual. This form is the homogeneous representation for the curvature form of the universal bundle over $\mathbf{P}_n$ with the standard metric on the frame bundle [except for sign; see equation (4.3) in Chap. III]. In particular, then, $\tilde{\mathbf{\Omega}}$ defines a d-closed differential form $\mathbf{\Omega}$ on $\mathbf{P}_n$ of type (1, 1). In terms of local coordinates in a particular coordinate system, for example,

$$w_j = \frac{\xi_j}{\xi_0}, \quad j = 1, \ldots, n,$$

we can write $\mathbf{\Omega}$ as

$$\mathbf{\Omega}(w) = \frac{i}{2} \frac{(1 + |w|^2) \sum_{\mu=1}^{n} dw_\mu \wedge d\bar{w}_\mu - \sum_{\mu,\nu=1}^{n} \bar{w}_\mu w_\nu dw_\mu \wedge d\bar{w}_\nu}{(1 + |w|^2)^2}$$

Thus the associated tensor

$$h = \left(\sum h_{\mu\nu}(w) dw_\mu \otimes d\bar{w}_\nu\right)(1 + |w|^2)^{-2}$$

has for coefficients (ignoring the positive denominator above)

$$h_{\mu\nu}(w) = (1 + |w|^2)\delta_{\mu\nu} - \bar{w}_\mu w_\nu, \quad \mu, \nu = 1, \ldots, n.$$

It is easy to see that $\tilde{h} = [h_{\mu\nu}]$ is Hermitian symmetric and positive definite. In fact, suppose that $u \in \mathbf{C}^n$. Then

$$\begin{aligned} {}^t\bar{u}\tilde{h}u = \sum_{\mu,\nu} h_{\mu\nu} u_\mu \bar{u}_\nu &= \sum_{\mu,\nu} (1 + |w|^2)\delta_{\mu\nu} u_\mu \bar{u}_\nu - \Big(\sum_\mu \bar{w}_\mu u_\mu\Big)\Big(\sum_\nu w_\nu \bar{u}_\nu\Big) \\ &= |u|^2 + |u|^2|w|^2 - (\bar{w}, \bar{u})(w, u), \end{aligned}$$

letting (,) denote the standard inner product in $\mathbf{C}^n$. Hence by Schwarz's inequality we have

$${}^t\bar{u}\tilde{h}u \geq |u|^2,$$

and hence $\tilde{h}$ is positive definite. It then follows that h defines a Hermitian metric on $\mathbf{P}_n$ (which is called the *Fubini-Study metric* classically). Since $\mathbf{\Omega}$ is a closed (1, 1)-form on $\mathbf{P}_n$, as noted earlier, we see that h is, in fact, a Kähler metric. This Kähler metric is invariant with respect to transformations of $\mathbf{P}_n$ induced by unitary transformations of $\mathbf{C}^{n+1} - \{0\}$ onto itself, a property which will not concern us too much but which is important from the point of view of homogeneous spaces.

The next proposition combined with the above basic examples gives many additional examples of Kähler manifolds.

Proposition 4.6: Let X be a Kähler manifold with Kähler metric h and let M be a complex submanifold of X. Then h induces a Kähler metric on M, and with this metric M becomes, therefore, a Kähler manifold.

Proof: Let $j: M \to X$ be the injection mapping. Then $h_M = j^*h$ defines a metric on M, and $j^*\mathbf{\Omega} = \mathbf{\Omega}_M$ is the associated fundamental form to h_M on M. Since $d\,\mathbf{\Omega}_M = j^*d\mathbf{\Omega} = 0$, it is clear that $\mathbf{\Omega}_M$ is also a Kähler fundamental form.

Q.E.D.

In terms of the differential operators d, ∂, and $\bar\partial$ on a Hermitian manifold, we can define the following Laplacian operators,

$$\Delta = dd^* + d^*d$$
$$\Box = \partial\partial^* + \partial^*\partial$$
$$\overline{\Box} = \bar\partial\bar\partial^* + \bar\partial^*\bar\partial,$$

of which the first and last will play an important role in our study of Kähler manifolds later in this chapter. Note that $\overline{\Box}$ is the complex conjugate of the operator $\Box$, thus justifying the notation. What can we say about the relation between the Laplacians Δ and $\overline{\Box}$? In general, not too much,† but on Kähler manifolds there is a striking relationship. Recall that an operator $P: \mathcal{E}^*(X) \to \mathcal{E}^*(X)$ is said to be real if $\overline{P\varphi} = P\bar\varphi$, i.e., $P = \bar P$.

Theorem 4.7: Let X be a Kähler manifold.‡ If the differential operators $d, d^*, \bar\partial, \bar\partial^*, \partial, \partial^*, \Box, \overline{\Box}$, and Δ are defined with respect to the Kähler metric on X, then Δ commutes with $*, d$, and L, and

$$\Delta = 2\Box = 2\overline{\Box}.$$

In particular,

(a) $\Box$ and $\overline{\Box}$ are real operators.
(b) $\Delta|_{\mathcal{E}^{p,q}}: \mathcal{E}^{p,q} \to \mathcal{E}^{p,q}$.

Remark: Neither (a) nor (b) of the above theorem are true in general and these properties will imply topological restrictions on Kähler manifolds, as we shall see in the next section.

To prove Theorem 4.7, we shall first develop some consequences of the representation theory from Sec. 3 as applied to the study of the interaction of the operators d and ∂ and their adjoints. The operators L and L^* will be used as auxiliary tools in this work§ and we shall also use the concept of a *primitive differential form* on a Hermitian complex manifold X. We shall say that $\varphi \in \mathcal{E}^p(X)$ is *primitive* if $L^*\varphi = 0$, and we shall denote by $\mathcal{E}^p_o(X)$ the vector space of primitive p-forms. All the results of Sec. 1 concerning primitive forms on an Hermitian vector space then apply to the primitive differential forms.

We also define the operators

$$d_c = J^{-1}dJ = wJdJ$$
$$d_c^* = J^{-1}d^*J = wJd^*J,$$

†There is a relationship which involves the *torsion tensor;* cf. Chern [2] or Goldberg [1].

‡Note that we do not necessarily assume compactness here. In the noncompact case, we assume that the formal adjoints are given by $\bar\partial^* = -\bar{*}\bar\partial\bar{*}$, etc. (cf. Propositions 2.2 and 2.3), which would be the formal L^2-adjoints for forms with compact support on an open manifold.

§Note that $L^* = w{*}L{*}$ (cf. Sec. 1) can be shown to be identical with the L^2-formal adjoint of the linear operator L [for the *Hodge metric* on $\wedge^*T^*(X)$] in the same way that $\bar\partial^* = -\bar{*}\bar\partial\bar{*}$ is derived as in Proposition 2.3.

a twisted conjugate to d. These are real operators which are useful in applications involving integration and Stokes' theorem, and this is one reason for introducing them. For instance, if we let d_c act on a function φ, we have

$$\begin{aligned} d_c\varphi &= wJdJ\varphi \\ &= (-1)J(\partial\varphi + \bar{\partial}\varphi) \\ &= (-1)(i\partial\varphi - i\bar{\partial}\varphi) \\ &= -i(\partial - \bar{\partial})\varphi, \end{aligned}$$

and we could use this last expression for d_c as a definition. From $d_c = -i(\partial - \bar{\partial})$, it follows immediately that

$$dd_c = 2i\partial\bar{\partial}, \tag{4.1}$$

which is a real operator of type (1, 1) acting on differential forms in $\mathcal{E}^*(X)$.

We now have an important theorem concerning the commutators of these various operators.

Theorem 4.8: Let X be a Kähler manifold. Then

(a) $[L, d] = 0, [L^*, d^*] = 0.$
(b) $[L, d^*] = d_c, [L^*, d] = -d_c^*.$

Proof: Part (a) is simple and follows from the fact that the fundamental form $\mathbf{\Omega}$ on X is closed, the basic Kähler assumption. The second part of (a) is the adjoint form of the first part. Similarly, the second part of (b) is the adjoint statement of the first part, and the first statement holds if and only if the second statement holds. Let us show then that

$$L^*d - dL^* = -J^{-1}d^*J.$$

Letting $L^* = \Lambda$ as before, we see by Proposition 2.3 that

$$d^* = (-1)^{p+1} * d *^{-1}, \quad \text{acting on } p\text{-forms.}$$

Now let φ be a p-form on X; then we find that, from Lemma 3.15,

$$\begin{aligned} \#\varphi &= i^{-p^2+n} J * \varphi, \\ \#^{-1}\varphi &= i^{(2n-p)^2-n} *^{-1} J^{-1}\varphi = i^{p^2-n} *^{-1} J^{-1}\varphi. \end{aligned}$$

Therefore we see that

$$\begin{aligned} \#d\#^{-1}\varphi &= i^{-(2n-p+1)^2+n} i^{p^2-n} J * d *^{-1} J^{-1}\varphi \\ &= i^{-1} i^{2p} J * d *^{-1} J^{-1}\varphi \\ &= iJ[(-1)^{p+1} * d *^{-1}]J^{-1}\varphi \qquad (4.2) \\ &= iJd^*J^{-1}\varphi. \end{aligned}$$

Now let

$$\begin{aligned} d_t &= \exp[it\alpha(X + Y)] \circ d \circ \exp[-it\alpha(X + Y)] \\ &= \exp[it(\Lambda + L)] \circ d \circ \exp[-it(\Lambda + L)]. \end{aligned}$$

Just as in the proof of Lemma 3.14, we have

$$d_t = \sum_{k=0}^{\infty}(1/k!)ad^k[it(\Lambda + L)]d$$

and, since $ad(L)d = 0$, by the Kähler hypothesis,

$$d_t = \sum_{k=0}^{\infty} a_k(t)ad^k(\Lambda)d, \tag{4.3}$$

where $a_k(t)$ are real-analytic functions of t. Now $d_{\pi/2}$ is an operator of degree -1, which implies that

$$d_{\pi/2} = a_1(\pi/2)ad(\Lambda)d, \tag{4.4}$$

since all other terms in the expansion (4.3) are operators of degree $+1, -3, -5, \ldots,$ etc., and the expansion clearly has only a finite number of nonzero terms. But then it follows from (4.2) and (4.4) that iJd^*J^{-1} is proportional to $ad(\Lambda)d$, and hence that $ad^k(\Lambda)d = 0$ for $k \geq 2$, since Λ commutes with d^* and J. Thus

$$d_t = a_0(t)d + a_1(t)ad(\Lambda)d,$$

and just as in the proof of Lemma 3.14, we conclude that

$$d_t = (\cos t)d + i(\sin t)ad(\Lambda)d,$$

and now the theorem follows by letting $t = \pi/2$ and observing that $Jd^*J^{-1} = -J^{-1}d^*J$.

Q.E.D.

Corollary 4.9: Let X be a Kähler manifold. Then

$$[L, d_c] = 0, \quad [L^*, d_c^*] = 0, \quad [L, d_c^*] = -d, \quad \text{and} \quad [L^*, d_c] = d^*.$$

Proof: This follows easily from Theorem 4.8, since the operator J commutes with the real operators, L, L^*, and so $(d_c)_c = -d$ and $(d_c^*)_c = -d^*$.

Q.E.D.

Considering the bidegree structure of the differential forms, we obtain the following corollary to Theorem 4.8.

Corollary 4.10: Let X be a Kähler manifold. Then

$$\begin{aligned} &[L, \partial] = [L, \bar{\partial}] = [L^*, \partial^*] = [L^*, \bar{\partial}^*] = 0 \\ &[L, \partial^*] = i\bar{\partial}, [L, \bar{\partial}^*] = -i\partial \\ &[L^*, \partial] = i\bar{\partial}^*, [L^*, \bar{\partial}] = -i\partial^*. \end{aligned} \tag{4.5}$$

$$\begin{aligned} d^*d_c &= -d_cd^* = d^*Ld^* = -d_cL^*d_c \\ dd_c^* &= -d_c^*d = d_c^*Ld_c^* = -dL^*d \\ \partial\bar{\partial}^* &= -\bar{\partial}^*\partial = -i\bar{\partial}^*L\bar{\partial}^* = -i\partial L^*\partial \\ \bar{\partial}\partial^* &= -\partial^*\bar{\partial} = i\partial^*L\partial^* = i\bar{\partial}L^*\bar{\partial}. \end{aligned} \tag{4.6}$$

Proof: Equations (4.5) follow from Theorem 4.8 by comparing bidegrees and using the fact that $d_c = i(\bar\partial - \partial)$.

To obtain (4.6) we use, for example, $d^* = [L^*, d_c]$ as follows:

$$d^* d_c = L^* d_c d_c - d_c L^* d_c = -d_c L^* d_c$$

$$-d_c d^* = -d_c L^* d_c + d_c d_c L^* = -d_c L^* d_c,$$

and so

$$d^* d_c = -d_c L^* d_c = -d_c d^*.$$

Similarly for the others, setting $d_c = [L, d^*]$, etc.

Q.E.D.

Using the above results, we are now in a position to prove Theorem 4.7 concerning the Laplacians on a Kähler manifold.

Proof of Theorem 4.7: It is clear from the definition of d^* and $\mathbf{\Delta}$ that $\mathbf{\Delta}$ commutes with d and $*$. So we have to see that $L\mathbf{\Delta} - \mathbf{\Delta} L$ vanishes. We have

$$\begin{aligned}\mathbf{\Delta} L - L\mathbf{\Delta} &= dd^*L + d^*dL - Ldd^* - Ld^*d \\ &= dd^*L + d^*Ld - dLd^* - Ld^*d \\ &= -d[L, d^*] - [L, d^*]d,\end{aligned}$$

and substituting, from Theorem 4.8, we obtain

$$\mathbf{\Delta} L - L\mathbf{\Delta} = -dd_c - d_c d.$$

It follows from (4.1) that $dd_c = -d_c d$, since $\partial\bar\partial + \bar\partial\partial = 0$; thus we obtain $\mathbf{\Delta} L - L\mathbf{\Delta} = 0$.

To prove the relationship between $\mathbf{\Delta}$ and the other Laplacians, we write, using Corollary 4.9,

$$\begin{aligned}\mathbf{\Delta} &= dd^* + d^*d = d[L^*, d_c] + [L^*, d_c]d \\ &= dL^*d_c - dd_cL^* + L^*d_cd - d_cL^*d.\end{aligned}$$

Note that all the information about the metric in the operator $\mathbf{\Delta}$ is contained in the operator L^*, since d and d_c depend only on the differentiable and complex structure, respectively. Multiply on the left by J^{-1} and on the right by J; we obtain

$$\mathbf{\Delta}_c = -d_cL^*d + d_cdL^* - L^*dd_c + dL^*d_c.$$

But since $d_c d = -dd_c$, we have that $\mathbf{\Delta} = \mathbf{\Delta}_c$, in a trivial manner.

We now write (noting that $2\partial = d + id_c$, etc.)

$$\begin{aligned}4(\partial\partial^* + \partial^*\partial) &= (d + id_c)\left(d^* - id_c^*\right) \\ &\quad + \left(d^* - id_c^*\right)(d + id_c) \\ &= (dd^* + d^*d) + \left(d_cd_c^* + d_c^*d_c\right) \\ &\quad + i(d_cd^* + d^*d_c) - i\left(dd_c^* + d_c^*d\right).\end{aligned}$$

By (4.6) in Corollary 4.10, we see that the last two parentheses vanish. We also have

$$\mathbf{\Delta}_c = J^{-1}\mathbf{\Delta}J = J^{-1}dd^*J + J^{-1}d^*dJ$$
$$= d_c d_c^* + d_c^* d_c.$$

Therefore we have

$$4\Box = \mathbf{\Delta} + \mathbf{\Delta}_c + 0$$
$$= 2\mathbf{\Delta}.$$

Thus, $2\Box = \mathbf{\Delta}$. The other assertion is proved in a similar manner. The fact that $\mathbf{\Delta}$ is of bidegree (0, 0) follows now trivially from the fact that $\Box$ is of bidegree (0, 0). Similarly, since $\mathbf{\Delta}$ is a real operator, $\Box$ and $\overline{\Box}$ must also be real operators.

Q.E.D.

Corollary 4.11: On a Kähler manifold, the operator $\mathbf{\Delta}$ commutes with J, $L^*, d, \partial, \bar{\partial}, \partial^*$, and d^*.

Since L^* commutes with $\mathbf{\Delta}$ on a Kähler manifold, we have an analogue to Theorem 3.12. On a Kähler manifold X, $\mathbf{\Delta}$-harmonic differential forms are the same as, by Theorem 4.7, $\Box$-harmonic or $\overline{\Box}$-harmonic forms, and we shall say simply *harmonic forms* on X, to be denoted by $\mathcal{H}^r(X)$ and $\mathcal{H}^{p,q}(X)$ as before. We shall denote by $\mathcal{H}_0^r(X)$ and $\mathcal{H}_0^{p,q}(X)$ the *primitive harmonic r-forms* and *(p, q)-forms*, respectively; i.e., $\mathcal{H}_0^r(X)$ is the kernel of the mapping L^*: $\mathcal{H}^r(X) \to \mathcal{H}^{r-2}(X)$ and $\mathcal{H}_0^{p,q}(X)$ is the kernel of the mapping L^*: $\mathcal{H}^{p,q}(X) \to \mathcal{H}^{p-1,q-1}(X)$. These maps are well defined since L^* commutes with $\mathbf{\Delta}$.

Corollary 4.12: On a compact Kähler manifold X there are direct sum decompositions:

$$\mathcal{H}^r(X) = \sum_{s \geq (r-n)^+} L^s \mathcal{H}_0^{r-2s}(X)$$

$$\mathcal{H}^{p,q}(X) = \sum_{s \geq (p+q-n)^+} L^s \mathcal{H}_0^{p-s,q-s}(X).$$

This result follows immediately from the primitive decomposition theorem (Theorem 3.12) and the fact that $\mathbf{\Delta}$ commutes with L and L^*.

Our last corollary to the Lefschetz decomposition theorem is the following result, also due to Lefschetz.

Corollary 4.13: Let X be a compact Kähler manifold, then

$$L^{n-p} = e(\mathbf{\Omega}^{n-p}): H^p(X, \mathbf{C}) \longrightarrow H^{2n-p}(X, \mathbf{C})$$

is an isomorphism, where $\mathbf{\Omega}$ is the Kähler form on X.

Remark: This implies the Poincaré duality theorem (Theorem 2.5) in this context, and is referred to in algebraic geometry as the "strong Lefschetz theorem" (cf., Grothendieck [1]).

Proof: This is an immediate consequence of part (c) of the Lefschetz decomposition theorem (Theorem 3.12), where we represent the cohomology groups by harmonic forms as in Corollary 4.12.

Q.E.D.

Remark: The basic result of this section is Theorem 4.7, and we shall develop its consequences in the next section. The derivation of this result was based on Theorem 4.8 and its corollaries, and this depended in turn on the representation theory of Sec. 3. However, the statement of Theorem 4.7 does not involve the representation theory, and there are alternative methods of deriving Theorem 4.8 (from which then follows Theorem 4.7) which do not involve this concept. One basic approach is the following one. Suppose that

$$\mathbf{\Omega} = \frac{i}{2}\sum h_{\mu v}(z) dz_\mu \wedge d\bar{z}_v$$

is the fundamental form on a Kähler manifold for z near 0 in some appropriate coordinate system. By a linear change of coordinates, one can obtain easily that the matrix $h(z) = [h_{\mu v}(z)]$ is the identity at $z = 0$

$$h_{\mu v}(0) = \delta_{\mu v}$$

or

$$h(z) = I + O(|z|).$$

By using the fact that $d\mathbf{\Omega} = 0$, one finds easily that the coefficient matrix satisfies the differential equations

$$\begin{aligned} &\frac{\partial h_{\mu v}}{\partial z_\lambda}(z) = \frac{\partial h_{\lambda v}}{\partial z_\mu}(z), \qquad \mu, v, \lambda = 1, \ldots, n \\ (4.7) \qquad &\frac{\partial h_{\mu v}}{\partial \bar{z}_\lambda}(z) = \frac{\partial h_{\mu\lambda}}{\partial \bar{z}_v}(z), \qquad \mu, v, \lambda = 1, \ldots, n. \end{aligned}$$

By making a new (quadratic) change of variables of the form

$$\tilde{z}_\mu = z_\mu + \frac{1}{2}\sum_{\alpha,\beta} A^\mu_{\alpha\beta} z_\alpha z_\beta,$$

where $[A^\mu_{\alpha\beta}]$ is a symmetric (in α, β) complex matrix (for fixed μ), one can choose the coefficients $A^\mu_{\alpha\beta}$ [by using the differential equations (4.7)] so that

$$A^\mu_{\alpha\beta} = -\frac{\partial h_{\beta\mu}}{\partial z_\alpha}(0),$$

and it will follow that

$$h(z) = I + O(|z|^2);$$

i.e., all the linear terms in the Taylor expansion of h at 0 vanish. Such a coordinate system is called a *geodesic coordinate system*. At the point 0, one can derive Theorem 4.8 by ignoring the higher-order terms, since in the commutator only first-order derivations of L and L^* will appear. Then one is reduced to proving the commutator relations in $\mathbf{C}^n$ with the canonical Kähler metric as in Example 4.3. This is not difficult but will involve a sort of combinatoric multilinear algebra similar to that developed in Sec. 1 in going back and forth between the real and complex structures.

One can also prove the Lefschetz decomposition theorem for differential forms on a Kähler manifold independent of the representation theory of $\mathfrak{sl}(2, \mathbf{C})$, per se, and then use this to prove the basic Kähler identities. This is the approach followed by Weil [1].

5. The Hodge Decomposition Theorem on Compact Kähler Manifolds

In this section we shall derive the Hodge decomposition theorem for Kähler manifolds and give various applications. Let X be a compact complex manifold. Then we have the de Rham groups on X, $\{H^r(X, \mathbf{C})\}$, represented by d-closed differential forms with complex coefficients, and the Dolbeault groups on $X, \{H^{p,q}(X)\}$, represented by $\bar\partial$-closed (p, q)-forms (Sec. 3 in Chap. II). We have seen that these vector spaces are finite dimensional (Sec. 5 in Chap. IV). Moreover, there is a spectral sequence relating them (Fröhlicher [1]). However, in general, if φ is a $\bar\partial$-closed (p, q)-form on X, then φ need not be d-closed, and, conversely, if ψ is a d-closed r-form on X and $\psi = \psi^{r,0} + \psi^{r-1,1} + \cdots + \psi^{0,r}$ are the bihomogeneous components of ψ, then the components $\psi^{p,q}$ need not be $\bar\partial$-closed. On manifolds of Kähler type, however, such relations are valid, as we see in the following decomposition theorem of Hodge (as amplified by Kodaira).

Theorem 5.1: Let X be a compact complex manifold of Kähler type. Then there is a direct sum decomposition†

$$H^r(X, \mathbf{C}) = \sum_{p+q=r} H^{p,q}(X), \tag{5.1}$$

and, moreover,

$$\bar{H}^{p,q}(X) = H^{q,p}(X). \tag{5.2}$$

Proof: We shall show that

$$\mathcal{H}^r(X) = \sum_{p+q=r} \mathcal{H}^{p,q}(X),$$

and then (5.1) follows immediately. Suppose that $\varphi \in \mathcal{H}^r(X)$. Then $\mathbf{\Delta}\varphi = 0$, but $2\overline{\square} = \mathbf{\Delta}$, by Theorem 4.7, and hence $\overline{\square}\varphi = 0$. But $\varphi = \varphi^{r,0} + \cdots + \varphi^{0,r}$ (writing out the bihomogeneous terms), and, moreover,

$$\overline{\square}\varphi = \overline{\square}\varphi^{r,0} + \cdots + \overline{\square}\varphi^{0,r}.$$

†Strictly speaking, there is an isomorphism $H^r(X, C) \cong \sum_{p+q=r} H^{p,q}(X)$, and it is easy to verify that the isomorphism is independent of the choice of the metric. We shall normally identify $H^{p,q}(X)$ with its image in $H^r(X, \mathbf{C})$ under this isomorphism. When both the Dolbeault groups and the de Rham groups are represented by harmonic forms for the same Kähler metric, then we have strict equality, as we see in the proof of the theorem.

Since $\overline{\square}$ preserves bidegree, we see that $\overline{\square}\varphi = 0$ implies that $\overline{\square}\varphi^{r,0} = \cdots = \overline{\square}\varphi^{0,r} = 0$, and therefore there is a mapping

$$\tau\colon \mathcal{H}^r(X) \longrightarrow \sum_{p+q=r} \mathcal{H}^{p,q}(X)$$

given by $\varphi \longrightarrow (\varphi^{r,0}, \ldots, \varphi^{0,r})$. The mapping is clearly injective, and, moreover, if $\varphi \in \mathcal{H}^{p,q}(X)$, then $\overline{\square}\varphi = \frac{1}{2}\mathbf{\Delta}\varphi = 0$, which implies that $\varphi \in \mathcal{H}^{p+q}(X)$, and thus τ is surjective, proving (4.1).

Assertion (5.2) follows immediately from the fact that $\overline{\square}$ is real (Theorem 4.7) and that conjugation is an isomorphism from $\mathcal{E}^{p,q}(X)$ to $\mathcal{E}^{q,p}(X)$.

Q.E.D.

Remark: One can also prove Theorem 5.1 by showing that the spectral sequence relating the Dolbeault and de Rham groups degenerates at the E_1 term (see Fröhlicher [1] and the appendix to Griffiths [4]). This proof also makes heavy use of the differential operators $\mathbf{\Delta}, \overline{\square}$ and the harmonic representation of the de Rham and Dolbeault groups. This approach, via spectral sequences, deserves mention because there are examples, namely K-3 surfaces† where one does not know (yet) whether they are Kähler in general or not.‡ However, one can show by other means that the spectral sequence degenerates, and one still obtains a Hodge decomposition, and this in turn is useful in the study of the moduli problem for K-3 surfaces.

As a consequence of the Hodge decomposition theorem, we have the following relations for the Betti numbers and Hodge numbers of a Kähler manifold. Recall that we set (see Sec. 5 in Chap. IV)

$$b_r(X) = \dim_{\mathbf{C}} H^r(X, \mathbf{C}),\, h^{p,q}(X) = \dim_{\mathbf{C}} H^{p,q}(X).$$

Corollary 5.2: Let X be a compact Kähler manifold. Then

(a) $b_r(X) = \sum_{p+q=r} h^{p,q}(X)$.
(b) $h^{p,q}(X) = h^{q,p}(X)$.
(c) $b_q(X)$ is even for q odd.
(d) $h^{1,0}(X) = \frac{1}{2}b_1(X)$ is a topological invariant.

These results are a simple consequence of the preceding theorem. We shall see shortly that there are examples of compact complex manifolds X which violate property (c), and hence such manifolds are not of Kähler type. Thus Corollary 5.2 places topological restrictions on a compact complex manifold admitting a Kähler metric. We already know that any such manifold always admits a Hermitian metric.

†A K-3 surface is a compact complex manifold X of complex dimension 2 such that (a) $H^1(X, \mathcal{O}_X) = 0$ and (b) $\wedge^2 T^*(X) = K$, the *canonical bundle* of X, is trivial; see, e.g., Kodaira [3] and Šafarevič [1], Chap. 9.

‡In 1983 Yum-Tong Siu showed that K-3 surfaces are Kähler (Y.-T. Siu, "Every K-3 Surface is Kähler," *Invent Math.* **73** (1983), no. 1, pp. 139–150.

The simplest example of a non-Kähler compact complex manifold is given by a *Hopf surface*, which we shall now construct. First, we recall one of the basic ways of constructing compact complex manifolds in general, namely, "dividing a given manifold by a group of automorphisms," an example being the complex tori considered in Sec. 4. Let X be a complex manifold and let $\mathbf{\Gamma}$ be a subgroup of the group of *automorphisms* of X (an *automorphism of* X is a biholomorphic self-mapping of X onto itself). We say that $\mathbf{\Gamma}$ is *properly discontinuous* if for any two compact sets $K_1, K_2 \subset X, \gamma(K_1) \cap K_2 \neq \varnothing$, for only a finite number of elements $\gamma \in \mathbf{\Gamma}$. The group $\mathbf{\Gamma}$ is said to have *no fixed points* if for each $\gamma \in \mathbf{\Gamma} - \{e\}$ (e $=$ identity in $\mathbf{\Gamma}$) $\gamma(x) \neq x$ for all $x \in X$. We let $X/\mathbf{\Gamma}$ be the set of equivalence classes with respect to the action of the group $\mathbf{\Gamma}$; i.e., x and $y \in X$ are equivalent (with respect to $\mathbf{\Gamma}$) if $x = \gamma(y)$ for some $\gamma \in \mathbf{\Gamma}$. Let $X/\mathbf{\Gamma}$ have the natural quotient topology given as follows: A basis for the open sets in $X/\mathbf{\Gamma}$ is given by the projection of the open sets in X under the natural projection mapping $\pi: X \to X/\mathbf{\Gamma}$.

Proposition 5.3: Let X be a complex manifold and let $\mathbf{\Gamma}$ be a properly discontinuous group of automorphisms of X without fixed points. Then $X/\mathbf{\Gamma}$ is a Hausdorff topological space which can be given uniquely a complex structure, so that the natural projection mapping $\pi: X \to X/\mathbf{\Gamma}$ is a holomorphic mapping, which is locally biholomorphic.

Proof: Let N be any compact neighborhood of a point $x_0 \in X$. Then there exists only finitely many elements $\gamma \in \mathbf{\Gamma}$ so that $\gamma(x_0) \in N$. This follows immediately from the definition of properly discontinuous, letting K_1 and $K_2 = N$. Thus for each point $x_0 \in X$ there exists an open neighborhood N_0 so that $\gamma(N_0) \cap N_0 = \varnothing$ for all $\gamma \in \mathbf{\Gamma} - \{e\}$. Then, clearly, $\gamma(N_0)$ will be a neighborhood of $\gamma(x_0)$ with the same property; i.e., $\gamma(N_0)$ is the only translate of N_0 by $\mathbf{\Gamma}$ that meets $\gamma(N_0)$. Let $y_0 = \pi(x_0)$. Then, clearly, $W_0 = \pi(N_0) = \pi(\cup_{\gamma\in\mathbf{\Gamma}}\gamma(N_0))$ is a neighborhood of y_0. If $y_1 \neq y_0$ is a second point in $X/\mathbf{\Gamma}$, then letting x_1 be any point in $\pi^{-1}(y_1)$, we can find a neighborhood N_1 of x_1 so that: (a) $\gamma(X_1) \notin N_1$ for all $\gamma \in \mathbf{\Gamma} - \{e\}$ and (b), $N_1 \cap \gamma(N_0) = \varnothing$, for all $\gamma \in \mathbf{\Gamma}$. Thus $\pi(\cup_{\gamma\in\mathbf{\Gamma}}\gamma(N_1))$ is an open neighborhood W_1 of y_1 which does not intersect W_0, and hence $X/\mathbf{\Gamma}$ is Hausdorff. We can use these neighborhoods as coordinate charts near the point y_0. Namely, N_0 is homeomorphic to W_0 under π since $\pi|_{N_0}$ is one-to-one, open, and continuous. Moreover, if W_0 and W_1 are two such coordinate systems near y_0 and y_1 and $W_0 \cap W_1 \neq \varnothing$, then there exists a $\gamma \in \mathbf{\Gamma}$ so that the corresponding $\gamma(N_0) \cap N_1 \neq \varnothing$, and thus the overlap transformation will be of the form

$$\gamma|_{N_0\cap\gamma^{-1}(N_1)}: N_0 \cap \gamma^{-1}(N_1) \longrightarrow \gamma(N_0) \cap N_1,$$

which is a biholomorphic mapping. Hence we have a complex structure, and the mapping π is clearly holomorphic and locally biholomorphic. The uniqueness is easy to verify, and we omit the proof.

Q.E.D.

Classical examples of complex manifolds constructed in this manner are

(a) Riemann surfaces of genus $g = 1$ (elliptic curves), where $X = \mathbf{C}$ and $\mathbf{\Gamma}$ is a two-dimensional lattice in $\mathbf{C}$ generated by two periods independent over $\mathbf{R}$.

(b) Complex tori (see Example 4.4). (a) is a complex torus of one complex dimension.

(c) Riemann surfaces of genus $g > 1$, where $X =$ unit disc in $\mathbf{C}$ and $\mathbf{\Gamma}$ is a properly discontinuous fixed point free subgroup of the group of automorphisms of X (which are all fractional-linear transformations of the unit disc onto itself, i.e., Möbius transformations).

Remark: If we omitted the assumption that $\mathbf{\Gamma}$ had no fixed points in Proposition 5.3, then $X/\mathbf{\Gamma}$ can still be given a complex structure as a complex space (a generalization of a complex manifold) with singularities at the image of the fixed points (see Cartan [1] for a proof of this).

To construct an example of a Hopf surface, we proceed as follows. Consider the 3-sphere S^3 defined by $\{z = (z_1, z_2) \in \mathbf{C}^2\colon |z_1|^2 + |z_2|^2 = 1\}$, and then we observe that there is a diffeomorphism

$$f\colon S^3 \times \mathbf{R} \xrightarrow{\cong} \mathbf{C}^2 - \{0\}$$

given by

$$f(z_1, z_2, t) = (e^t z_1, e^t z_2)$$

for $(z_1, z_2) \in S^3 \subset \mathbf{C}^2, t \in \mathbf{R}$ (i.e., we are shrinking and expanding S^3 in $\mathbf{C}^2$ by the parameter t exponentially). The infinite cyclic group $\mathbf{Z}$ acts on $S^3 \times \mathbf{R}$ in a natural manner, namely,

$$(z_1, z_2, t) \longrightarrow (z_1, z_2, t + m), \quad \text{for } m \in \mathbf{Z},$$

and it is clear that the quotient space under this action (defined as above) $(S^3 \times \mathbf{R})/\mathbf{Z}$ is diffeomorphic to $S^3 \times S^1$. Under the diffeomorphism f we can transfer the action of $\mathbf{Z}$ on $S^3 \times \mathbf{R}$ to an action of $\mathbf{Z}$ on $\mathbf{C}^2 - \{0\}$. Namely,

$$(z_1, z_2, m) \to (e^m z_1, e^m z_2)$$

for $(z_1, z_2) \in \mathbf{C}^2 - \{0\}$ and $m \in \mathbf{Z}$. Moreover, for a fixed $m \in \mathbf{Z}$, the mapping above is an automorphism of $\mathbf{C}^2 - \{0\}$. Thus the action of $\mathbf{Z}$ on $\mathbf{C}^2 - \{0\}$ is the action of a subgroup $\mathbf{\Gamma}$ of $\operatorname{Aut}(\mathbf{C}^2 - \{0\})$, which, it is easy to check, is properly discontinuous without fixed points (the orbit of a point under $\mathbf{\Gamma}$ is a discrete sequence of points with limits at 0 and ∞). Since the action of the groups $\mathbf{Z}$ and $\mathbf{\Gamma}$ commutes with the diffeomorphism, we have the commutative diagram

$$\begin{array}{ccc}
S^3 \times \mathbf{R} & \xrightarrow{f} & \mathbf{C}^2 - \{0\} \\
\downarrow & & \downarrow \\
(S^3 \times \mathbf{R})/\mathbf{Z} & \xrightarrow{f} & (\mathbf{C}^2 - \{0\})/\mathbf{\Gamma} \\
\| \wr & & \| \wr \\
S^3 \times S^1 & & \mathbf{X},
\end{array}$$

where the vertical arrows are the natural projections. By Proposition 5.3 we see that X is a complex manifold which is diffeomorphic under $\tilde{f}$ to $S^3 \times S^1$ (and this is compact). An integral basis for the homology of $S^3 \times S^1$ is given by the factors S^1, S^3 in those dimensions, and we have the Betti numbers

$$b_0(X) = b_1(X) = b_3(X) = b_4(X) = 1$$

$$b_2(X) = 0.$$

In particular, $b_1(X) = 1$, and hence X cannot be Kähler, since odd degree Betti numbers must be even on Kähler manifolds. A deep result of Kodaira [4] asserts that any compact complex manifold which is homeomorphic to $S^1 \times S^3$ is of the form $(\mathbf{C}^2 - \{0\})/\mathbf{\Gamma}$ for some appropriate $\mathbf{\Gamma}$ chosen in a manner similar to that of our example. Such manifolds are called *Hopf surfaces*.

We would like to give one last important example of Kähler manifolds.

Theorem 5.4: Every complex manifold X of complex dimension 1 (a *Riemann surface*) is of Kähler type.

Proof: Let g be an arbitrary Hermitian metric on X. Then it suffices to show that this metric is indeed a Kähler metric. But this is trivial, since the associated fundamental form $\mathbf{\Omega}$ is of type (1, 1) and therefore of total degree 2 on X. Since X has two real dimensions, it follows that $d\mathbf{\Omega} = 0$, since there are no forms of higher degree.

Q.E.D.

Suppose that X is a compact Riemann surface. Then we have, by the Hodge decomposition theorem for Kähler manifolds,

$$H^1(X, \mathbf{C}) = H^{1,0}(X) \oplus H^{0,1}(X).$$

Moreover, $h^{1,0}(X) = h^{0,1}(X)$, and hence $2h^{1,0}(X) = b_1(X)$. Thus $h^{1,0}(X)$ is a topological invariant of X, called the *genus* of the Riemann surface, usually denoted by g.

6. The Hodge-Riemann Bilinear Relations on a Kähler Manifold

In this section we want to study the structure of the de Rham groups on a Kähler manifold. If X is a Kähler manifold, then the fundamental form $\mathbf{\Omega}$ on X determines the *Lefschetz decomposition*,

$$H^r(X, \mathbf{C}) = \sum_{s \geq (r-n)^+} L^s H_o^{r-2s}(X, \mathbf{C}), \tag{6.1}$$

where $H_o^r(X, \mathbf{C})$ is the vector space of primitive cohomology classes of degree r. This follows immediately from the harmonic forms representation of the de Rham group and Corollary 4.12. Since we represent the cohomology ring $H^*(X, C)$ by differential forms, we shall write $\xi \wedge \eta$ for the product of two cohomology classes, where we mean by this the following: If $\varphi, \psi \in Z^*(X, \mathbf{C})$, the d-closed differential forms, and $[\varphi]$, $[\psi]$ are the classes of φ and ψ in $H^*(X, \mathbf{C})$, then $[\varphi] \wedge [\psi]$ is defined by $[\varphi \wedge \psi]$, and it is easy to verify that this cohomology product is well defined and, moreover, satisfies $\xi \wedge \eta = (-1)^{\deg \xi \cdot \deg \eta} \eta \wedge \xi$. If $\mathbf{\Omega}$ is the fundamental form on X, let $\omega = [\mathbf{\Omega}]$; then we define

$$L: H^*(X, \mathbf{C}) \to H^*(X, \mathbf{C})$$

by $L(\xi) = \omega \wedge \xi$. Thus the Kähler structure on X determines the linear mapping L on cohomology. However, the mapping L depends only on the class ω and not on the differential form representing it (nor on the metric inducing $\mathbf{\Omega}$); any cohomologous differential form would give the same result. The existence of a Kähler metric therefore implies the existence of a linear mapping $L: H^*(X, \mathbf{C}) \to H^*(X, \mathbf{C})$, which is *real*, i.e., L is actually defined on $H^*(X, \mathbf{R})$, and, moreover, the above Lefschetz decomposition (5.1) holds. The *primitive cohomology classes* $H_o^r(X, \mathbf{C}) \subset H^r(X, \mathbf{C})$ are those satisfying $L^{n-r+1}\xi = 0$, as before. The point we wish to make here is that the existence of L and of the decomposition (6.1) is a topological necessity that a (say, differentiable or topological) manifold admit a Kähler complex structure. This is analogous to and related to the requirement that odd degree Betti numbers must be even for Kähler manifolds.

Suppose that such an L exists on a compact oriented differentiable manifold of real dimension $2n$, i.e.,

$$L: H^r(X, \mathbf{R}) \to H^{r+2}(X, \mathbf{R}), \quad r = 0, \dots, 2n-2$$

and

$$H^r(X, \mathbf{R}) = \sum_{s \geq (r-n)^+} L^s H_o^{r-2s}(X, \mathbf{R}),$$

where $H_o^p(X, \mathbf{R})$ is the kernel of the mapping

$$L^{n-p+1}: H^p(X, \mathbf{R}) \to H^{2n-p+2}(X, \mathbf{R}), \quad p \leq n,$$

and L is extended to the complexification by linearity. We want to introduce a bilinear form on $H^r(X, \mathbf{R})$ as follows: For $\xi, \eta \in H^r(X, \mathbf{R})$, we let

$$Q(\xi, \eta) = \sum_{s \geq (r-n)^+} (-1)^{[r(r+1)/2]+s} \int_X L^{n-r+2s}(\xi_s \wedge \eta_s), \tag{6.2}$$

where $\xi = \sum L^s \xi_s$ and $\eta = \sum L^s \eta_s$ are the primitive decompositions of ξ and η, respectively. In the case where X is a Kähler manifold, the quadratic form above is well defined by the fundamental form $\mathbf{\Omega}$. However, we do not assume for the present that X is Kähler to emphasize the topological nature of the quadratic form Q above. Such a quadratic form was first introduced by Lefschetz in the context of a projective algebraic variety and then reinterpreted in the same context (for a projective algebraic manifold) by Hodge for de Rham cohomology represented by harmonic differential forms. The quadratic form Q is a sort of intersection matrix for cycles in X, and the signs reflect the decomposition induced by L. As we shall see, Q will have many important properties and applications, but first we want to discuss it from an intuitive and geometric point of view. Suppose that $\dim_{\mathbf{R}} X = 2$. Then we have, for $H^1(X, \mathbf{R})$,

$$Q(\xi, \eta) = -\int_X \xi \wedge \eta,$$

since $H^1(X, \mathbf{R}) = H_0^1(\mathbf{X}, \mathbf{R})$ and $n = r = 1$. Thus, if $\{\xi_\alpha\}$ is a real basis for $H^1(X, \mathbf{R})$ and if we let $\{\hat{\xi}_\alpha\}$ be a dual basis for $H^1(X, \mathbf{R})* \cong H_1(\mathbf{X}, \mathbf{R})$, then $\{\hat{\xi}_\alpha\}$ can be represented by geometric 1-cycles on X, which in turn can be represented by an algebraic sum of oriented closed curves $\mathbf{\Gamma}_\alpha$ on X. Then the matrix $Q(\xi_\alpha, \xi_\beta) = Q_{\alpha\beta}$ can be represented by (and is the same as) the *intersection matrix* $(\mathbf{\Gamma}_\alpha \cdot \mathbf{\Gamma}_\beta)$, which is defined by $\mathbf{\Gamma}_\alpha \cdot \mathbf{\Gamma}_\beta =$ the algebraic sum of the number of intersections of $\mathbf{\Gamma}_\alpha$ with $\mathbf{\Gamma}_\beta$, assuming that they are in general position, meeting only in a finite number of points. The sign of the intersection number is given by whether the local orientation of the intersecting curves agrees or disagrees with the orientation of X. This was, in fact, the context in which Lefschetz worked (see Lefschetz [1] or Hodge [1], where higher-dimensional intersections are also considered). The interaction between the two points of view is very important (especially in algebraic geometry), but in this book we shall restrict ourselves primarily to a discussion of the cohomology groups $H^*(X, \mathbf{C})$, defined by differential forms, and deduce what we can from the existence of a Kähler metric and other considerations.

Suppose now that X is a compact Kähler manifold with fundamental form $\mathbf{\Omega}$ and that we have the Lefschetz decomposition as given by (6.1) and the quadratic form Q defined by (6.2), which we extend to $H^r(X, \mathbf{C})$ by complex-linearity. Since X is a Kähler manifold, there is a bigrading on $H^*(X, \mathbf{C})$ induced by the complex structure; i.e.,

$$H^r(X, \mathbf{C}) = \sum_{p+q=r} H^{p,q}(X),$$

given by the Hodge decomposition, Theorem 5.1. The linear operator $J = \sum_{p,q} i^{p-q} \Pi_{p,q}$ is well defined on $H^r(X, \mathbf{C})$, where $\Pi_{p,q}$ denotes projections onto $H^{p,q}(X)$ (*cf.* Sec. 1). Then we have the following theorem.

Theorem 6.1: Let X be a compact Kähler manifold with fundamental form $\mathbf{\Omega}$ and with the associated quadratic form Q defined by (5.2). Then Q is a nondegenerate real bilinear form with the following properties: If ξ and $\eta \in H^r(X, \mathbf{C})$, then

(a) $Q(\xi, \eta) = (-1)^r Q(\eta, \xi)$.
(b) $Q(J\xi, J\eta) = Q(\xi, \eta)$.
(c) $Q(\xi, J\eta) = Q(\eta, J\xi)$.
(d) $Q(\xi, J\bar{\xi}) > 0$, if $\xi \neq 0$.

Proof: Property (a) is obvious from the definition of Q. Property (d) has as a consequence that the quadratic form Q is nondegenerate, since $Q(\xi, J\bar{\eta})$ is the composition of the bilinear form Q with two isomorphisms of $H^r(X, \mathbf{C})$ onto itself. In a matrix representation of this composition we would have the product of the matrices, and since the composition is a positive definite Hermitian symmetric form, it must have a nonzero determinant. Thus Q must have a nonzero determinant with respect to some basis and hence is nonsingular.

To show property (d), we note that we can rewrite

$$Q(\xi, J\eta) = \sum_{s \geq (r-n)^+} c_s \int_X L^s \xi_s \wedge *L^s \eta_s,$$

where the $\{c_s\}$ are positive constants. This follows from Theorem 3.16. Namely, in this case we have (recall that degree $\eta_s = r - 2s$)

$$*L^s \eta_s = (-1)^{[r(r+1)/2]+s} c_{r,s} L^{n-r+s} J\eta_s,$$

where $c_{r,s}$ is a positive constant. Thus we obtain, with $c_s > 0$,

$$Q(\xi, J\bar{\xi}) = \sum_{s \geq (r-n)^+} c_s \int_X L^s \xi_s \wedge *L^s \bar{\xi}_s,$$

and this is > 0 since $\xi \neq 0$ implies at least one of the $L^s \xi_s \neq 0$ and hence the sum is positive, by the positive definite nature of the Hodge inner product.

The proofs for properties (b) and (c) are similar and will be omitted.

Q.E.D.

Property (a) in Theorem 6.1 tells us that Q is either symmetric or skew-symmetric depending on whether Q is acting on cohomology of even or odd degree. It is well known from linear algebra that there are canonical forms for such quadratic forms. Namely, for r odd, there exists a basis $\{\xi_\alpha\}$ for $H^r(X, \mathbf{R})$ so that if we let $Q(\xi_\alpha, \xi_\beta) = Q_{\alpha\beta}$, then the matrix $[Q_{\alpha\beta}]$ has the form

$$[Q_{\alpha\beta}] = \begin{bmatrix} 0 & I_g \\ -I_g & 0 \end{bmatrix}, \tag{6.3}$$

where $g = \frac{1}{2} b_r(X)$ and I_g is the $g \times g$ identity matrix [note that it is necessary that $b_r(X)$ be even in this case]. Similarly, if r is even, then there is a basis $\{\xi_\alpha\}$ of $H^r(X, \mathbf{R})$ so that

$$[Q_{\alpha\beta}] = \begin{bmatrix} I_h & 0 \\ 0 & -I_k \end{bmatrix}, \tag{6.4}$$

and $h - k$ is the *signature* of the quadratic form Q.

Our next results will show that the subspaces of $H^r(X, \mathbf{C})$ on which Q is positive or negative definite are very much related to the bigrading of $H^r(X, \mathbf{C})$ given by the Hodge decomposition. First, however, we want to discuss the distinction between primitive and nonprimitive cohomology classes. We shall be interested primarily in the de Rham groups $H^r(X, \mathbf{C})$ for $r \leq n$, since by Poincaré duality the vector spaces $H^{2n-r}(X, \mathbf{C})$ for $r < n$, are conjugate-linearly isomorphic to $H^r(X, \mathbf{C})$ and, in effect, do not contain any new information.

Let us compute the primitive cohomology of a simple space, $X = \mathbf{P}_n(\mathbf{C})$. We have seen before that

$$b_0(\mathbf{P}_n) = \cdots = b_{2n}(\mathbf{P}_n) = 1$$

$$b_1(\mathbf{P}_n) = \cdots = b_{2n-1}(\mathbf{P}_n) = 0,$$

as is most easily seen by a cell decomposition, and the generators of the homology groups are given by $0 = \mathbf{P}_0 \subset \cdots \subset \mathbf{P}_n$. The cohomology groups $H^*(\mathbf{P}_n, \mathbf{C})$ have as basis elements $1, \omega, \omega^2, \ldots, \omega^n$, where $\mathbf{\Omega}$ is the fundamental form of $\mathbf{P}_n$ and $[\mathbf{\Omega}] = \omega$ is the class of $\mathbf{\Omega}$ in $H^2(\mathbf{P}_n, \mathbf{C})$; i.e.,

$$H*(X, \mathbf{C}) = \mathbf{C} \oplus \mathbf{C}\omega \oplus \mathbf{C}\omega^2 \oplus \cdots \oplus \mathbf{C}\omega^n,$$

where $\mathbf{C}\omega^m$ represents the complex vector space spanned by the (m, m) class ω^m in $H^{2m}(X, \mathbf{C})$. We claim that the only primitive cohomology classes in $H^*(\mathbf{P}_n, \mathbf{C})$ are the constants, i.e., $H_o^0(\mathbf{P}_n, \mathbf{C}) \cong \mathbf{C}$, $H_o^{2m}(\mathbf{P}_n, \mathbf{C}) = 0, m = 1, \ldots, n$. This follows from the fact that ω is *not* primitive, since

$$\omega^{n-2r+1} \wedge \omega^r = \omega^{n-r+1} \neq 0 \quad \text{if} \quad r \geq 1.$$

Thus, in a very easy case, all of the cohomology is determined by primitive cohomology (the constants) and the fundamental form. In general, on a compact Kähler manifold a nonprimitive cohomology class ξ is of the form

$$\xi = \xi_0 + \omega \wedge \xi_1 + \cdots + \omega^m \wedge \xi_m,$$

where the ξ_j are primitive cohomology classes and ω is the fundamental class, and some $\xi_j \neq 0$ for $j > 0$. How large is $H*_o(X, \mathbf{C})$ in general? Let $b_o^j = \dim_{\mathbf{C}} H_o^j(X, \mathbf{C})$. Then we have the following proposition.

Proposition 6.2: Let X be a compact Kähler manifold. Then

$$b_o^r(X) = \dim H_o^r(X, \mathbf{C}) = b_r(X) - b_{r-2}(X)$$

for $r \leq n$.

Before we give the proof, we note that for projective space we get the right answer, since $b_r - b_{r-2} = 0$ for $r \geq 1$. Similarly, another simple example (which follows from Proposition 6.2) would be cohomology of degree 2 on a Kähler manifold X, and we see that in this case $b_o^0 = b_0, b_o^1 = b_1$, and $b_o^2 = b_2 - b_0 = b_2 - 1$. Moreover, if ω is the fundamental class on X, then ω is of type (1, 1), and hence we have

$$\begin{aligned} H^2(X, \mathbf{C}) &= H^{2,0}(X) \oplus H^{1,1}(X) \oplus H^{0,2}(X) \\ &= H_o^{2,0}(X) \oplus H_o^{1,1}(X) \oplus \mathbf{C}\omega \oplus H_o^{0,2}(X), \end{aligned}$$

noting that, by dimension considerations, we have $H^{2,0}(X) = H_o^{2,0}(X)$ and $H^{0,2}(X) = H_o^{0,2}(X)$; i.e., all of the nonprimitive cohomology is in the middle and is one-dimensional.

Geometrically, what this means is the following. If X is a smooth complex submanifold of $\mathbf{P}_n$ (and hence Kähler), then there are many cycles on X of the form $X \cap \mathbf{P}_j, j = 0, 1, \ldots, n-1$, where $\mathbf{P}_0 \subset \cdots \subset \mathbf{P}_n$ is the cell decomposition of $\mathbf{P}_n$ (assuming that the intersecting manifold X and $\mathbf{P}_j$ are in general position). This determines *part* of the homology of X; the remainder of the homology, which is not so determined, is the *primitive* part (or as Lefschetz called them, *effective* cycles). In the case of a complex surface $X \subset \mathbf{P}_3$, then, $X \cap \mathbf{P}_2$ is (generically) a real two-dimensional closed submanifold, which is a cycle in $H_2(X, \mathbf{Z})$, which corresponds in cohomology to $\omega \in H^2(X, \mathbf{Z})$ (since ω is, in this case, an integral cohomology class). Again, we shall not formally prove this correspondence; we merely mention it as motivation for the discussion at hand.

Proof of Proposition 6.2: The proposition is clearly true for $r = 0, 1$, and so we shall prove it by induction for general r. Suppose that $b_o^q = b_q - b_{q-2}$ for $q = 0, \ldots, r-1$. Then let $\{\xi_i^{(j)}\}$ be a basis for $H_o^{r-2j}(X, \mathbf{C})$, $i = 1, \ldots, b_{r-2j} - b_{r-2j-2}$, $j = 1, \ldots, [r/2]$($b_q = 0$, for $q < 0$, by definition), and consider the set $\{L^j \xi_i^{(j)}\}$ of classes in $H^r(X, \mathbf{C})$. We claim that these vectors are linearly independent. Suppose that

$$\sum_{ij} \alpha_{ij} L^j \xi_i^{(j)} = 0, \quad \alpha_{ij} \in \mathbf{C},$$

Then we have

$$0 = \sum_j L^j \Big(\sum_i \alpha_{ij} \xi_i^{(j)}\Big),$$

and by the uniqueness of primitive decomposition, we obtain $\sum_i \alpha_{ij} \xi_i^{(j)} = 0$, $j = 1, \ldots, [r/2]$. By the linear independence of the $\{\xi_i^{(j)}\}$ in $H_o^{r-2j}(X, \mathbf{C})$, we see that $\alpha_{ij} = 0$ for all i and j. We claim now that none of the vectors of the form $L^j \xi_i^{(j)}$ can be primitive in $H^r(X, \mathbf{C})$. To show this, suppose that $\xi \in H_o^{r-2j}(X, \mathbf{C})$, and, moreover, suppose that $L^j \xi$ is primitive; i.e., $L^{n-r+1}(L^j \xi) = 0$. Then it follows from Theorem 3.12 that ξ must be zero. Suppose that $\{\eta_1, \ldots, \eta_m\}$ is a basis for $H_o^r(X, \mathbf{C})$. Then it follows from the above remark that the vectors $\{\eta_1, \ldots, \eta_m, L^j \xi_i^{(j)}\}$ are linearly independent in $H^r(X, \mathbf{C})$. By the primitive decomposition theorem, they clearly span $H^r(X, \mathbf{C})$, and hence

$$\begin{aligned} b_o^r = m &= b_r - \{(b_{r-2} - b_{r-4}) + (b_{r-4} - b_{r-6}) + \cdots\} \\ &= b_r - b_{r-2}. \end{aligned}$$

Q.E.D.

It is interesting to note that although the primitive cohomology is defined via the fundamental class ω, the dimensions $b_o^j(X)$ are topological invariants of X and independent of the fundamental class ω (of course, for $j \le n$).†

We would now like to discuss the restriction of the quadratic form Q for a compact Kähler manifold X to subspaces of $H^r(X, \mathbf{C})$. For reasons which will become apparent, we shall want to consider Q restricted to the primitive cohomology $H_0^r(X, \mathbf{C})$. We have the following important theorem, due to

†The same proof shows that for the Hodge numbers $h^{p,q}$ we have $h_o^{p,q} = h^{p,q} - h^{p-1,q-1}$.

Hodge, which generalizes a theorem of Riemann for the case $n = r = 1$ (in which case primitive cohomology coincides with cohomology).

To simplify the notation we let $P^n(X, \mathbf{C}) = H_o^n(X, \mathbf{C})$ and $P^{p,q}(X) = H_o^{p,q}(X, \mathbf{C})$ denote primitive cohomology, and by definition $P_n(X, \mathbf{C})$, etc., will be the dual *primitive homology groups* (the effective cycles of Lefschetz).

Theorem 6.3: Let X be a compact Kähler manifold, let $P^r(X, \mathbf{C}) = \sum_{p+q=r} P^{p,q}(X)$ be the primitive cohomology on $X, r = 0, \ldots, 2n$, and let Q be the quadratic form on $P^r(X, \mathbf{C})$ given by (6.2). Then

(a) $Q(P^{r-q,q}, P^{s,r-s}) = 0 (q \neq s)$.
(b) $i^{-r}(-1)^q Q(P^{r-q,q}, \bar{P}^{r-q,q}) > 0$.

Here (a) means $Q(\xi, \eta) = 0$ for $\xi \in P^{r-q,q}$ and $\eta \in P^{s,r-s}$, and (b) means that

$$i^{-r}(-1)^q Q(\xi, \bar{\xi}) > 0, \quad \text{for all nonzero } \xi \in P^{r-q,q}.$$

Proof: First we observe that Q restricted to $P^r(X, \mathbf{C})$ has a simpler form, namely,

$$Q(\xi, \eta) = (-1)^{r(r+1)/2} \int_X L^{n-r} \xi \wedge \eta, \quad \xi, \eta \in P^r(X, \mathbf{C}) \tag{6.5}$$

and as in the proof of Theorem 6.1, we have

$$*\eta = (-1)^{r(r+1)/2} c_0^{-1} L^{n-r} J\eta, \quad c_0 > 0,$$

as given by Theorem 3.16. Substituting in, we find that

$$Q(\xi, \eta) = c_0 i^{b-a} \int_X \xi \wedge *\eta$$

if $\eta \in P^{a,b}$. Now, for part (a), suppose that $\xi \in P^{r-q,q}$ and $\eta \in P^{s,r-s}$, $q \neq s$. Then we have

$$Q(\xi, \eta) = c_0 i^{r-2s} \int_X \xi \wedge \bar{*}\bar{\eta},$$

and ξ and $\bar{\eta}$ have different bidegrees, by assumption, and so, by Proposition 2.2, $Q(\xi, \eta) = 0$. Similarly, if $\xi \in P^{r-q,q}$ and $\xi \neq 0$, then we see that $i^{2q-r} = i^{-r}(-1)^q$, and thus

$$i^{-r}(-1)^q Q(\xi, \bar{\xi}) = c_0 \int_X \xi \wedge *\bar{\xi} > 0.$$

Q.E.D.

We shall call the relations in Theorem 5.3 (a) and (b) the *Hodge-Riemann bilinear relations*. These play an important role in the study of the moduli of algebraic manifolds (cf. Griffiths [1], [3]). They are the natural generalization of the Riemann period matrix of a Riemann surface or of an abelian variety (cf. Sec. VI.4). These topics will be discussed briefly in the remainder of this section in connection with the general moduli problem for compact complex manifolds. The reason we restrict our attention to primitive cohomology in

Theorem 6.3 is that the corresponding quadratic form in (b) for the full Dolbeault group $H^{p,q}(X)$ contained in the full de Rham group $H^r(X, \mathbf{C})$ is an Hermitian symmetric form which is nondegenerate, but it is no longer positive definite, in general (cf. Hodge [1]). Since the primitive cohomology generates the full cohomology by means of the fundamental class ω, there is no essential loss of information.

Remark: If X is a compact Kähler manifold of *even* complex dimension $2m$, then one can use the above type of considerations to show that the *signature* of the underlying topological manifold, which is the same as the signature of the quadratic form

$$A(\xi, \eta) = \int_X \xi \wedge \eta, \quad \xi, \eta \in H^{2m}(X, \mathbf{R}),$$

can be computed in terms of the Hodge numbers $h^{p,q}(X)$. More precisely, one has

$$\sigma(X) = \sum_{p,q} (-1)^p h^{p,q}(X) = \sum_{p=q(2)} (-1)^p h^{p,q}(X),$$

where $\sigma(X)$, the signature of X, is the difference between the number of positive and negative eigenvalues of the (symmetric, nondegenerate) quadratic form A, and, as is well known in algebraic topology, is a topological invariant of such a real $4m$-dimensional oriented topological manifold (see, e.g., Hirzebruch [1]). For more details see Weil [1], p. 78.

Let X be a compact Kähler manifold and consider the Hodge decomposition of the primitive cohomology group of degree r,

$$P^r(X, \mathbf{C}) = \sum_{p+q=r} P^{p,q}(X).$$

Then we have the subspace relation

$$P^{p,q}(X) \subset P^r(X, \mathbf{C}),$$

and we note that Theorem 6.3 imposes restrictions that subspaces be of this form. Let $\varphi = \{\varphi^1, \ldots, \varphi^h\}$ be a basis for $P^{p,q}(X)$, where $h = h_o^{p,q}(X)$, and let $\hat{\gamma} = \{\hat{\gamma}_1, \ldots, \hat{\gamma}_b\}$ be a basis for $P^r(X, \mathbf{R})$ with dual (real) basis $\gamma = \{\gamma_1, \ldots, \gamma_b\}$ for $P_r(X, \mathbf{R})$. For instance, we can choose the basis $\hat{\gamma}$ so that Q in terms of this basis has the canonical form (6.3) or (6.4) depending on the parity of r, but this is not necessary for our discussion here. We can express φ^α in terms of the basis $\hat{\gamma}$, namely,

$$\varphi^\alpha = \sum_{\sigma=1}^{b} \omega_{\alpha\sigma} \hat{\gamma}_\sigma$$

and we can integrate this relationship over the cycles $\{\gamma_\rho\}$, obtaining

$$\int_{\gamma_\rho} \varphi^\alpha = \sum_\sigma \omega_{\alpha\sigma} \int_{\gamma_\rho} \hat{\gamma}_\sigma = \omega_{\alpha\rho}$$

since γ and $\hat{\gamma}$ are dual bases and the duality pairing is given (via de Rham's theorem) by integration of differential forms over cycles. Thus we have a matrix

$$\mathbf{\Omega} = (\omega_{\alpha\rho}) = \begin{bmatrix} \int_{\gamma_1} \varphi^1 \cdots \int_{\gamma_b} \varphi^1 \\ \cdot \quad\quad\quad \cdot \\ \cdot \quad\quad\quad \cdot \\ \cdot \quad\quad\quad \cdot \\ \int_{\gamma_1} \varphi^h \cdots \int_{\gamma_b} \varphi^h \end{bmatrix},$$

which we call the *period matrix* of the differential forms $\{\varphi^\alpha\}$ with respect to the cycles $\{\gamma_\rho\}$. It is clear that $\mathbf{\Omega}$ is an $h \times b$ matrix of maximal rank. We can now express the Hodge-Riemann bilinear relations in terms of this matrix representation for the subspace relation $P^{p,q} \subset H^r$. Namely, we see that

$$\begin{aligned} Q(\varphi^\alpha, \varphi^\beta) &= Q\Big(\sum_\sigma \omega_{\alpha\sigma}\hat{\gamma}_\sigma, \sum_\tau \omega_{\beta\tau}\hat{\gamma}_\tau\Big) \\ &= \sum_{\sigma,\tau} \omega_{\alpha\sigma} Q_{\sigma\tau}\omega_{\beta\tau} = 0, \end{aligned}$$

and, similarly,

$$i^{-r}(-1)^q Q(\varphi^\alpha, \bar{\varphi}^\beta) = i^{-r}(-1)^q \sum \omega_{\alpha\sigma} Q_{\sigma\tau}\bar{\omega}_{\beta\tau} > 0,$$

which can be expressed in the form (letting Q denote the matrix $[Q_{\sigma\tau}]$)

(a) $\mathbf{\Omega} Q^t \mathbf{\Omega} = 0$.
(b) $i^{-r}(-1)^q \mathbf{\Omega} Q^t \bar{\mathbf{\Omega}} > 0$.

The bilinear relations above were first written down in this form by Riemann for periods of holomorphic 1-forms (abelian differentials) on a Riemann surface (Riemann [1]). If we make a change of basis for $P^{p,q}(X)$, then we get another period matrix $\tilde{\mathbf{\Omega}}$ which is related to the original $\mathbf{\Omega}$ by the relation $\tilde{\mathbf{\Omega}} = A\mathbf{\Omega}$ for $A \in GL(h, \mathbf{C})$.

If we consider the Grassmannian manifold

$$G_h(P^r(X, \mathbf{C})),$$

then the subspace relation $P^{p,q}(X) \subset P^r(X, \mathbf{C})$ defines a point in the above Grassmannian manifold. We thus have the association

$$X \longrightarrow \mathbf{\Phi}(X) = (P^{p,q}(X) \subset P^r(X, \mathbf{C}) \in G_h(P^r(X, \mathbf{C})),$$

where $\mathbf{\Phi}(X)$ is, by definition, the associated point in the Grassmannian, given by the subspace relation. We call $\mathbf{\Phi}$ the *period mapping* since the image point $\mathbf{\Phi}(X)$ can be represented by periods of integrals as above. The choice of basis $\{\hat{\gamma}_1, \ldots, \hat{\gamma}_b\}$ gives us

$$G_h(P^r(X, \mathbf{C})) \cong G_h(\mathbf{C}^b) = G_{h,b}(\mathbf{C}),$$

and the choice of basis $\{\varphi_1, \ldots, \varphi_h\}$ gives us an $h \times b$ matrix (the period matrix) $\mathbf{\Omega} \in M_{h,b}(\mathbf{C})$, which is mapped onto the corresponding point in the Grassmannian via the canonical projection mapping

$$M_{h,b}(\mathbf{C}) \xrightarrow{\pi} G_{h,b}(\mathbf{C})$$

(see Chap. 1). The invariant description of the period mapping given above is due to Griffiths. If the complex structure on X is allowed to vary in some manner (for a fixed cup product operator L on a fixed topological manifold X_{top}), then the subspace $P^{p,q}(X) \subset P^r(X, \mathbf{C})$ will vary, although the primitive de Rham group remains fixed. Thus the variation of the Hodge group $P^{p,q}$ in P^r is a reflection of the variation of the complex structure on the underlying topological manifold X_{top}.† We refer to this generally as a *variation of Hodge structure*, and Griffiths has introduced a formulation for making this variation of Hodge structure precise and in many instances a true measure of the variation of complex structures (see Griffiths [1], where he introduces the period mapping, and his survey article [3], which contains an up-to-date bibliography of the very active work in this field as well as a long list of conjectures and problems).

We shall introduce here what we shall call a *Griffiths domain*, which is a classifying space for Hodge structures and which is chosen in such a manner that an a priori holomorphic variation of complex structures induces a holomorphic mapping into the Griffiths domain (a subset of an appropriate Grassmannian-type domain manifold generalizing the classical upper half-plane and Siegel's upper half-space).

Let X be a Kähler manifold as above and let

$$P^r(X) = \sum_{p+q=r} P^{p,q}(X)$$

be the Hodge decomposition for primitive cohomology. Then we define

$$F^s(X) = P^{r,0}(X) + \cdots + P^{r-s,s}(X), \quad s \leq r,$$

and we see that

$$F^1 \subset F^2 \subset \cdots \subset F^r = P^r$$

and we call $\{F^s\}$ the *Hodge filtration* of the primitive de Rham group P^r.‡ Then let $f^s = \dim_{\mathbf{C}} F^s$, $\sigma = [(r-1)/2]$, and $f = (f^0, \ldots, f^\sigma)$ ([ ] denotes greatest integer). We consider the flag manifold $F(f, W)$, where $W = P^r(X, \mathbf{C})$; i.e., a point in $F(f, W)$ (called a *flag*) is by definition a sequence of subspaces

$$F^0 \subset F^1 \subset \cdots \subset F^\sigma \subset W,$$

where

$$\dim_{\mathbf{C}} F^j = f^j.$$

Thus $F(f, W)$ is a natural generalization of a Grassmannian $G_h(W)$, which is the flag manifold for $\sigma = 0$ (which is the case if $r = 1$, for instance). The detailed construction of a flag manifold is analogous to that of a Grassmannian, and we omit any details here. Now, to a Kähler manifold X we can

†The above discussion works equally well for nonprimitive cohomology, i.e., considering $H^{p,q}(X) \subset H^r(X, \mathbf{C})$ as a point in a different Grassmannian. The period relations which will play a role later are defined only for primitive cohomology, and hence the restriction. However, by the Lefschetz decomposition theorem, there is no loss of information.

‡One can also define the Hodge filtration of the full de Rham group in the same manner.

associate the integers $f^0, \ldots, f^\sigma$ coming from the Hodge filtration, and there is then a mapping defined,

$$X \longrightarrow F^0(X) \subset F^1(X) \subset \cdots \subset F^\sigma(X) \subset P^r(X, \mathbf{C}),$$

which we then write as

$$\Phi(X) \in F(f, W).$$

This is Griffiths' period mapping (Griffiths [1]).

Let $X \xrightarrow{\pi} T$ be a proper surjective holomorphic mapping of maximal rank from a complex manifold X to a complex manifold T. Then $X \xrightarrow{\pi} T$ is called a *complex-analytic family* of compact complex manifolds. Let $X_t = \pi^{-1}(t)$. Then X_t is the *fibre over* t, or the compact complex submanifold of X corresponding to the parameter $t \in T$. A basic fact about such families is the following proposition asserting that they are locally differentiably trivial.

Proposition 6.4: If $t_o \in T$, then there exists a neighborhood U of t_o in T and a fibre preserving diffeomorphism

$$f\colon X_{t_0} \times U \longrightarrow \pi^{-1}(U). \tag{6.6}$$

Proof: This is a local problem in the parameter space T, and so let T be an open set in $\mathbf{C}^k$ and let $t_0 = 0$ be the origin assumed to be in T. Then we have coordinates $(t_1, \ldots, t_k)$ for points in T, and by the implicit function theorem, if $p \in X_0 = \pi^{-1}(0)$, it follows that we can find a neighborhood U_p and a biholomorphic mapping

$$\psi_p\colon U_p \longrightarrow U'_p \underset{\text{open}}{\subset} \mathbf{C}^n \times \mathbf{C}^k,$$

with

$$\psi_p|_{U_p \cap X_t} \xrightarrow{\cong} U'_p \cap \mathbf{C}^n \times \{t\};$$

i.e., the fibres of the family in this coordinate system are given by [$t =$ constant], where $(z, t) \in U'_p$, $z \in \mathbf{C}^n$, $t \in \mathbf{C}^k$. In other words, near p, the family is holomorphically trivial (= to a product family). We can find a finite covering $\{U_\alpha\}$ of a neighborhood of X_0 in X by such coordinate systems, and we denote the coordinates for U_α by (z^α, t). The transition functions from (z^α, t) coordinates to (z^β, t) coordinates are of the form

$$\begin{bmatrix} f_{\alpha\beta}(z, t) & 0 \\ 0 & 1 \end{bmatrix},$$

where $f_{\alpha\beta}(z, t)$ is an $n \times n$ complex matrix of holomorphic functions. By using a partition of unity we can piece together the usual Euclidean metric in each coordinate system to obtain a global Hermitian metric h, which, expressed in one of the above coordinate systems, has the form (in real coordinates)

$$h = \sum g_{ij}(x, s) dx_i \otimes dx_j + \sum h_{i\nu}(x, s) dx_i \otimes ds_\nu + \cdots,$$

where

$$z_j = x_j + ix_{j+n}, \quad j = 1, \ldots, n$$

$$t_j = s_j + is_{j+k}, \quad j = 1, \ldots, k,$$

and where (g_{ij}) is a real positive definite matrix and $z_j = z_j^\alpha$ (dropping the notational dependence on α). Consider a curve of the form (in U^α)

$$\gamma_{p,t}(\tau) = (z(\tau), \tau t),$$

depending on the parameters (p, t), where $p = (z, 0)$ is a point on $X_0 \cap U_\alpha$. We require that

(a) $\gamma_{p,t}(0) = (z, 0)$.
(b) The curve $\gamma_{p,t}$ be orthogonal to $X_{\tau t}$ with respect to the metric h at $\gamma_{p,t}(\tau)$.

Note that the nature of the parameterization and the coordinate system ensures us that the curve intersects $X_{\tau t}$ precisely at the point $\gamma_{p,t}(\tau)$. Condition (b) can be rewritten as the system of ordinary differential equations

$$\sum_{j=1}^{2n} g_{ij}(x(\tau), \tau s)x_j'(\tau) + \sum_{\nu=1}^{2k} h_{i\nu}(x(\tau), \tau s)s_\nu = 0, \qquad i = 1, \ldots, 2n.$$

It follows that this nonlinear system of equations satisfies a Lipschitz condition (it is quasilinear) such that the standard existence, uniqueness, and parameter dependence theorems for ordinary differential equations hold, and thus there is a unique curve associated to each parameter point (p, t), and we define

$$f(p, t) = \gamma_{p,t}(1)$$

and obtain a mapping

$$f\colon X_0 \times T \longrightarrow X,$$

which is (for $|t|$ small) an injective differentiable mapping. Moreover, the differential of this mapping at points of X_0 is readily seen to be invertible, and thus the mapping

$$f\colon X_0 \times \{|t| < \epsilon\} \longrightarrow X|_{|t|<\epsilon}$$

is a diffeomorphism for ϵ sufficiently small.

Q.E.D.

Remark: The above result clearly does not depend on the complex structures.

Proposition 6.4 tells us, in particular, that all the fibres X_t for t near t_0 are diffeomorphic. Then we can consider $f^{-1}(X_t)$ as inducing possibly different complex structures on the same differentiable manifold $(X_{t_0})_{\text{diff}}$. This is the point of view of deformation theory, introduced in the general context by Kodaira and Spencer in 1958 and begun by Riemann in his study of the number of moduli necessary to parameterize the different complex structures on a Riemann surface. The recent book by Morrow and Kodaira [1] gives a good introduction to deformation theory along with many examples, and we refer the reader to this reference as well as the original papers of Kodaira and Spencer [1, 2]. One of Griffiths' objects in introducing the period mapping above was to obtain a representation for the variation of complex structure (in the sense of deformation theory) in terms of the variation of Hodge structure. To describe this mapping, we need some auxiliary results from deformation theory, which we shall now describe.

Proposition 6.5: Let $X \xrightarrow{\pi} T$ be a complex-analytic family and let $h_t^{p,q} = h_t^{p,q}(X_t)$. Then $h_t^{p,q}$ is an upper semicontinuous function of the parameter t; moreover, $h_t^{p,q} \leq h_{t_0}^{p,q}$, $t_0 \in T$ and t near t_0.

Proof: This is a local result. Let $T \subset \mathbf{C}^k$ and $t_0 = 0 \in \mathbf{C}^k$. We first use (6.6) to get a diffeomorphism

$$f_t \colon X_t \longrightarrow X_0,$$

which induces a differentiable vector bundle isomorphism,

$$f_t^* \colon \wedge^r T^*(X_0) \longrightarrow \wedge^r T^*(X_t).$$

The almost complex structure J_t acting on $T(X_t)$ induces an almost complex structure $\tilde{J}_t$ on $T(X_0)$, via f_t, and hence a projection

$$\Pi_{p,q,t} \colon \wedge^* T^*(X_0) \longrightarrow \wedge_t^{p,q} T^*(X_0),$$

which is maximal rank for $t = 0$ and thus for t near 0. Therefore the diagram

$$\begin{array}{ccc} \wedge^r T^*(X_0) & \longrightarrow & \wedge_t^{p,q} T^*(X_0) \cong \wedge^{p,q} T^*(X_t) \\ \uparrow & \overset{\mu}{\underset{\cong}{\nearrow}} & \\ \wedge^{p,q} T^*(X_0) & & \end{array}$$

induces an isomorphism μ for t sufficiently small. Thus we have the operator $\bar{\partial}$ on X_t acting on the complex $\wedge^{p,q} T^*(X_t)$, induces via μ, the complex

$$\longrightarrow \mathcal{E}^{p,q}(X_0) \xrightarrow{\bar{\partial}_t} \mathcal{E}^{p,q+1}(X_0) \longrightarrow,$$

where $\bar{\partial}_0 = \bar{\partial}$ and the operator $\bar{\partial}_t$ depends continuously on the parameter t. The proposition now follows from Theorem 4.13 and Sec. 5 in Chap. IV.

Q.E.D.

Corollary 6.6: Suppose that $X \xrightarrow{\pi} T$ is a complex-analytic family such that T is connected and X_t is Kähler for $t \in T$. Then $h_t^{p,q} = h_{t_0}^{p,q}$ for some fixed $t_0 \in T$; i.e., $h_t^{p,q}$ is constant on T.

Proof: By Corollary 4.2 we know that $\sum_{p+q=r} h_t^{p,q} = b_{r,t}$, but since all the fibres are diffeomorphic, $b_{r,t} = b_{r,t_0} = b_r$. Thus for $|t - t_0| < \delta$, we have $h_t^{p,q} \leq h_{t_0}^{p,q}$, and therefore

$$b_r = \sum_{p+q=r} h_t^{p,q} \leq \sum_{p+q=r} h_{t_0}^{p,q} = b_r.$$

If for some $p, q, h_t^{p,q} < h_{t_0}^{p,q}$, for $|t - t_0| < \delta$, then we would have a contradiction.

Q.E.D.

If now $X \xrightarrow{\pi} T$ is a complex-analytic family of Kähler manifolds (e.g., a family of projective algebraic submanifolds, parameterized by varying the coefficients of the defining homogeneous equations) and T is connected, then for all fibres X_t in the family we have the same Hodge numbers $h^{p,q}$ and hence the same primitive Hodge numbers $h_0^{p,q}$, and finally the same Hodge filtration numbers $f^s, 0 \le s \le \sigma = [r/2]$. Thus for this family we may define the flag manifold

$$F(f^0, \ldots, f^\sigma, W), \quad W = P^r(X_{t_0}, \mathbf{C}),$$

and we see that the mapping

$$\Phi: T \longrightarrow F(f, W) \tag{6.7}$$

given by

$$\Phi(t) = \Phi(X_t) = [F^0(X_t) \subset \ldots \subset F^\sigma(X_t) \subset W]$$

is well defined.

Theorem 6.7 (Griffiths): The period mapping (6.7) is a holomorphic mapping.

Remark: The proof of this theorem depends principally on the Kodaira-Spencer deformation theory formalism (Kodaira and Spencer [1]), which we do not develop here (see e.g., Morrow and Kodaira [1]). In fact, Griffiths shows many more properties of the period mapping such as the nature of the curvature of certain natural metrics restricted to $\Phi(T)$, or that $\Phi(T)$ is a locally closed analytic subvariety of $F(f, W)$, etc. (see Griffiths [2, 6]). He also gives conditions (verifiable in many examples) such that if $\Phi(t_1) \neq \Phi(t_2)$, then the two complex manifolds X_{t_1} and X_{t_2} are not biholomorphically equivalent. In other words, the period mapping is a description (sometimes complete) of the variation of the complex structure.

If Q is the fundamental quadratic form defined on $P^r(X, \mathbf{C})$ (6.5), then let

$$X \subset F(f^0, \ldots, f^\sigma, W) = F(f, W)$$

be defined by the set of flags in $F(f, W)$ satisfying the *first bilinear condition*

$$Q(F^s/F^{s-1}, F^s/F^{s-1}) = 0, \tag{6.8}$$

where F^s/F^{s-1} is defined to be a subspace of $F^s \subset W$ by defining

$$F^s/F^{s-1} = \{v \in F^s\colon Q(v, F^{s-1}) = 0\}$$

(note that Q is nondegenerate). Then let $D \subset X$ be the set of flags in $F(f, W)$ satisfying in addition to (6.8) the *second bilinear condition*

$$i^{-r}(-1)^s Q(F^s/F^{s-1}, \bar{F}^s/\bar{F}^{s-1}) > 0. \tag{6.9}$$

One proves that X is a compact projective algebraic manifold and that D is an open subset of X. Both are homogeneous spaces, with natural invariant metrics. We call such a domain D a *Griffiths domain*.† Because of Theorem 6.3, it follows that

$$\Phi(T) \subset D \subset X \subset F(f, W).$$

Moreover, there is a natural fibering of D (because of the homogeneous structure) as a real-analytic family of compact complex submanifolds of D, as in Example 6.8 below (possibly zero-dimensional, as in the classical case), and Griffiths obtains an *infinitesimal period relation* which asserts that the mapping Φ is transversal to the fibres in the real-analytic fibering mentioned above.

We mention two examples of Griffiths domains.

Example 6.8: Let $r = 1$. Then $\sigma = 0, F^0 = H^{1,0}(X)$, and the flag manifold $F(f, W)$ becomes

$$F(f, W) = G_{h,2h}(\mathbf{C}),$$

and letting Q be in standard form (6.3),

$$Q = \begin{bmatrix} 0 & I_h \\ -I_h & 0 \end{bmatrix},$$

we see that X and D are defined in terms of the "homogeneous coordinates" for $G_{h,2h}(\mathbf{C})$,

$$X = \{\boldsymbol{\Omega} \in M_{h,2h}(\mathbf{C}) \colon \boldsymbol{\Omega} Q\, {}^t\boldsymbol{\Omega} = 0\}$$

$$D = \{\boldsymbol{\Omega} \in M_{h,2h}(\mathbf{C}) \colon \boldsymbol{\Omega} Q\, {}^t\boldsymbol{\Omega} = 0, -i\boldsymbol{\Omega} Q\, {}^t\bar{\boldsymbol{\Omega}} > 0\}.$$

This Griffiths domain D is biholomorphically equivalent to Siegel's upper half-space (see Griffiths [1]),

$$D_s = \{Z \in \mathfrak{M}_{h,h}(\mathbf{C}) \colon \mathbf{Z} = {}^t\mathbf{Z}, \operatorname{Im} \mathbf{Z} > 0\},$$

which is itself a generalization of the classic upper half-plane ($h = 1$) (see Siegel [1]). D can also be expressed in the homogeneous space form

$$D = Sp(h)/U(h)$$

where $Sp(h)$ is the real symplectic group and $U(h)$ is the unitary group and is a classical bounded symmetric domain (see Helgason [1]).

Example 6.9: If $r = 2$, then we have the relationship

$$F^0 = P^{2,0} \subset P^2$$

(note that $P^{2,0} = H^{2,0}$), and, moreover,

$$\dim\ F^0 = \dim\ H^{2,0},$$

†Griffiths called these domains *period matrix domains* (Griffiths [1]) and *classifying spaces for Hodge structures* (Griffiths [3]).

and

$$i^2 Q|_{H^{2,0}+H^{0,2}} \text{ is positive definite}$$

and

$$i^2 Q|_{P^{1,1}} \text{ is negative definite.}$$

Therefore we have that

$$i^2 Q = \begin{bmatrix} I_{2h} & 0 \\ 0 & -I_k \end{bmatrix},$$

where $k = \dim P^2 - 2h$. D then has the homogeneous representation

$$D = SO(2h, k)/(U(h) \times SO(k)),$$

and we note that the maximal compact subgroup of the noncompact real group $SO(2h, k)$ is $SO(2h) \times SO(k)$. Thus we have a natural fibering

$$\begin{array}{l} D = SO(2h, k)/(U(h) \times SO(h)) \\ \downarrow \qquad\qquad\quad \downarrow \\ M = SO(2h, k)/(SO(2h) \times SO(k)), \end{array}$$

and it so happens that the fibres of this mapping are compact complex submanifolds of positive dimension when $h \neq 1$.

The reader is referred to Griffiths' papers in the References for a further discussion of the period mapping and its relation to the study of the variation of complex structure on a given (usually projective algebraic) manifold.

The discussion and analytic behavior of the period mapping into a Griffiths domain is contained in Griffiths [1, 3 and 6], while the geometry of a Griffiths domain itself is discussed in Griffiths and Schmid [1], Schmid [1], and Wells [1, 2], Wells–Wolf [1]. The relation of the periods of harmonic forms on an algebraic hypersurface V of $\mathbf{P}_n$ and the rational forms on $\mathbf{P}_n - V$ with poles of various orders along V is studied in Griffiths [5] along with some interesting applications to algebraic geometry.

CHAPTER VI

KODAIRA'S PROJECTIVE EMBEDDING THEOREM

In this chapter we are going to prove a famous theorem due to Kodaira, which gives a characterization of which compact complex manifolds admit an embedding into complex projective space. In Sec. 1 we shall define Hodge manifolds as those which carry an integral (1, 1) form which is positive definite in local coordinates. We then give various examples of such manifolds. Kodaira's theorem asserts that a compact complex manifold is projective algebraic if and only if it is a Hodge manifold. This is a very useful theorem, as we shall see, since it is often easy to verify the criterion. Chow's theorem asserts that projective algebraic manifolds are indeed *algebraic*, i.e., defined by the zeros of homogeneous polynomials. Thus the combination of these two theorems allows one to reduce problems of analysis to ones of algebra (cf. Serre's famous paper [2] in which this program of comparison is carried out in great detail).

In Sec. 2 we shall use the Hodge theory developed in the previous two chapters to prove Kodaira's *vanishing theorem*, which plays a role in compact complex manifold theory similar to that of Theorem B of Cartan in Stein manifold theory (see Gunning and Rossi [1]).

In Sec. 3 we shall introduce the concept of a quadratic transform of a complex manifold at a given point (the Hopf blowup) and study the behavior of metrics on holomorphic line bundles under pullbacks with respect to a quadratic transform. In Sec 4. we shall bring together the tools of Secs. 2 and 3 (which depended in turn on the work in the previous chapters) to prove Kodaira's embedding theorem.

1. Hodge Manifolds

In this section we want to consider a restricted class of Kähler manifolds defined by a certain topological (integrality) condition. If X is a compact complex manifold, then a d-closed differential form φ on X is said to be *integral* if its cohomology class in the de Rham group, $[\varphi] \in H^*(X, \mathbf{C})$, is in the image of the natural mapping:

$$H^*(X, \mathbf{Z}) \longrightarrow H^*(X, \mathbf{C}).$$

Let h be a Kähler metric on a complex manifold of Kähler type and let $\mathbf{\Omega}$ be the associated fundamental form.

Definition 1.1: If $\mathbf{\Omega}$ is an integral differential form, then $\mathbf{\Omega}$ is called a *Hodge form* on X, and h is called a *Hodge metric*. A manifold of Kähler type is called a *Hodge manifold* if it admits a Hodge metric.

This terminology was first used by A. Weil. The main theorem of this chapter (due to Kodaira [2]) is that a compact complex manifold is Hodge if and only if it is projective algebraic. First we shall see that there are many examples of Hodge manifolds, some of which are not at all obviously projective algebraic, and in passing we shall note that the Hodge condition is often easy to verify in practice.

Let E be a holomorphic line bundle over a complex manifold X. Then we let

$$E^{\mu} = \underbrace{E \otimes \cdots \otimes E}_{\mu \text{ factors}}$$

and

$$E^{-\mu} = (E^*)^{\mu},$$

for any positive integer μ. We let $E^0 = X \times \mathbf{C}$, the trivial line bundle over X, which is isomorphic to $E^{\mu} \otimes E^{-\mu}$ for all positive μ, as is easy to see. If $\{g_{\alpha\beta}\}$ is a set of transition functions for E with respect to some locally finite set of trivializations, then $\{g_{\alpha\beta}^{\mu}\}$ is a set of transition functions for E^{μ} for all integers μ. This is a simple fact, whose verification we leave to the reader (cf. Sec. 2 in Chap. I). In various examples below we shall use this principle to compare different line bundles on the same space, by comparing appropriate transition functions on the same open covering. If X is of complex dimension n, then we let

$$K_X = \wedge^n T^*(X)$$

be the *canonical line bundle of* X. It follows that

$$\mathcal{O}_X(K_X) = \mathcal{O}_X(\wedge^n T^*(X)) = \mathbf{\Omega}_X^n$$

the sheaf of holomorphic n-forms on X. For simplicity we denote the canonical line bundle simply by K whenever X is fixed in a given discussion.

We now present a list of examples of Hodge manifolds.

Example 1.2: Let X be a compact projective algebraic manifold. Then X is a submanifold of $\mathbf{P}_N$ for some N. Let $\mathbf{\Omega}$ be the fundamental form associated with the Fubini-Study metric on $\mathbf{P}_N$ (see Example V.4.5). Since $\mathbf{\Omega}$ is the negative of the Chern form for the universal bundle $U_{1,N+1} \to \mathbf{P}_N$, it follows that $\mathbf{\Omega}$ is a Hodge form on $\mathbf{P}_N$ (see Propositions III.4.3 and III.4.6). The restriction of $\mathbf{\Omega}$ (as a differential form) to X will also be a Hodge form, and hence X is a Hodge manifold. In general, by the same principle, a complex submanifold of a Hodge manifold is again a Hodge manifold.

Example 1.3: Let X be a compact complex manifold which is an unramified covering of a Hodge manifold Y; i.e., there is a holomorphic mapping $X \xrightarrow{\pi} Y$ such that $\pi^{-1}(p)$ is discrete and π is a local biholomorphism at each point $x \in X$. Then X is a Hodge manifold. To see this, simply let $\mathbf{\Omega}$ be a Hodge form on Y and then $\pi^*\mathbf{\Omega}$ will be a Hodge form on X. Similarly, if $X \xrightarrow{f} Y$ is an immersion, then $f^*\mathbf{\Omega}$ will give a Hodge manifold structure to X.

Example 1.4: Let X be a compact connected Riemann surface. Then X is a Hodge manifold. Namely, since $\dim_{\mathbf{R}} X = 2$, we have by Poincaré duality that $\mathbf{C} \cong H^0(X, \mathbf{C}) \cong H^2(X, \mathbf{C})$, and, moreover, $H^2(X, \mathbf{C}) = H^{1,1}(X)$. Let $\tilde{\mathbf{\Omega}}$ be the fundamental form on X associated with a Hermitian metric. Then $\tilde{\mathbf{\Omega}}$ is a closed form [of type (1, 1)] which is a basis element for the one-dimensional de Rham group $H^2(X, \mathbf{C})$. Let $c = \int_X \tilde{\mathbf{\Omega}}$, and then $\mathbf{\Omega} = c^{-1}\tilde{\mathbf{\Omega}}$ will be an integral positive form on X of type (1, 1). Hence X is Hodge. This example generalizes to the assertion that any Kähler manifold X with the property that $\dim_{\mathbf{C}} H^{1,1}(X) = 1$ is necessarily Hodge. This follows from the fact that multiplication by an appropriate constant will make the Kähler form on X integral, as above in the Riemann surface case (one has to also make an appropriate choice of basis for the integral 2-cycles).

Example 1.5: Let D be a bounded domain in $\mathbf{C}^n$ and let $\mathbf{\Gamma}$ be a fixed point free properly discontinuous subgroup of the group of biholomorphisms of D onto itself [$= \text{Aut}(D)$] with the property that $X = D/\mathbf{\Gamma}$ is compact (cf. Proposition V.5.3). Then X is a Hodge manifold. Let $\mathbf{\Omega}_D$ be the fundamental form associated with the Bergman metric h_D on D (see, e.g., Bergman [1], Helgason [1], or Weil [1]). The Bergman metric has the very useful property that it is invariant under the action of $\text{Aut}(D)$ and hence under the action of any subgroup $\mathbf{\Gamma}$. Thus h_D induces a metric h on X, which has associated with it a fundamental form $\mathbf{\Omega}$ which is of type (1,1) and positive definite. Moreover, since (for a particular normalization)

$$\mathbf{\Omega}_D(z) = \partial\bar{\partial} \log k_D(z), \tag{1.1}$$

where $k_D(z)$ is the Bergman kernel function for the domain D, it follows that $\mathbf{\Omega}$ is Kähler. What remains to be shown is that the Bergman metric form (1.1) above induces an integral form $\mathbf{\Omega}$ on X. To do this, we shall show that $(i/2\pi)\mathbf{\Omega}$ is, in fact, the Chern form of the canonical bundle over X and hence belongs to an integral cohomology class in $H^2(X, \mathbf{Z})$. We shall need the property that $k_D(z) > 0$ for all $z \in D$ [which is almost self-evident from the fact that

$$k_D(z) = \sum_v |\varphi_v(z)|^2$$

for an orthonormal basis $\{\varphi_v\}$ for the Hilbert space $L^2(D) \cap \mathcal{O}(D)$], and we shall also need the property that

$$k_D(\gamma(z)) = k_D(z) \left|\det \frac{\partial \gamma(z)}{\partial z}\right|^{-2}, \quad \gamma \in \mathrm{Aut}(D), \tag{1.2}$$

where $\partial\gamma/\partial z$ is the Jacobian matrix of the biholomorphism γ (see Bergman [1] or Weil [1]). Suppose that $\{U_\alpha\}$ is a covering of X by a finite number of coordinate patches, with ψ_α: $U_\alpha \to D$ being the coordinate functions (which can be taken as local inverses for the projection π: $D \to X = D/\mathbf{\Gamma}$). Then the transition function $\gamma_{\alpha\beta} = \psi_\alpha \circ \psi_\beta^{-1} \in \mathbf{\Gamma}$ and is defined on all of D. It then follows that $\{(\partial\gamma_{\alpha\beta}/\partial z)(\psi_\beta(p))\}$ are the transition functions for $T(X)$ and that $g_{\alpha\beta}(z) = \det(\partial\gamma_{\alpha\beta}/\partial z)(\psi_\beta(p))$ are the transition functions for $\wedge^n T(X)$. Thus the functions $\{g_{\alpha\beta}^{-1}\}$ are the transition functions for the canonical line bundle $K_X = \wedge^n T^*(X)$. Let $k^\alpha = k_D \circ \psi_\alpha$ be positive functions defined on $U_\alpha \subset X$. Then it is easy to check from (1.2) that

$$k^\alpha(p) = \left|\det \frac{\partial \gamma_{\alpha\beta}(z)}{\partial z}\right|^2 k^\beta(p),$$

where $z = \psi_\beta(p)$. This shows that the $\{k^\alpha\}$ transform like a metric for K and thus define a metric on K_X. By the results in Chap. III, we see that

$$\begin{aligned} c_1(K_X) &= \frac{i}{2\pi}\bar{\partial} k_\alpha^{-1} \partial k_\alpha = \frac{-i}{2\pi}\partial\bar{\partial}\log k_\alpha \\ &= -\frac{i}{2\pi}\partial\bar{\partial}\log k(z) \qquad \text{(in the coordinates of } D), \end{aligned}$$

but this is (except for sign) the fundamental form associated with the Bergman metric and thus the induced Bergman metric is a Hodge metric. Therefore, X is a Hodge manifold.

Remark: Note that the above example is quite different from the example of a Hopf surface given in Sec. 5 of Chap. V, since the Hopf surface was defined as a quotient space $D/\mathbf{\Gamma}$, where D was *not* a bounded domain, and by the results above it cannot be biholomorphically equivalent to one. Being biholomorphically equivalent to a bounded domain is rather crucial for the Bergman kernel theory to apply.

Example 1.6: Consider a complex torus X, as in Example V.4.4, with $2n$ independent periods $\{\omega_1, \ldots, \omega_{2n}\}$ in $\mathbf{C}^n$, and let

$$\mathbf{\Omega} = \begin{bmatrix} \omega_{11} & \cdots & \omega_{1,2n} \\ \cdot & & \cdot \\ \cdot & & \cdot \\ \cdot & & \cdot \\ \omega_{n1} & \cdots & \omega_{n,2n} \end{bmatrix}$$

be the matrix of periods. Suppose that there exists a nonsingular integral skew-symmetric matrix Q of rank $2n$ such that

$$(1.3)\qquad \begin{aligned} &\text{(I)} \qquad \mathbf{\Omega} A^t \mathbf{\Omega} = 0 \\ &\text{(II)} \quad -i \mathbf{\Omega} A^t \bar{\mathbf{\Omega}} = M > 0 \quad \text{(positive definite)}, \end{aligned}$$

where $A = Q^{-1}$. Then we say that $\mathbf{\Omega}$ is a *Riemann matrix* (cf. Conforto [1] and Siegel [1]). Consider the matrix $P = \begin{bmatrix} \mathbf{\Omega} \\ \bar{\mathbf{\Omega}} \end{bmatrix}$, called the *big period matrix*. Then it follows from the conditions above that P is nonsingular. Namely, consider the product [using the relations (1.3) above] and, noting that ${}^tM = \bar{M}$,

$$(1.4)\qquad PA^t\bar{P} = \begin{bmatrix} \mathbf{\Omega} \\ \bar{\mathbf{\Omega}} \end{bmatrix} A[{}^t\bar{\mathbf{\Omega}}, {}^t\mathbf{\Omega}] = \begin{bmatrix} iM & 0 \\ 0 & -i\bar{M} \end{bmatrix},$$

which is nonsingular, since $M > 0$, and hence P is nonsingular. Thus we find that, by taking the inverse of (1.4),

$$Q = \frac{i}{2}[{}^t\mathbf{\Omega} H \bar{\mathbf{\Omega}} - {}^t\bar{\mathbf{\Omega}} \bar{H} \mathbf{\Omega}],$$

where we let ${}^tH = 2M^{-1}$, which is also positive definite, and thus we find that

$$(1.5)\qquad Q_{\alpha\beta} = \frac{i}{2} \sum_{\mu,\nu} h_{\mu\nu}(\omega_{\mu\alpha}\bar{\omega}_{\nu\beta} - \omega_{\mu\beta}\bar{\omega}_{\nu\alpha}).$$

Conversely, if the periods $\{\omega_\alpha\}$ satisfy (1.5) for some Hermitian positive definite matrix H, where Q is a skew-symmetric nondegenerate matrix with integer coefficients, then $\mathbf{\Omega}$ is a Riemann matrix. Let

$$\omega = \frac{i}{2} \sum h_{\mu\nu} dz_\mu \wedge d\bar{z}_\nu$$

be the fundamental form for a Hermitian metric for X defined by the constant positive definite matrix H. The integral homology group $H_2(X, \mathbf{Z})$ is generated by the integral 2-cycles $\{C_{\alpha\beta}\}$, defined by the parametric representation

$$C_{\alpha\beta} = \{s\omega_\alpha + t\omega_\beta \colon 0 \leq s, t \leq 1\},$$

where ω_α, ω_β are given periods, $1 \leq \alpha < \beta \leq 2n$. Then the period of ω over the 2-cycles is given by

$$\int_{C_{\alpha\beta}} \omega = \frac{i}{2} \sum_{\mu,\nu} h_{\mu\nu}(\omega_{\mu\alpha}\bar{\omega}_{\nu\beta} - \omega_{\mu\beta}\bar{\omega}_{\nu\alpha}).$$

This is easy to verify and consists of evaluating the integral of $dz_\mu \wedge d\bar{z}_\nu$ over the real two-dimensional parallelogram determined by the two vectors ω_α and ω_β in $\mathbf{C}^n$. Thus ω is a Hodge form for the torus X.

In the other direction, suppose that we know that a torus X admits an embedding into some projective space $\mathbf{P}_N$. Then the standard Kähler form on $\mathbf{P}_N$ induces a Hodge form ω on X and in the coordinates of $\mathbf{C}^n$,

$$(1.6)\qquad \omega = \frac{i}{2} \sum h_{\mu\nu} dz_\mu \wedge d\bar{z}_\nu,$$

where the functions $h_{\mu\nu}$ are not necessarily constant, as we had above. However,

$$\int_{C_{\alpha\beta}} \omega \in \mathbf{Z},$$

and we can replace $h_{\alpha\beta}$ in (1.5) by the mean value

$$\hat{h}_{\alpha\beta} = \mu(X)^{-1} \int_X h_{\alpha\beta} d\mu,$$

where μ is the invariant measure on the torus induced by Lebesgue measure in $\mathbf{C}^n$. One can then verify that the resulting form $\hat{\omega}$ will satisfy the condition (1.5) and will have a positive definite coefficient matrix. Thus the existence of a Hodge form on X implies that the period matrix is a Riemann matrix.

An example of a complex torus not satisfying Riemann's condition is given by the period matrix ($n = 2$)

$$\mathbf{\Omega} = \begin{bmatrix} 1 & 0 & \sqrt{-2} & \sqrt{-5} \\ 0 & 1 & \sqrt{-3} & \sqrt{-7} \end{bmatrix}. \tag{1.7}$$

Namely, suppose that there existed a matrix A with rational coefficients such that

$$\mathbf{\Omega} A^t \mathbf{\Omega} = 0. \tag{1.8}$$

Then the element in the first row and second column of (1.8) is given by

$$\begin{aligned} a_{12} + a_{13}\sqrt{-3} + a_{14}\sqrt{-7} - a_{23}\sqrt{-2} - a_{24}\sqrt{-5} \\ + a_{34}(\sqrt{14} - \sqrt{15}) = 0, \end{aligned}$$

from which it follows easily that

$$a_{12} = a_{13} = a_{14} = a_{23} = a_{24} = a_{34} = 0,$$

since A was assumed to have rational entries. Since A is skew-symmetric, it follows that A cannot be nonsingular, which contradicts the assumption of $\mathbf{\Omega}$ being a Riemann matrix. Thus this particular complex torus cannot be projective algebraic. One can show, in fact, that the complex torus defined by the period matrix $\mathbf{\Omega}$ in (1.7) does not admit any nonconstant meromorphic function (cf. Siegel [1], pp. 104–106), which also implies that X is not embeddable in any projective space.

2. Kodaira's Vanishing Theorem

The vanishing theorem of Kodaira plays a role in the theory of compact complex manifolds analogous to the well-known Theorem B of Stein manifold theory (due to Cartan and Serre; see, e.g., Gunning and Rossi [1]). The basic difference is that on a compact complex manifold X, the cohomology groups $H^q(X, \mathcal{O}(E))$, $q \geq 1$, do not need to vanish for all holomorphic vector bundles E, which would be the case for Stein manifolds. There are basic obstructions, due to the compactness.

We shall now formulate the vanishing theorems for line bundles. A differential form φ of type (1, 1) on a complex manifold is said to be *positive* if, in local coordinates at any point p,

$$\varphi = i \sum_{\mu,\nu} \varphi_{\mu\nu}(z) dz_\mu \wedge d\bar{z}_\nu,$$

and the matrix $[\varphi_{\mu\nu}(z)]$ is a positive definite Hermitian symmetric matrix for each fixed point z near p. Notationally, we denote this condition by $\varphi > 0$.

Definition 2.1: Let $E \to X$ be a holomorphic line bundle and let $c_1(E)$ be the first Chern class of E considered as an element of the de Rham group $H^2(X, \mathbf{R})$. Then E is said to be *positive* if there is a real closed differential form ψ of type (1, 1) such that $\psi \in c_1(E)$ and ψ is a positive differential form. E is said to be *negative* if E^* is positive.

For computational ease we prove the following proposition.

Proposition 2.2: Let $E \longrightarrow X$ be a holomorphic line bundle over a compact complex manifold X. Then E is positive if and only if there is a Hermitian metric h on E such that $i\mathbf{\Theta}_E$ is a positive differential form, where $\mathbf{\Theta}_E$ is the curvature of E with respect to the canonical connection induced by h.

Proof: It is obvious from the differential-geometric definition of $c_1(E)$ that $i\mathbf{\Theta}_E$ positive for some metric h will imply that E is positive. Conversely, suppose that E is positive and that $\varphi \in c_1(E)$, where φ is a positive differential form. Let h be any metric on E, and then with respect to a local frame f we have $[h = h(f)]$

$$\varphi_0 = \frac{i}{2\pi} \bar{\partial}\partial \log h \in c_1(E),$$

and hence

$$\varphi - \varphi_0 = d\eta, \quad \eta \in \mathcal{E}^1(X).$$

Moreover, the differential form φ is a Kähler form on X, and X becomes a Kähler manifold when equipped with the associated Kähler metric. Then we may apply the harmonic theory, and let H be the harmonic projection onto $\mathcal{H}^*(X)$, and let G be the Green's operator associated with the d-Laplacian $\mathbf{\Delta} = 2\square = 2\bar{\square}$. Then we note that

$$\eta = H\eta + \mathbf{\Delta} G\eta,$$

and hence

$$d\eta = dH\eta + d\mathbf{\Delta} G\eta = \mathbf{\Delta} G d\eta,$$

since $dH = 0$ and d commutes with both $\mathbf{\Delta}$ and G.† It follows that

$$d\eta = 2\bar{\partial}\bar{\partial}^* G d\eta + 2\bar{\partial}^* G \bar{\partial} d\eta,$$

and we claim that $\bar{\partial} d\eta = 0$ and $\partial d\eta = 0$. This follows from the fact that

†The operators ∂ and $\bar{\partial}$ also commute with $G = \frac{1}{2} G_\square = \frac{1}{2} G_{\bar{\square}}$, (cf. Theorem IV.5.2).

$\bar\partial\varphi = \bar\partial\varphi_0 = 0$, and $\partial\varphi = \partial\varphi_0 = 0$, since φ can locally be written in the form $\varphi = \partial\bar\partial u$, for some C^∞ function u (Lemma II.2.15), and φ_0 is already of this form. Thus

$$d\eta = 2\bar\partial\bar\partial^* G d\eta,$$

and we can use the Kähler identity

$$i\bar\partial^* = L^*\partial - \partial L^*$$

(Corollary V.4.10), obtaining

$$d\eta = 2i\bar\partial\partial L^* G d\eta,$$

since $\partial G = G\partial$ and $\partial d\eta = 0$. Therefore we set $r = 2L^* G d\eta$, and by letting $h' = h \cdot e^{2\pi r}$ be a new metric for E, we obtain

$$\begin{aligned}\frac{i}{2\pi}\bar\partial\partial \log h' &= \frac{i}{2\pi}\bar\partial\partial \log h + i\bar\partial\partial r\\ &= \varphi_0 + d\eta\\ &= \varphi.\end{aligned}$$

Q.E.D.

Example 2.3: Let $X = \mathbf{P}_n$, and consider the following three basic line bundles over $\mathbf{P}_n$:

(a) The *hyperplane section bundle:* $H \longrightarrow \mathbf{P}_n$.
(b) The *universal bundle:* $U \longrightarrow \mathbf{P}_n (U = U_{1,n+1})$.
(c) The *canonical bundle:* $K = \wedge^n T^*(\mathbf{P}_n) \longrightarrow \mathbf{P}_n$.

Here H is the line bundle associated to the divisor of a hyperplane in $\mathbf{P}_n$, e.g., $[t_0 = 0]$, in the homogeneous coordinates $[t_0, \ldots, t_n]$. Then the divisor is defined by $\{t_0/t_\alpha\}$ in $U_\alpha = \{t_\alpha \neq 0\}$, and the line bundle H has transition functions (cf. (III.4.9))

$$\begin{aligned}h_{\alpha\beta} &= \left(\frac{t_0}{t_\alpha}\right)\left(\frac{t_0}{t_\beta}\right)^{-1} \quad \text{in } U_\alpha \cap U_\beta\\ &= \frac{t_\beta}{t_\alpha}.\end{aligned}$$

The universal bundle (Example I.2.6) has transition functions

$$u_{\alpha\beta} = \frac{t_\alpha}{t_\beta} \quad \text{in } U_\alpha \cap U_\beta,$$

and thus $H^* = U$. Let us now compute the transition functions for the canonical bundle K on $\mathbf{P}_n$. If we let $\zeta_j^\beta = t_j/t_\beta$, $j \neq \beta$, the usual coordinates in U_β, then a basis for $K|_{U_\beta}$ is given by the n-form

$$\Phi_\beta = (-1)^\beta d\zeta_0^\beta \wedge \cdots \wedge d\zeta_{\beta-1}^\beta \wedge d\zeta_{\beta+1}^\beta \wedge \ldots \wedge d\zeta_n^\beta.$$

Since

$$\zeta_j^\beta = \frac{t_j}{t_\beta} = \frac{t_j}{t_\alpha}\cdot\frac{t_\alpha}{t_\beta},$$

we have

$$\zeta_j^\beta = \zeta_j^\alpha \cdot (\zeta_\beta^\alpha)^{-1}$$

in $U_\alpha \cap U_\beta$ which is the (nonlinear) change of coordinates for $\mathbf{P}_n$ from U_α to U_β. Thus we obtain easily, by substituting into the above form $\mathbf{\Phi}_\alpha$,

$$\begin{aligned}\mathbf{\Phi}_\alpha &= (\zeta_\beta^\alpha)^{n+1}(-1)^\beta d\zeta_0^\beta \wedge \cdots \wedge d\zeta_{\beta-1}^\beta \wedge d\zeta_{\beta+1}^\beta \wedge \cdots \wedge d\zeta_n^\beta \\ &= (\zeta_\beta^\alpha)^{n+1}\mathbf{\Phi}_\beta\end{aligned}$$

Now we see that these transition functions for the frames $\{\mathbf{\Phi}_\alpha\}$ induce transition functions $\{k_{\alpha\beta}\}$ for the canonical bundle K which are given by

$$k_{\alpha\beta}([t_0, \ldots, t_n]) = \left(\frac{t_\alpha}{t_\beta}\right)^{n+1}.$$

We note that the choice of the minus sign in the trivializing sections was necessary for the transition functions for K to be comparable to the transition functions for U and H. Thus $K = \wedge^n T^*(\mathbf{P}_n) = U^{n+1} = (H^*)^{n+1}$. Moreover, the universal bundle $U \to \mathbf{P}_n$ has the curvature form given in (III.2.10), which is the negative of the positive differential form

$$\mathbf{\Omega} = \frac{i}{2}\frac{|t|^2 \sum\limits_{\mu=0}^{n} dt_\mu \wedge d\bar{t}_\mu - \sum\limits_{\mu,\nu=0}^{n} \bar{t}_\mu t_\nu dt_\mu \wedge d\bar{t}_\nu}{|t|^4},$$

expressed in homogeneous coordinates. Namely, $\mathbf{\Omega}$ is the canonical Kähler form on $\mathbf{P}_n$ associated with the Fubini-Study metric (see Example V.4.5). Thus H^*, U, and K are *negative* line bundles over $\mathbf{P}_n$, and the hyperplane section bundle $H \longrightarrow \mathbf{P}_n$ is *positive*. These are the primary examples of positive and negative line bundles.

Remark: It follows from the Hodge decomposition theorem that $H^1(\mathbf{P}_n, \mathcal{O}) = H^2(\mathbf{P}_n, \mathcal{O}) = 0$. Namely,

$$H^1(\mathbf{P}_n, \mathbf{C}) = H^{1,0}(\mathbf{P}_n) \oplus H^{0,1}(\mathbf{P}_n),$$

and $H^1(\mathbf{P}_n, \mathbf{C}) = 0$, by the cell decomposition of $\mathbf{P}_n$. Also,

$$\mathbf{C} \cong H^2(\mathbf{P}_n, \mathbf{C}) = H^{2,0}(\mathbf{P}_n) \oplus H^{1,1}(\mathbf{P}_n) \oplus H^{0,2}(\mathbf{P}_n),$$

and since $H^{1,1}(\mathbf{P}_n) = \mathbf{C}[\mathbf{\Omega}]$, where $\mathbf{\Omega}$ is the fundamental form on $\mathbf{P}_n$, it follows that $H^2(\mathbf{P}_n, \mathcal{O} \cong H^{0,2}(\mathbf{P}_n) = 0$.† Now consider the short exact sequence

$$0 \longrightarrow \mathbf{Z} \longrightarrow \mathcal{O} \longrightarrow \mathcal{O}^* \longrightarrow 0$$

on $\mathbf{P}_n$ and the induced cohomology sequence

$$H^1(\mathbf{P}_n, \mathcal{O}) \longrightarrow H^1(\mathbf{P}_n, \mathcal{O}^*) \xrightarrow{c_1} H^2(\mathbf{P}_n, \mathbf{Z}) \longrightarrow H^2(\mathbf{P}_n, \mathcal{O}),$$

which gives us, since $H^1(\mathbf{P}_n, \mathcal{O}) = H^2(\mathbf{P}_n, \mathcal{O}) = 0$,

$$0 \longrightarrow H^1(\mathbf{P}_n, \mathcal{O}^*) \xrightarrow{c_1} H^2(\mathbf{P}_n, \mathbf{Z}) \longrightarrow 0.$$
$$\qquad\qquad\qquad\qquad \| \wr$$
$$\qquad\qquad\qquad\qquad \mathbf{Z}$$

†In the same manner, one obtains that $H^q(\mathbf{P}_n, \mathbf{\Omega}^p) = 0$, $p \neq q$, $H^p(\mathbf{P}_n, \mathbf{\Omega}^p) \cong H^{2p}(\mathbf{P}_n, \mathbf{C})$, which are special cases of a vanishing theorem due to Bott [1].

Let $\mathbf{P}_1 \subset \mathbf{P}_n$ be a generator for $H^2(\mathbf{P}_n, Z)$, and then we see that if we consider powers of the hyperplane section bundle H^m, we obtain

$$c_1(H^m)(\mathbf{P}_1) = m.$$

Namely, by the properties of Chern classes,†

$$c_1(H) = c_1(U^*) = -c_1(U) \quad \text{and} \quad c_1(U)(\mathbf{P}_1) = -1.$$

Since c_1 is an isomorphism of abelian groups, it follows that every holomorphic line bundle $L \longrightarrow \mathbf{P}_n$ (in particular U and K in the above example) is a power of the hyperplane section bundle, $L = H^m$, and $c_1(L)(\mathbf{P}_1) = m$. We use here the fact that $c_1(L) = c_1(H^m) = c_1(H \otimes \cdots \otimes H) = c_1(H) + \cdots + c_1(H)$ (cf. the proof of Theorem III.3.6). In particular, we obtain from Example 2.3 that $c_1(K_{\mathbf{P}_n})(\mathbf{P}_n) = -(n+1)$. Thus the holomorphic line bundles on $\mathbf{P}_n$ are completely classified in this manner by their Chern classes.

We now state the basic vanishing theorem due originally to Kodaira [1].

Theorem 2.4: Suppose that X is a compact complex manifold.

(a) Let $E \longrightarrow X$ be a holomorphic line bundle with the property that $E \otimes K^*$ is a positive line bundle. Then

$$H^q(X, \mathcal{O}(E)) = 0, \quad q > 0.$$

(b) Let $E \longrightarrow X$ be a negative line bundle. Then

$$H^q(X, \mathbf{\Omega}^p(E)) = 0, \quad p + q < n.$$

Remark: Kodaira's theorems were first proved in Kodaira [1] ((a) and $p = 0$ in (b)) and were generalized later by Nakano [1] to the case we have given here. There are various generalizations of these types of results for vector bundles which are not as precise as the above theorems but which have numerous applications. See, e.g., Grauert [2], Griffiths [2], Nakano [1], Hartshorne [1], and Grauert and Riemenschneider [1].

To prove the above theorem we want to derive some fundamental inequalities due to Nakano. First suppose that X is a Kähler manifold with a fundamental form $\mathbf{\Omega}$ associated to the Kähler metric. Then the operators L and L^* are well-defined endomorphisms of $\mathcal{E}^*(X)$. Suppose that $E \longrightarrow X$ is a holomorphic vector bundle over X. Then we want to show that L and L^* extend in a natural manner to endomorphisms of $\mathcal{E}^*(X, E)$ (differential forms with coefficients in E). If $\xi \in \mathcal{E}^p(X, E)$, then for a choice of a local holomorphic frame f for E in an open set $U \subset X$, we see that

$$\xi(f) = \begin{bmatrix} \xi^1(f) \\ \cdot \\ \cdot \\ \cdot \\ \xi^\rho(f) \end{bmatrix},$$

†Compare the proof of Proposition III.4.3, where $c_1(U)(\mathbf{P}_1) = \int_{\mathbf{P}_1} c_1(U) = -1$.

where $\xi^j(f) \in \mathcal{E}^p(U)$. Moreover, if g is a holomorphic change of frame, then we have the compatibility condition that

$$g^{-1}\xi(f) = \xi(gf)$$

(see Sec. 2 of Chap. III), where the matrix g^{-1} of functions is multiplied with the vector $\xi(f)$ of differential forms. We now let $*$ be the Hodge operator defined with respect to the Kähler metric on X, and $*$ acts naturally on vector-valued forms by setting

$$*\xi(f) = \begin{bmatrix} *\xi^1(f) \\ \cdot \\ \cdot \\ \cdot \\ *\xi^\rho(f) \end{bmatrix}$$

and noting that, since $*$ is $\mathbf{C}$-linear,

$$*\xi(gf) = *g^{-1}\xi(gf) = g^{-1}*\xi(f),$$

and hence $*\xi(f)$ satisfies the compatibility conditions and defines a global element in $\mathcal{E}^p(X, E)$. This is true of any zeroth order differential operator (which is a homomorphism of the underlying vector bundles). Thus

$$L\colon \mathcal{E}^p(X, E) \longrightarrow \mathcal{E}^{p+2}(X, E)$$

is well defined by letting

$$L(\xi(f)) = (L\xi^j(f)), \quad j = 1, \ldots, \rho,$$

and hence $L^* = w{*}L{*}$ is also defined. Of course, exterior differentiation d does not extend to vector-valued forms, and we have to introduce a connection on E in order to define *covariant differentiation* on E, a generalization of exterior differentiation. Namely, as in Chap. III, we let

$$D = d + \theta,$$

where θ is the connection defined by

$$\theta = h^{-1}\partial h \quad \text{(with respect to a local holomorphic frame)}$$

if h is the metric. Moreover,

$$D = D' + D'',$$

where

$$D' = \partial + \theta$$
$$D'' = \bar{\partial}$$

are the splitting of the covariant differentiation into types. With respect to the Hodge inner product on $\mathcal{E}^*(X, E)$, we have the L^2-adjoints of the above differential operators, computed as in Proposition V.2.3:

(2.1) $$(D')^*_E = -{*}\bar{\partial}{*} = \partial^*$$

(2.2) $$(D'')^*_E = -{*}\partial{*} + w{*}\theta{*} \quad \text{on } r\text{-forms}$$
$$= \bar{\partial}^* + w{*}\theta{*}.$$

Note that in making this computation the Hodge inner product can be represented with respect to a local holomorphic frame

$$(\xi, \eta) = (-1)^p \int \bar{*}^t \eta \bullet h \bullet \xi, \quad \xi, \eta \in \mathcal{E}^p(X, E),$$

where $\xi = \xi(f), \eta = \eta(f)$, and $h = h(f)$ are vectors and matrices, respectively, and the multiplication inside the integral is matrix multiplication. Also it suffices to compute the adjoint (which we know is a differential operator by Proposition IV.2.8), to assume that ξ and η have support where the holomorphic frame is defined. The crucial factor for our later use is that the adjoint $(D')^*_E$ does not depend on the Hermitian metric of the fibres of E, and, in particular, is the more classical scalar adjoint ∂^* acting in $\mathcal{E}^*(X)$. The adjoint of $\bar{\partial}$ (a scalar operator) is no longer scalar, however, Then we can conclude that, by Corollary V.4.10, since the scalar operator adjoints are with respect to a scalar metric,

$$\bar{\partial} L^* - L^* \bar{\partial} = i\partial^* = i(D')^*_E. \tag{2.3}$$

Under these circumstances we have the following inequality due to Nakano [1].

Proposition 2.5: Let $\xi \in \mathcal{H}^{p,q}(E)$. Then

(a) $(i/2)(\mathbf{\Theta} \wedge L^* \xi, \xi) \leq 0$.
(b) $(i/2)(L^* \mathbf{\Theta} \wedge \xi, \xi) \geq 0$.

In both (a) and (b) $\mathbf{\Theta}$ $(= \bar{\partial} h^{-1} \partial h)$ is the curvature form for the metric h on the holomorphic vector bundle E.

Proof: We recall that (Proposition III.1.9), as an operator,

$$D^2 = (d + \theta)^2 = \mathbf{\Theta},$$

and thus

$$\mathbf{\Theta} \wedge \eta = D^2 \eta = (D' \bar{\partial} + \bar{\partial} D')\eta$$

for $\eta \in \mathcal{E}^*(X, E)$ (noting that $(D')^2 = 0$, because of type). Hence we have

$$i(\partial^* \xi, \partial^* \xi) = ([\bar{\partial} L^* - L^* \bar{\partial}]\xi, \partial^* \xi)$$

by (2.3), and since ξ is harmonic, we have $\bar{\partial}\xi = \bar{\partial}^*_E \xi = 0$, and thus

$$\begin{aligned} i(\partial^* \xi, \partial^* \xi) &= (\bar{\partial} L^* \xi, \partial^* \xi) \\ &= (L^* \xi, [\bar{\partial}^*_E \partial^* + \partial^* \bar{\partial}^*_E]\xi), \end{aligned}$$

since $\bar{\partial}^*_E \xi = 0$. Then, taking adjoints, we get

$$\begin{aligned} i(\partial^* \xi, \partial^* \xi) &= ([D' \bar{\partial} + \bar{\partial} D'] L^* \xi, \xi) \\ &= (\mathbf{\Theta} \wedge L^* \xi, \xi), \end{aligned}$$

which immediately gives part (a). Part (b) is proved in a similar manner. Q.E.D.

It is now a simple matter to derive Kodaira's vanishing theorem.

Proof of Theorem 2.4: Suppose that E is a negative line bundle. Then a fundamental form for a Kähler metric on X is given by $\mathbf{\Omega} = -(i/2)\mathbf{\Theta}$ [noting that $\mathbf{\Theta}$ is a closed form of type (1, 1), whose coefficient matrix is negative definite]. Then subtract part (b) from part (a) in Proposition 2.5, and we obtain [noting that $-(i/2)\mathbf{\Theta}\wedge$ gets replaced by L]

$$([L^*L - LL^*]\xi, \xi) \leq 0.$$

Recalling from Proposition V.1.1(c) that

$$(L^*L - LL^*)\xi = (n - p - q)\xi,$$

we have immediately

$$(n - p - q)(\xi, \xi) \leq 0$$

if $\xi \in \mathcal{H}^{p,q}(E)$, and thus part (b) of Theorem 2.4 follows, by using the results in Example IV.5.7. Part (a) follows from part (b) by Serre duality (Theorem V.2.7). Namely, if $E \otimes K^*$ is positive, then $(E \otimes K^*)^* = K \otimes E^*$ is negative. We then have

$$H^q(X, \mathcal{O}(E)) = H^q(X, \mathcal{O}(K \otimes K^* \otimes E) = H^q(X, \mathbf{\Omega}^n(K^* \otimes E)),$$

which is dual to $H^{n-q}(X, \mathcal{O}(K \otimes E^*))$, which vanishes for $q > 0$, by part (b).

Q.E.D.

3. Quadratic Transformations

In this section we are going to study the behavior of positive line bundles under quadratic transformations. Let X be a complex manifold and suppose that $p \in X$. Then we want to define the *quadratic transform* of the manifold X at the point p. Let U be a coordinate neighborhood of the point p, with coordinates $z = (z_1, \ldots, z_n)$, where $z = 0$ corresponds to the point p. Consider the product $U \times \mathbf{P}_{n-1}$, where we assume that $(t_1, \ldots, t_n)$ are homogeneous coordinates for $\mathbf{P}_{n-1}$. Then let

$$W = \{(z, t) \in U \times \mathbf{P}_{n-1}\colon t_\alpha z_\beta - t_\beta z_\alpha = 0, \alpha, \beta = 1, \ldots, n\}, \tag{3.1}$$

which is a submanifold of $U \times \mathbf{P}_{n-1}$. Then there is a holomorphic projection $\pi\colon W \longrightarrow U$ given by $\pi(z, t) = z$. Moreover, π has the following properties, as is easy to verify:

$$\begin{aligned} &\pi^{-1}(0) = S = \{0\} \times \mathbf{P}_{n-1} \cong \mathbf{P}_{n-1} \\ &\pi|_{W-S}\colon W - S \longrightarrow U - \{0\} \quad \text{is a biholomorphism.} \end{aligned} \tag{3.2}$$

We define $\tilde{X} = Q_p(X)$, *the quadratic transform of X at p*, by letting

$$\tilde{X} = \begin{cases} W, x \in U \\ X - U, & x \in X - U. \end{cases}$$

This process if often referred to as *blowing up X at the point p*. We may also denote the manifold $\tilde{X}$ by $Q_p(X)$ to indicate the dependence on the point p, and the projection will be denoted by $\pi_p\colon Q_p(X) \longrightarrow X$.

We recall from Sec. 4 in Chap. III that a divisor D in a complex manifold X determines an associated holomorphic line bundle $L(D) \to X$, which is unique up to holomorphic equivalence of line bundles. Let $\tilde{X} \xrightarrow{\pi} X$ be the quadratic transform of X at the point p and let $S = \pi^{-1}(p)$. Then $S \cong \mathbf{P}_{n-1}$ and is an $(n-1)$-dimensional compact hypersurface embedded in $\tilde{X}$. As such it is a divisor in $\tilde{X}$ and determines a line bundle $L(S) \to \tilde{X}$, which we shall simply denote by L. Moreover, since $S \cong \mathbf{P}_{n-1}$, there is a canonical line bundle, the *hyperplane section bundle* $H \to S$ (cf. Example 2.3), which is the line bundle determined by the divisor corresponding to a fixed linear hyperplane (e.g., $[t_1 = 0] \subset \mathbf{P}_{n-1}$), all such line bundles being isomorphic. Let σ denote the projection $\sigma \colon W \to \mathbf{P}_{n-1}, \sigma(z, t) = t$, and let $L|_W$ denote the restriction of the line bundle $L \to \tilde{X}$ to $W \subset \tilde{X}$. Then we have the following proposition.

Proposition 3.1: $L|_W = \sigma^* H^*$.

Proof: Let U be a coordinate neighborhood of p in X, and represent $\tilde{X}$ near $\pi^{-1}(U)$ by $W \subset U \times \mathbf{P}_{n-1}$, with coordinates $(z_1, \ldots, z_n) \in U$, $[t_1, \ldots, t_n] \in \mathbf{P}_{n-1}$. Then S is defined by $z_1 = \cdots = z_n = 0$ in the product space $U \times \mathbf{P}_{n-1}$. Now the hyperplane $[t_1 = 0]$ is defined by the equations $[(t_1/t_\alpha) = 0]$ in $V_\alpha \subset \mathbf{P}_{n-1}$, where $V_\alpha = \{[t_1, \ldots, t_n] \colon t_\alpha \neq 0\}$ is a coordinate patch for $\mathbf{P}_{n-1}$. Therefore $H \to S$ is the line bundle given by the transition functions

$$h_{\alpha\beta} = \left(\frac{t_1}{t_\alpha}\right) \cdot \left(\frac{t_1}{t_\beta}\right)^{-1} = \frac{t_\beta}{t_\alpha} \quad \text{in } V_\alpha \cap V_\beta,$$

and $\sigma^* H$ has the same transition functions in $(U \times V_\alpha \cap V_\beta) \cap W$. Now $S \cap (U \times V_\alpha) \cap W$ is defined by the single equation $[z_\alpha = 0]$, as is easily checked, using the defining relation for W. Thus the line bundle L associated to the divisor $S \subset W$ has the transition functions

$$g_{\alpha\beta}(z, t) = \frac{z_\alpha}{z_\beta} \quad \text{in } (U \times V_\alpha \cap V_\beta) \cap W.$$

It follows that $g_{\alpha\beta} = h_{\alpha\beta}^{-1}$ and thus $L|_W = \sigma^* H^*$.

Q.E.D.

We now want to study the differential-geometric behavior of a line bundle on X when lifted to a quadratic transformation of X at some point p. First we look at the behavior of the canonical bundles. Let X be a compact complex manifold, which will remain fixed in the following discussion, and $L_p \to Q_p X := Q_p(X)$ is the line bundle given in Proposition 3.1.

Lemma 3.2: $K_{Q_p X} = \pi_p^* K_X \otimes L_p^{n-1}$.

Proof: First we note that $(z_1, t_2/t_1, \ldots, t_n/t_1)$ are holomorphic coordinates for $U \times V_1 \cap W$ (using the same coordinates as above). Hence

$$f_1 = dz_1 \wedge d\left(\frac{t_2}{t_1}\right) \wedge \cdots \wedge d\left(\frac{t_n}{t_1}\right)$$

is a holomorphic frame for $K_{\tilde{X}}$ over this open set, letting $\tilde{X} = Q_pX$. Moreover, one obtains easily that

$$f_1 = z_1^{1-n} dz_1 \wedge \cdots \wedge dz_n,$$

using the defining equations for W. More generally, we have that

$$f_\alpha = z_\alpha^{1-n} dz_1 \wedge \cdots \wedge dz_n$$

is a frame for $K_{\tilde{X}}$ over $U \times V_\alpha \cap W$, and hence transition functions for the line bundle $K_{\tilde{X}}|_W$ are given by

$$f_\beta = g_{\alpha\beta} f_\alpha, \tag{3.3}$$

since the local frames $\{f_\alpha\}$ define a system of trivializations, which then gives (3.3). It then follows that $g_{\alpha\beta} = (z_\alpha/z_\beta)^{n-1}$, which implies that

$$K_{\tilde{X}}|_W = L^{n-1}|_W \cong L^{n-1} \otimes \pi_p^* K_X|_W$$

since K_X is trivial on U. Also, $L|_{\tilde{X}-W}$ is trivial, and π_p is biholomorphic on $\tilde{X} - W$. Hence

$$K_{\tilde{X}}|_{\tilde{X}-W} \cong K_{\tilde{X}} \otimes L^{n-1}|_{\tilde{X}-W}.$$

Thus

$$K_{\tilde{X}} \cong L^{n-1} \otimes \pi_p^* K_X. \tag{3.4}$$

Q.E.D.

Let $p \in X$ and let $L_p \to Q_pX$ be the line bundle corresponding to the divisor $\pi_p^{-1}(p) \subset X$. If $q \neq p$ is another point on X, then it is clear that $Q_qQ_pX \cong Q_pQ_qX$, since blowing up at the points p and q are local and independent operations. Let $\pi_{p,q}$: $Q_pQ_qX \to X$ be the composite projection and and let $L_{p,q}$ be the line bundle corresponding to the divisor $\pi_{p,q}^{-1}(\{p\} \cup \{q\})$.

Proposition 3.3: Let $E \to X$ be a positive holomorphic line bundle. There exists an integer $\mu_0 > 0$ such that if $\mu \geq \mu_0$, then for any points $p, q \in X$, $p \neq q$,

(a) $\pi_p^* E^\mu \otimes L_p^* \otimes K_{Q_pX}^*$,
(b) $\pi_p^* E^\mu \otimes (L_p^*)^2 \otimes K_{Q_pX}^*$, and
(c) $\pi_{p,q}^* E^\mu \otimes L_{p,q}^* \otimes K_{Q_pQ_qX}^*$

are positive holomorphic line bundles.

Proof: To prove the above proposition, we shall construct a metric on each of the above line bundles whose curvature form multiplied by i is positive. We shall first look at a special case. Suppose that $p \in X$ is fixed, and let $Q_pX = \tilde{X}$, as before. The basic fact that we shall be using is that if F and G are Hermitian line bundles over a complex manifold Y, then, denoting the curvature by $\boldsymbol{\Theta}$,

$$\boldsymbol{\Theta}_{F\otimes G} = \boldsymbol{\Theta}_F + \boldsymbol{\Theta}_G.$$

This is easy to see, since if $\rho_\alpha = |g_{\beta\alpha}|^2\rho_\beta$ and $r_\alpha = |h_{\beta\alpha}|^2 r_\beta$ are local transition functions for metrics $\{\rho_\alpha\}$ and $\{r_\alpha\}$ for F and G, respectively (cf. Chap. III), then

$$\Theta_F = \bar{\partial}\partial \log \rho_\alpha,$$

$$\Theta_G = \bar{\partial}\partial \log r_\alpha,$$

but $\{g_{\alpha\beta} \bullet h_{\alpha\beta}\}$ are the transition functions for $F \otimes G$, and thus $\{\rho_\alpha \bullet r_\alpha\}$ defines a metric for $F \otimes G$, since

$$\rho_\alpha r_\alpha = |g_{\beta\alpha}|^2 |h_{\beta\alpha}|^2 \rho_\beta r_\beta.$$

Thus

$$\begin{aligned}\Theta_{F\otimes G} &= \bar{\partial}\partial \log\,(\rho_\alpha r_\alpha)\\ &= \Theta_F + \Theta_G.\end{aligned}$$

We have, then, using the given metric on E and letting $\pi = \pi_p$,

$$\Theta_{\pi^* E^\mu} = \pi^*\Theta_{E^\mu} = \mu\pi^*\Theta_E.$$

We now need to construct appropriate metrics on L_p and $K_{\tilde{X}}$.

First we consider L_p. Suppose that U is a coordinate neighborhood near p, with coordinates $(z_1, \ldots, z_n)$, that $\mathbf{P}_{n-1}$ has homogeneous coordinates $[t_1, \ldots, t_n]$ as before, and that $W \subset U \times \mathbf{P}_{n-1}$ is the local representation for $\tilde{X}$ near $\pi_p^{-1}(p)$ as given by (3.1). Let U' be an open subset of U such that $0 \in U' \subset\subset U$, and let $\rho \in \mathcal{D}(U)$ be chosen so that $\rho \geq 0$ in U and $\rho \equiv 1$ on U'. Let

$$\Theta_H = \bar{\partial}\partial \log \frac{|t_\alpha|^2}{|t_1|^2 + \cdots + |t_n|^2} \qquad (\text{in } V_\alpha \subset \mathbf{P}_{n-1})$$

be the curvature of the hyperplane section bundle $H \to \mathbf{P}_{n-1}$, with respect to the natural metric h_0 (see Example III.2.4 for the construction of this metric for $U_{1,n-1} = H^*$). In particular $(i/2)\Theta_H$ is the fundamental form associated with the standard Kähler metric on $\mathbf{P}_{n-1}$. Since $L|_W = \sigma^*H^*$, we can equip $L^*|_W$ with the metric $h_1 = \sigma^* h_0$. Now $L^*|_{\tilde{X}-U'}$ is trivial, and we can equip it with a constant metric h_2. Then, letting ρ be chosen as above, we see that

$$h = \rho h_1 + (1-\rho)h_2$$

defines a metric on $L^* \to \tilde{X}$ and that, moreover, $h = h_1$ in $W' = U' \times \mathbf{P}_{n-1} \cap W$. Thus

$$\Theta_{L^*} = \Theta_{\sigma^* H} \qquad \text{in } W'$$

$$\Theta_{L^*} \equiv 0 \qquad \text{in } \tilde{X} - W.$$

We now let K_X be equipped with an arbitrary Hermitian metric, and then we have from Lemma 3.2 (letting L be equipped with the dual metric)

$$\Theta_{K_{\tilde{X}}} = \Theta_{\pi^* K_X} + (n-1)\Theta_L.$$

Therefore it follows that

$$\Theta_{\pi^*E^\mu\otimes L^*\otimes K_{\tilde{X}}^*} = \mu\Theta_{\pi^*E} + n\Theta_{L^*} + \Theta_{\pi^*K_{X^*}^*}$$

Consider the sum

$$\mu\Theta_{\pi^*E} + \Theta_{\sigma^*H}$$

as differential forms in $U' \times \mathbf{P}_{n-1}$, with the coordinates (z, t) as before. Then Θ_{π^*E} depends only on the z-variable and Θ_{σ^*H} depends only on the t-variable, and the coefficient matrix of each is positive definite in each of the respective directions, so their sum is a positive differential form† in $U' \times \mathbf{P}_{n-1}$, and the restriction to W is likewise positive. Moreover, Θ_{π^*E} is positive on $U - U'$, so there exists a $\mu_1(p)$ such that $\mu > \mu_1(p)$ implies that

$$[\mu\Theta_{\pi^*E}] + \Theta_{L^*} > 0 \tag{3.5}$$

on all of $\tilde{X}$.

Let μ_2 be chosen such that

$$\mu_2\Theta_E + \Theta_{K_X^*} > 0,$$

which is possible since E is positive and since X is compact. Thus we see that there is a $\mu_0(p)$ such that

$$\mu\Theta_{\pi_p^*E} + n\Theta_{L^*} + \Theta_{\pi_p^*K_X^*} > 0 \tag{3.6}$$

if $\mu > \mu_0(p)$. Namely, let $\mu_0(p) = \mu_2 + n\mu_1(p)$ and note that

$$\mu_2\Theta_{\pi^*E} + \Theta_{\pi^*K_X^*}$$

is positive everywhere on $\tilde{X}$ except at points of S, where it is positive semidefinite (in the obvious sense).

Suppose that $q \in U'$. Then we claim that if $\mu \geq \mu_1(p)$, then the estimate (3.5) will hold for points q near p, namely,

$$\mu\Theta_{\pi_q^*E} + \Theta_{L^*} > 0$$

for all $x \in X$. This is a simple continuity argument which is easily seen by expressing the equations for the quadratic transform at q in terms of local coordinates centered at p, namely,

$$W_q = \{(z, t) \in U \times \mathbf{P}_{n-1}\colon (z_i - q_i)t_j = (z_j - q_j)t_i\},$$

where $q = (q_1, \ldots, q_n)$ and $p = (0, \ldots, 0)$.

By covering X with a finite number of such neighborhoods, we find that there is a μ_0 such that (3.6) holds for all points $p \in X$, if $\mu \geq \mu_0$, and this concludes the proof of part (a). Parts (b) and (c) are proved in exactly the same manner. In (b) we put the same metric on L_p near the point p as above, and $(L_p^*)^2$ will have the same positivity properties as L_p^*, compensating for the lack of positivity of $\pi_p^*E^\mu$ on $\pi_p^{-1}(p)$. In part (c) one has the same local constructions near each of the two distinct points p and q. The continuity

†In this argument we ignore the factor of i and mean by > 0 that the coefficient matrix of the (1, 1) form is positive definite.

and compactness arguments go through in exactly the same manner, and we leave further details to the reader.

Q.E.D.

4. Kodaira's Embedding Theorem

After the preliminary preparations of the previous sections we are now prepared to prove Kodaira's embedding theorem for Hodge manifolds. This theorem was conjectured by Hodge [2] and proved by Kodaira [2].

Theorem 4.1: Let X be a compact Hodge manifold. Then X is a projective algebraic manifold.

Remark: (a) As a consequence of the Kodaira embedding theorem, each of the examples of Hodge manifolds in Sec. 1 admits a projective algebraic embedding. In particular, any compact Riemann surface is projective algebraic (a well-known classical result), and a complex torus admits a projective embedding if and only if the periods defining the torus give rise to a Riemann matrix. Such tori are called *abelian varieties* and can also be characterized by the fact that a complex torus X is an abelian variety if and only if there are n algebraically independent nonconstant meromorphic functions on X, where $n = \dim_{\mathbf{C}} X$ (cf. Siegel [1]).

(b) It follows immediately from Theorem 4.1 that any compact complex manifold X which admits a positive line bundle $L \to X$ is projective algebraic (and conversely). Namely, in this case, $c_1(L)$ will have a Hodge form as a representative, and thus X will be projective algebraic. This is a very useful version of the theorem, and in this form the theorem has been generalized by Grauert [2] to include the case where X admits singularities. Grauert's proof can be found in Gunning and Rossi [1], and it depends on the finiteness theorem for strongly pseudoconvex manifolds and spaces.

Proof: By hypothesis, there is a Hodge form $\mathbf{\Omega}$ on X. By Proposition III.4.6, it follows that there is a holomorphic line bundle $E \to X$ such that $\mathbf{\Omega}$ is a representative for $c_1(E)$. Hence, E is a positive holomorphic line bundle. Let μ_0 be given by Proposition 3.3, let $\mu \geq \mu_0$, and set $F = E^\mu$. Consider the vector space of holomorphic sections $\mathcal{O}(X, F) = \mathbf{\Gamma}(F)$, for short, which is finite dimensional by Theorem IV.5.2. Our object is to show that there is an embedding of X into $\mathbf{P}(\mathbf{\Gamma}(F))$. We shall prove this by a sequence of lemmas, which will reduce the embedding problem and hence the proof of Theorem 4.1 to the vanishing theorem proved in Sec. 2. First we have some preliminary considerations.

Consider the subsheaf of $\mathcal{O} = \mathcal{O}_X$ consisting of germs of holomorphic functions which vanish at p and q; call it m_{pq}. If $p = q$, then we mean by this the holomorphic functions which vanish to second order at p, and we denote it simply by $m_p^2(= m_{pp})$, where m_p is the ideal sheaf of germs vanishing to first order at p. Then there is an exact sequence of sheaves

$$0 \longrightarrow m_{pq} \longrightarrow \mathcal{O} \longrightarrow \mathcal{O}/m_{pq} \longrightarrow 0,$$

and we can tensor this with the locally free sheaf $\mathcal{O}(F)$ (the sheaf of holomorphic sections of F), obtaining

$$0 \longrightarrow m_{pq} \otimes_{\mathcal{O}} \mathcal{O}(F) \longrightarrow \mathcal{O}(F) \longrightarrow \mathcal{O}/m_{pq} \otimes_{\mathcal{O}} \mathcal{O}(F) \longrightarrow 0. \tag{4.1}$$

We see that the quotient sheaf in this sequence becomes

$$\begin{array}{ll} \mathcal{O}_p/m_p^2 \otimes_{\mathbf{C}} F_p, & x = p = q \\ 0, & x \neq p \end{array} \tag{4.2}$$

if $p = q$ and

$$\begin{array}{ll} F_p, & x = p \\ F_q & x = q \\ 0, & x \neq p \text{ or } q \end{array} \tag{4.3}$$

if $p \neq q$, where we have used the fact that $\mathcal{O}_p/m_p \cong \mathbf{C}$, where m_p is the maximal ideal in the local ring $\mathcal{O}_p$.

Lemma 4.2: $\mathcal{O}_p/m_p^2 \cong \mathbf{C} \oplus \boldsymbol{T}_p^*(X)$, and the quotient mapping is represented by $f \in \mathcal{O}_p \to f(p) + df(p)$.

Proof: If $f \in \mathcal{O}_p$ is expanded in a power series near $z = p$ in local coordinates, we have

$$f(z) = \sum_{|\alpha| \geq 0} \frac{1}{\alpha!} D^\alpha f(p)(z-p)^\alpha,$$

using the standard multiindex notation (see Chap. IV). Then if [ ] denotes equivalence classes in $\mathcal{O}/m_p^2$, we see that

$$[f]_p = [f(p) + \sum_{|\alpha|=1} D^\alpha f(p)(z-p)^\alpha],$$

since the higher-order terms $\in m_p^2$. Then define the mapping

$$\mathcal{O}_p \xrightarrow{\psi} \mathbf{C} \oplus T_p^*(X)$$

by $\psi(f) = [f(p), df(p)]$, and it is easy to check that ψ factors through the quotient mapping

$$\begin{array}{ccc} \mathcal{O}_p & \xrightarrow{\psi} & \mathbf{C} \oplus T_p^*(X) \\ & \searrow \quad \tilde{\psi} \nearrow & \\ & \mathcal{O}_p/m_p^2 & \end{array}$$

and $\tilde{\psi}$ is an isomorphism.

Q.E.D.

Consider now the sequence (4.1) and the induced mapping on global sections

$$\mathcal{O}(X, F) \xrightarrow{r} \mathcal{O}_p/m_p^2 \otimes F_p \quad (4.4)$$
$$\wr\|$$
$$[\mathbf{C} \oplus T_p^*(X)] \otimes F_p.$$

If f is a local frame for F near p and if $\xi \in \mathcal{O}(X, F)$, then

$$r(\xi(f)) = (\xi(f)(p), d\xi(f)(p)) \in \mathbf{C} \oplus T_p^*(X),$$

noting that $F_p \cong \mathbf{C}$, by the choice of a frame f. Suppose that the map r in (4.4) is *surjective*. Then we can find sections $\{\xi_0, \xi_1, \ldots, \xi_n\}$, $\xi_j \in \mathbf{\Gamma}(X, F)$, such that

$$\begin{aligned} \xi_0(p) &= 1 \\ \xi_j(p) &= 0, \; j = 1, \ldots, n \\ d\xi_j(p) &= dz_j \qquad \text{(in local coordinates).} \end{aligned} \quad (4.5)$$

This means that the global sections $\xi_1, \ldots, \xi_n$, expressed in terms of the frame f, give local coordinates for X, in particular, $d\xi_1(f)\wedge \cdots \wedge d\xi_n(f) \neq 0$. Moreover, $\xi_0(p) \neq 0$.

Similarly, suppose that the mapping

$$\mathcal{O}(X, F) \xrightarrow{s} F_p \oplus F_q \quad (4.6)$$

induced from the sheaf sequence (4.3) is surjective. Then we can find global sections ξ_1 and ξ_2 such that

$$\begin{aligned} \xi_1(p) &\neq 0, \qquad \xi_1(q) = 0 \\ \xi_2(p) &= 0, \qquad \xi_2(q) \neq 0. \end{aligned} \quad (4.7)$$

Lemma 4.3: If the mappings r and s in (4.4) and (4.6) are surjective for all points p and $q \in X$, then there exists a holomorphic embedding of X into $\mathbf{P}_m$, where $\dim_{\mathbf{C}}\mathcal{O}(X, F) = m + 1$.

Proof: Let $\varphi = \{\varphi_0, \ldots, \varphi_m\}$ be a basis for $\mathcal{O}(X, F)$. If f is a holomorphic frame for F at p, then $(\varphi_0(f)(x), \ldots, \varphi_m(f)(x)) \in \mathbf{C}^{m+1}$ for x near p. By assumption, (4.4) is surjective, and hence at least one of the basis elements φ_j is nonzero at p and hence in a neighborhood of p. Thus $[\varphi_0(f)(x), \ldots, \varphi_m(f)(x)]$ is a well-defined point in $\mathbf{P}_m$, for x near p, and is a holomorphic mapping as a function of the parameter x. If $\tilde{f}$ is another holomorphic frame at p, then it is easy to check that

$$\varphi_j(f)(x) = c(x)\varphi_j(\tilde{f})(x),$$

where c is holomorphic and nonvanishing near p, and thus we have a well-defined holomorphic mapping from X into $\mathbf{P}_m$ by

$$\mathbf{\Phi}_\varphi \colon X \longrightarrow \mathbf{P}_m,$$

with

$$\Phi_\varphi(x) = [\varphi_0(f)(x), \dots, \varphi_m(f)(x)].$$

Suppose that the basis φ is replaced by another basis, $\tilde{\varphi} = \{\tilde{\varphi}_j\}$, where

$$\tilde{\varphi}_i = \Sigma c_{ij}\varphi_j, \quad c_{ij} \in \mathbf{C},$$

and that the matrix $C = [c_{ij}]$ is nonsingular. Then consider the diagram

$$\begin{array}{ccc} & \overset{\Phi_\varphi}{\nearrow} & \mathbf{P}_m \\ X & & \downarrow C \\ & \underset{\Phi_{\tilde{\varphi}}}{\searrow} & \mathbf{P}_m, \end{array}$$

where C is the mapping on $\mathbf{P}_m$ defined by the action of the matrix C on the homogeneous coordinates. The diagram is then commutative, and since C is a biholomorphic mapping, it follows that the holomorphic mapping Φ_φ has maximal rank or is an embedding if and only if $\Phi_{\tilde{\varphi}}$ has the same property.

To complete the proof of the lemma, we see that to prove that Φ_φ has maximal rank at $p \in X$ it suffices to find a nice choice of basis φ which demonstrates this property. By hypothesis, the mapping r in (4.4) is surjective, and it follows that we can find sections $\xi_0, \xi_1, \dots, \xi_n \in \mathcal{O}(X, F)$ satisfying the conditions in (4.5). It is easy to verify that $\xi_0, \xi_1, \dots, \xi_n$ are linearly independent in the vector space $\mathcal{O}(X, F)$ and that we can extend them to a basis, $\tilde{\varphi}$. Then the mapping $\Phi_{\tilde{\varphi}}$ is defined, in terms of the frame used in (4.5), by

$$\Phi_{\tilde{\varphi}}(x) = [\xi_0(f)(x), \xi_1(f)(x), \dots, \xi_n(f)(x), \dots],$$

and using the local coordinates

$$(1, \zeta_1, \dots, \zeta_n, \dots)$$

in $\mathbf{P}_m$ and $(z_1, \dots, z_n)$ in X, we see that the Jacobian determinant

$$\frac{\partial(\zeta_1, \dots, \zeta_n)}{\partial(z_1, \dots, z_n)}$$

is given by the coefficient of

$$d\left(\frac{\xi_1(f)}{\xi_0(f)}\right) \wedge \cdots \wedge d\left(\frac{\xi_n(f)}{\xi_0(f)}\right)(p) = \xi_0(f)(p)^{-n} dz_1 \wedge \cdots \wedge dz_n,$$

which is nonzero. Thus $\Phi_{\tilde{\varphi}}$ and hence Φ_φ have maximal rank at p, and consequently Φ_φ is an immersion.

To see that Φ_φ is one to one, we let p and q be two distinct points on X and choose global sections ξ_1 and ξ_2 satisfying (4.7). Then ξ_1 and ξ_2 are clearly linearly independent and extend to a basis

$$\tilde{\varphi} = \{\xi_1, \xi_2, \dots\},$$

and it is clear that $\Phi_{\tilde{\varphi}}$ is one to one, and thus that Φ_φ is an embedding.

Q.E.D.

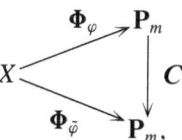

The following lemma will then complete the proof of Theorem 4.1.

Lemma 4.4: The mappings r and s in (4.4) and (4.6) are surjective for all points p and $q \in X$.

Proof: Consider first the mapping r in (4.4). Let $\tilde{X} = Q_pX$ and let $S = \pi_p^{-1}(p)$. Then let $\mathcal{J}_S$ be the ideal sheaf of the submanifold $S \subset \tilde{X}$, i.e., the sheaf of holomorphic functions on $\tilde{X}$ which vanish along S. Let $\mathcal{J}_S^2$ be the ideal sheaf of holomorphic functions on $\tilde{X}$ which vanish to second order along S, i.e., $f_x \in \mathcal{J}_{S,x}^2$ if $f_x = (g_x)^2$, where $g_x \in \mathcal{J}_{S,x}$. Let $\tilde{\mathcal{O}} = \mathcal{O}_{\tilde{X}}$ be the structure sheaf for $\tilde{X}$, let $\mathcal{O} = \mathcal{O}_X$ be the structure sheaf for X, and let $\tilde{F} = \pi^*F$. Then we have the exact sequence of sheaves over $\tilde{X}$ (tensor products are over the structure sheaves),

$$0 \longrightarrow \tilde{\mathcal{O}}(\tilde{F}) \otimes \mathcal{J}_S^2 \longrightarrow \tilde{\mathcal{O}}(\tilde{F}) \longrightarrow \tilde{\mathcal{O}}(\tilde{F}) \otimes \tilde{\mathcal{O}}/\mathcal{J}_S^2 \longrightarrow 0, \tag{4.8}$$

and the mapping π_p induces a commutative diagram

$$\begin{array}{ccccccccc} 0 & \longrightarrow & \tilde{\mathcal{O}}(\tilde{F}) \otimes \mathcal{J}_S^2 & \longrightarrow & \tilde{\mathcal{O}}(\tilde{F}) & \longrightarrow & \tilde{\mathcal{O}}(\tilde{F}) \otimes \tilde{\mathcal{O}}/\mathcal{J}_S^2 & \longrightarrow & 0 \\ & & \uparrow \pi_1^* & & \uparrow \pi^* & & \uparrow \pi_2^* & & \\ 0 & \longrightarrow & \mathcal{O}(F) \otimes m_p^2 & \longrightarrow & \mathcal{O}(F) & \longrightarrow & \mathcal{O}(F) \otimes \mathcal{O}/m_p^2 & \longrightarrow & 0 \end{array} \tag{4.9}$$

given by the topological pullback of the sheaves on X to sheaves on $\tilde{X}$, where π_1^* is the restriction of π^* to the subsheaf $\mathcal{O}(F) \otimes m_p^2$, and π_2^* is the induced map on quotients. We note that if $f \in \mathbf{\Gamma}(U, F)$ for some $U \subset X$, then f vanishes to second order at p if and only if $\pi^*f \in \mathbf{\Gamma}(\pi_p^{-1}(U), \tilde{F})$ vanishes to second order along S, where $S \subset \pi_p^{-1}(U)$, if $p \in U$. This only has to be verified for the structure sheaves, since F is trivial near p, and hence $\tilde{F}$ is trivial in a neighborhood of S. This is easy to do using the local coordinates (z, t) in W as in Sec. 3. Thus π_1^* and hence π_2^* are well defined mappings and one checks easily that π_2^* is injective. Moreover, we claim that there exist isomorphisms α and β making the following diagram commutative:

$$\begin{array}{ccccccc} 0 \longrightarrow & \Gamma(\tilde{X}, \tilde{\mathcal{O}}(\tilde{F}) \otimes \mathcal{J}_S^2) & \longrightarrow & \Gamma(\tilde{X}, \tilde{\mathcal{O}}(\tilde{F})) & \longrightarrow & \Gamma(\tilde{X}, \tilde{\mathcal{O}}(\tilde{F}) \otimes \tilde{\mathcal{O}}/\mathcal{J}_S^2) \\ & \alpha \downarrow \uparrow \pi_1^* & & \beta \downarrow \uparrow \pi^* & & \uparrow \pi_2^* \\ 0 \longrightarrow & \Gamma(X, \mathcal{O}(F) \otimes m_p^2) & \longrightarrow & \Gamma(X, \mathcal{O}(F)) & \xrightarrow{\ r\ } & \Gamma(X, \mathcal{O}(F) \otimes \mathcal{O}/m_p^2). \end{array} \tag{4.10}$$

If we can show that $H^1(\tilde{X}, \mathcal{O}(\tilde{F}) \otimes \mathcal{J}_S^2) = 0$, it follows from (4.10) that r must be surjective. First, we shall construct α and β. For $n = 1$ this is trivial, since $\tilde{X} = X$ and $\pi =$ identity. For $n > 1$ we shall need to use Hartogs' theorem, which asserts that a holomorphic function f defined on $U - \{0\}$, where U is a neighborhood of the origin in $\mathbf{C}^n, n > 1$, can be analytically continued to all of U.† We shall define β and see that the restriction of β to the subspace $\mathbf{\Gamma}(\tilde{X}, \tilde{\mathcal{O}}(\tilde{F}) \otimes \mathcal{J}_S^2)$ (which we shall call α) has the desired image. Namely, suppose that $\xi \in \mathbf{\Gamma}(\tilde{X}, \tilde{\mathcal{O}}(\tilde{F}))$. Then the projection π_p: $\tilde{X} \rightarrow X$ is biholomorphic on the complement of S, and let

$$\tilde{\beta}(\xi) = (\pi_p^{-1})^*\xi,$$

†For an elementary proof of this theorem, see Hörmander [2], Chap. II.

which is a well-defined element of $\mathbf{\Gamma}(X - \{p\}, \mathcal{O}(F))$. Then by Hartogs' theorem, there is a unique extension of $\tilde{\beta}(\xi)$ to a section of $\mathcal{O}(F)$ on X, which we call $\beta(\xi)$. Clearly, we have that $\beta^{-1} = \pi_p^*$ and hence β is an isomorphism. Moreover, as noted above $\beta^{-1}(\eta)$ will vanish to second order along S if and only if $\eta \in \mathbf{\Gamma}(X, \mathcal{O}(F) \otimes m_p^2)$.

It thus remains to show that

$$H^1(\tilde{X}, \tilde{\mathcal{O}}(\tilde{F}) \otimes \mathcal{J}_S^2) = 0. \tag{4.11}$$

To do this, we note that $\mathcal{J}_S$ is a locally free sheaf of rank 1, since it is the ideal sheaf of a divisor in $\tilde{X}$. Moreover, any locally free sheaf corresponds to the sheaf of sections of a vector bundle, and we see that in fact

$$\mathcal{J}_S \cong \tilde{\mathcal{O}}(L^*),$$

where L is the line bundle associated to the divisor $S \subset \tilde{X}$. This is easy to check by verifying that L^* and $\mathcal{J}_S$ have the same transition functions in terms of the coverings of S used in Sec. 3 in its coordinate representation as a subset of $W \subset U \times \mathbf{P}_{n-1}$. Moreover, one also has $\mathcal{J}_S^2 \cong \mathcal{O}((L^*)^2)$, and then we see that

$$H^1(\tilde{X}, \tilde{\mathcal{O}}(\tilde{F}) \otimes \mathcal{J}_S^2) = H^1(\tilde{X}, \tilde{\mathcal{O}}(\tilde{F} \otimes (L^*)^2)).$$

But, by hypothesis, $F = E^\mu$, where $\mu > \mu_0$, and by Proposition 3.3(b)

$$\tilde{F} \otimes (L^*)^2 \otimes K_{\tilde{X}}^* > 0,$$

and thus by Kodaira's vanishing theorem [Theorem 2.4(a)], we see that (4.11) holds.

To see that s in (4.6) is surjective, we let $S = \pi_{pq}^{-1}(\{p\} \cup \{q\})$, let $\mathcal{J}_S$ be the ideal sheaf of this divisor, let $\tilde{\mathcal{O}}$ be the structure sheaf for $Q_p\, Q_q\, X$, and let $\tilde{F} = \pi_{pq}^* F$. We then have the exact sequence

$$0 \longrightarrow \tilde{\mathcal{O}}(\tilde{F}) \otimes \mathcal{J}_S \longrightarrow \tilde{\mathcal{O}}(\tilde{F}) \longrightarrow \tilde{\mathcal{O}}(\tilde{F}) \otimes \tilde{\mathcal{O}}/\mathcal{J}_S \longrightarrow 0, \tag{4.12}$$

and there exists isomorphisms α and β such that the following diagram commutes:

$$\begin{array}{ccccccc} 0 \longrightarrow & \Gamma(\tilde{X}, \tilde{\mathcal{O}}(\tilde{F}) \otimes \mathcal{J}_S) & \longrightarrow & \Gamma(\tilde{X}, \tilde{\mathcal{O}}(\tilde{F})) & \longrightarrow & \Gamma(\tilde{X}, \tilde{\mathcal{O}}(\tilde{F}) \otimes \tilde{\mathcal{O}}/\mathcal{J}_S) \\ & \alpha \downarrow\uparrow \pi_{p,q}^* & & \beta \downarrow\uparrow \pi_{p,q}^* & s & \uparrow \pi_{p,q}^* \\ 0 \longrightarrow & \Gamma(X, \mathcal{O}(F) \otimes m_{pq}) & \longrightarrow & \Gamma(X, \mathcal{O}(F)) & \longrightarrow & \Gamma(X, \mathcal{O}(F) \otimes \mathcal{O}/m_{pq}). \end{array} \tag{4.13}$$

The isomorphisms α and β are constructed using Hartogs' theorem as before, and thus we see that the vanishing of $H^1(\tilde{X}, \tilde{\mathcal{O}}(\tilde{F}) \otimes \mathcal{J}_S)$ will ensure the surjectivity of s. But $\mathcal{J}_S \cong \tilde{\mathcal{O}}(L_{pq}^*)$, and it follows from Proposition 3.3(c) that $\tilde{F} \otimes L_{pq}^* \otimes K_{\tilde{X}}^* > 0$. Applying Kodaira's vanishing theorem again, we obtain the desired result.

Q.E.D.

Remark: Note that in the diagrams (4.10) and (4.13) we would be able to complete the proof if we knew that

$$H^1(X, \mathcal{O}(F) \otimes m_p^2) = H^1(X, \mathcal{O}(F) \otimes m_{pq}) = 0.$$

This is the approach taken by Grauert [2] and gives an alternative proof of the embedding theorem. Namely, Grauert proves the more general vanishing theorem: If E is a positive line bundle and $\mathcal{F}$ is any coherent analytic sheaf, then there is an integer $\mu_0 > 0$ so that

$$H^q(X, \mathcal{O}(E^\mu) \otimes \mathcal{F}) = 0$$

for $\mu \geq \mu_0$ and $q \geq 1$. This result is derived from the general theory of coherent analytic sheaves on pseudoconvex analytic spaces and involves, in particular, Grauert's solution to the Levi problem (Grauert [1]; see Gunning and Rossi [1] for this derivation). Moreover, one needs to know that the ideal sheaves m_p^2 and m_{pq} are coherent analytic sheaves (which for these *particular* ideals sheaves is not too difficult to prove). Kodaira's approach, which we have followed here, says, in effect, that if you blow up the points p appropriately, then the coherent sheaves m_p^2 and m_{pq} become locally free on the blown up complex manifold $\tilde{X}$, and then the theory of harmonic differential forms (which applies at this time only to locally free sheaves, i.e., vector bundles) can be applied to give the desired vanishing theorems. To prove Grauert's vanishing theorem via harmonic theory, it is necessary to *first* obtain a projective embedding; then, by finding a global projective resolution of the given coherent sheaf by locally free sheaves (which follows from the work of Serre [2]), one can deduce Grauert's result (see Griffiths [1] for this derivation).

A recent generalization of Kodaira's vanishing and embedding theorems to Moishezon spaces (generalizations of projective algebraic spaces) by Grauert and Riemenschneider [2] and Riemenschneider [1] has involved the approach used by Kodaira presented here, combined with the theory of coherent analytic sheaves on pseudoconvex spaces.

APPENDIX

BY OSCAR GARCÍA-PRADA*

MODULI SPACES AND GEOMETRIC STRUCTURES

1. Introduction

Moduli spaces arise naturally in classification problems in geometry. Typically, one has a collection of objects and an equivalence relation, and the problem is to describe the set of equivalence classes. Usually, there are discrete invariants that partition this set in a countable number of subsets. In most cases there exist continuous families of objects, and one would like to give the set of equivalence classes some geometric structure to reflect this fact. This is the object of the theory of moduli spaces.

The word *moduli* is due to Riemann†, who used it as a synonym for parameters when showing that the space of equivalence classes of Riemann surfaces of a given genus g depends on $3g-3$ complex numbers. After this, the concept of moduli has been used in geometry in a rather loose sense to measure variations of geometric structures of one kind or another, but it has not been however until the 1960s that one has been able to formulate moduli problems in precise terms and in some cases to obtain solutions to them.

The study of moduli spaces has a local and a global aspect. While the local theory progressed through the initial work of Kodaira–Spencer and the contributions of Kuranishi, and others, the global theory, with the proper definition of moduli spaces and how to construct them is due to Mumford [25] based on Grothendieck's theory of schemes.

In this appendix we are concerned with moduli spaces of holomorphic vector bundles and Higgs bundles over a Riemann surface, as well as related structures, like connections and representations of the fundamental group of the surface. These moduli spaces play a very prominent role in algebraic geometry, differential geometry, topology and theoretical physics. The emphasis here is not so much on the construction of these moduli spaces

*Instituto de Ciencias Matemáticas, Consejo Superior de Investigaciones Científicas, Serrano, 121, 28006 Madrid, Spain.

†"hängt ...von $3g-3$ stetig veränderliche Grössen ab, welche die Moduln dieser Klasse gennant werden sollen" [32].

as on the correspondences among them and the geometric structures that they carry. Our goal is to review some developments in the theory that have taken place mostly since the beginning of the 1980s, including some more recent progress. Naturally, due to space limitations, we do this in a rather scketchy way, avoiding many technical aspects and omitting most proofs, but trying to address the reader to the main bibliography in the subject.

In section 2 we study the moduli space of holomorphic vector bundles. After the classification of holomorphic vector bundles for genus 0 by Grothendieck [16] and genus 1 by Atiyah [2], vector bundles on higher genus Riemann surfaces have been studied extensively with the fundamental work of Mumford [25] and of Narasimhan and Seshadri [28], who defined the concept of a stable vector bundle and constructed the moduli spaces which classify these bundles. In their theorem Narasimhan and Seshadri [28] identified the moduli space of stable vector bundles over a compact Riemann surface with the moduli space of irreducible projective unitary representations of the fundamental group of the surface (or equivalently, unitary representations of the universal central extension of the fundamental group). Donaldson [10] gave another proof of this theorem, which we briefly scketch, following the gauge-theoretic point of view of Atiyah and Bott [4], where these representatations are identified with projectively flat connections.

A Higgs bundle over a Riemann surface is a pair consisting of a holomorphic vector bundle and an endomorphism of the bundle twisted with the canonical line bundle of the surface — the Higgs field. Higgs bundles were introduced by Hitchin [17] in the study of self-duality equations. They are the holomorphic objects corresponding to complex representations of the fundamental group which are not necessarily unitary. In fact when the Higgs field vanishes, one recovers unitary representations. To establish this correspondence requires two fundamental theorems: one proving the existence of a solution to Hitchin equations on a stable Higgs bundle [17, 34, 35], which we explain in Section 3, and another one, studied in Sections 4 and 5, proving the existence of a harmonic metric on a bundle equipped with a reductive flat conection [8, 14]. This correspondence is in some sense a non-abelian version of the classical Hodge theory.

In Section 6 we explain how Higgs bundles can also be used to study representations of the fundamental group of the surface and its universal central extension that preserve an indefinite Hermitian metric of signature (p, q) on $\mathbf{C}^n$, with $n = p+q$. This is equivalent to studying representations in the non-compact real form $\mathrm{U}(p, q)$ of the complex group $\mathrm{GL}(n, \mathbf{C})$. The representations in the compact form $\mathrm{U}(n)$ are those preserving a definite Hermitian metric and, as mentioned above, correspond to Higgs bundles with vanishing Higgs field, i.e., holomorphic vector bundles. We will see that the representations in $\mathrm{U}(p, q)$ are in correspondence with Higgs bundles of a particular structure, determined by the geometry of $\mathrm{U}(p, q)$.

All our moduli spaces carry various types of geometric structures. Naturally, the moduli spaces of vector bundles and Higgs bundles have complex structures induced by the complex structure of the Riemann surface. Moreover, they also have a symplectic structure, which is in fact compatible with the complex structure, endowing the moduli spaces with a Kähler structure. One way to see this, is to exhibit the moduli space as a symplectic and Kähler quotient. This point of view, initiated by Atiyah and Bott [4], is studied in Section 7, where we see that the gauge-theoretic equations defining the moduli space can be interpreted as the vanishing of the moment map for a symplectic action of the gauge group on the space of connections and Higgs fields. In the case of Higgs bundles, more is actually true. The complex structure of $GL(n, \mathbf{C})$ defines an additional complex structure on the moduli space, which combined with the one induced by the surface give rise to a hyperkähler structure that can be obtained as a hyperkähler quotient.

We finish in Section 8 by briefly studying a generalization to higher dimensional compact Kähler manifolds of some of the previous results. We first explain the Hitchin–Kobayashi correspondence between stable vector bundles and Hermitian–Einstein connections, proved by Donaldson [12, 13] and Uhlenbeck and Yau [38], and then finish with the higher dimensional version of the non-abelian Hodge theory correspondence, due to Simpson [34, 35] and Corlette [8].

We follow the notation of the book.

Acknowledgements. First of all I want to thank Ronny Wells for having invited me to write this appendix for the third edition of his book. I must confess that I have always been very embarrassed by the certainty that the appendix could never approach the quality and elegance of Ronny's text. I hope, however, that the reader finds the topics presented here stimulating enough to pursue further study of a subject whose basics are so beautifully treated in Ronny's book. I have a great debt to Nigel Hitchin and Simon Donaldson, from whom I have learned the subject treated here. Thanks are also due to Steve Bradlow and Peter Gothen, my collaborators in the original work on $U(p, q)$ reviewed in this appendix. I am very grateful to my students, Álvaro Antón, Marta Aparicio, Mario García, and Roberto Rubio for having gone through several preliminary versions of this appendix and having made many useful comments.

Much of the final preparation of this appendix has taken place while the author was visiting the IHES, whose hospitality and support is warmly thanked.

2. Vector Bundles on Riemann Surfaces

2.1. Moduli Spaces of Vector Bundles

Let X be a connected compact Riemann surface. As we have seen in Chap. III†, the group $H^1(X, \mathcal{O}^*)$ parametrizes isomorphism classes of holomorphic

†All references to chapters, sections and page numbers refer to the preceeding Chap. I–VI of this book

line bundles on X, where the group structure is given by the tensor product of line bundles. This group is called the *Picard group* of X. If we fix the first Chern class — which we will call *degree* from now on — to be d, then $\mathrm{Pic}^d(X)$ is the set of equivalence classes of line bundles of degree d. In particular, $J(X) := \mathrm{Pic}^0(X)$ is a subgroup of $H^1(X, \mathcal{O}^*)$, which is called the *Jacobian* of X. The group $J(X)$ has the structure of a complex abelian variety (see Remark (a) on page 234), whose dimension is equal to the genus of X, and it is very important in the study of the geometry of the Riemann surface [1].

The problem we will address now is that of classifying isomorphism classes of holomorphic vector bundles of arbitrary rank on a compact Riemann surface X. This is a much harder problem than that of line bundles, basically due to the fact that this set is no longer a group. As in the case of line bundles, C^∞ vector bundles are classified by the first Chern class, and hence an equivalence class of C^∞ vector bundles is determined by the rank n and the degree d.

For genus zero, Grothendieck [16] showed that every holomorphic vector bundle on $\mathbf{P}_1(\mathbf{C})$ is a direct sum of holomorphic line bundles, and it is hence determined by the degrees of these line bundles. The case of a genus one surface (a torus) was studied by Atiyah [2], who showed that the set of equivalence classes of indecomposable vector bundles with fixed rank and degree is isomorphic to the surface itself. We will then assume most of the time that the genus of X is greater than one.

We will look now at holomorphic vector bundles from the point of view of $\bar\partial$-operators, which will be more adapted to our approach. As we have seen in Sec. 3 of Chap. II, a holomorphic vector bundle E on a complex manifold X defines a $\mathbf{C}$-linear mapping

$$\bar\partial_E\colon \mathcal{E}^0(X, E) \longrightarrow \mathcal{E}^{0,1}(X, E),$$

which satisfies

$$\bar\partial_E(\varphi\xi) = \bar\partial\varphi\xi + \varphi\bar\partial_E\xi,$$

where $\varphi \in \mathcal{E}(X)$ and $\xi \in \mathcal{E}(X, E)$. This can be extended to a $\mathbf{C}$-linear mapping

$$\bar\partial_E\colon \mathcal{E}^{0,1}(X, E) \longrightarrow \mathcal{E}^{0,2}(X, E),$$

such that $\bar\partial_E^2 = 0$.

Conversely, if $\mathbb{E}$ is a smooth vector bundle and we have an operator $\bar\partial_E$ as above satisfying $\bar\partial_E^2 = 0$, we can find holomorphic transition functions by solving locally the equation $\bar\partial_E s = 0$ [19] and endow $\mathbb{E}$ with the structure of a holomorphic vector bundle. Of course, if we are on a Riemann surface, $\mathcal{E}^{0,2}(X) = 0$, and hence the condition $\bar\partial_E^2 = 0$ is always satisfied.

Let $\mathbb{E}$ be a smooth complex vector bundle of rank n and degree d over a Riemann surface X. Let $\mathcal{C}$ be the set of all $\bar\partial_E$ operators defined on $\mathbb{E}$. As we have mentioned above, this set is in bijection with the set of holomorphic structures on $\mathbb{E}$. A holomorphic bundle E can be then thought of as a pair $(\mathbb{E}, \overline{\partial}_E)$. The group $\mathcal{G}^c$ of automorphisms of the vector bundle $\mathbb{E}$ acts on $\mathcal{C}$ by the rule:

$$g \cdot \overline{\partial}_E = g\overline{\partial}_E g^{-1}, \quad \text{where} \quad g \in \mathcal{G}^c \quad \text{and} \quad \overline{\partial}_E \in \mathcal{C},$$

and the above correspondence can be made precise in the following.

Proposition 2.1: The quotient $\mathcal{C}/\mathcal{G}^c$ can be identified with the set of isomorphism classes of holomorphic vector bundles of rank n and degree d.

The set $\mathcal{C}/\mathcal{G}^c$ can be endowed with the C^∞ topology, although for technical reasons one should consider appropriate Sobolev spaces instead of the C^∞ topology. It turns out that this space is in general non-Hausdorff (see [28]), but a "good" space is obtained if we consider only (semi)stable holomorphic structures.

Let $E = (\mathbb{E}, \overline{\partial}_E)$ be a holomorphic vector bundle over a compact Riemann surface X. Let $\deg(E)$ be the *degree* of E (i.e., the first Chern class $c_1(E)$). The *slope* of E is defined as

$$\mu(E) = \deg E / \operatorname{rank} E.$$

The vector bundle E is *stable* if for every proper subbundle $F \subset E$

$$\mu(F) < \mu(E).$$

This concept arises naturally from Mumford's Geometric Invariant Theory (GIT, see [25]).

Let

$$\mathcal{C}^s = \{\overline{\partial}_E \in \mathcal{C} \ : \ E = (\mathbb{E}, \overline{\partial}_E) \quad \text{is stable}\}$$

The *moduli space of stable vector bundles* of rank n and degree d is defined as

$$M^s(n, d) = \mathcal{C}^s / \mathcal{G}^c.$$

Using analytic methods (see e.g. [20]) one can show the following.

Theorem 2.2: The moduli space of stable bundles $M^s(n, d)$ has the structure of a complex manifold of dimension $1 + n^2(g - 1)$, where g is the genus of X.

Notice that every line bundle is stable and $M^s(1, d)$ is simply the component of $\operatorname{Pic}(X)$ consisting of line bundles of degree d, in particular, $M^s(1, 0)$ is the Jacobian $J(X)$.

We can easily compute the tangent space of $M^s(n, d)$ at a point $[\overline{\partial}_E]$. To do this, we first observe that $\mathcal{C}$ is an affine space modelled on $\mathcal{E}^{0,1}(X, \operatorname{End}\mathbb{E})$, where $\operatorname{End}\mathbb{E} = \operatorname{Hom}(\mathbb{E}, \mathbb{E})$, and hence the tangent space of $\mathcal{C}$ at $\overline{\partial}_E$ is isomorphic to $\mathcal{E}^{0,1}(X, \operatorname{End} \mathbb{E})$. Notice that in higher dimensions, the equation $\overline{\partial}_E^2 = 0$ is not automatically satisfied, and it is not difficult to show that the tangent space to the space of $\overline{\partial}_E$ operators satisfying $\overline{\partial}_E^2 = 0$ is the kernel of the map

$$\overline{\partial}_E \colon \mathcal{E}^{0,1}(X, \operatorname{End} \mathbb{E}) \longrightarrow \mathcal{E}^{0,2}(X, \operatorname{End} \mathbb{E}).$$

Here we are abusing notation since the $\overline{\partial}$-operator associated to $\operatorname{End} \mathbb{E} = \mathbb{E} \otimes \mathbb{E}^*$ is $\overline{\partial}_{E\otimes E^*} = \overline{\partial}_E \otimes I_{E^*} + I_E \otimes \overline{\partial}_{E^*}$, and $\overline{\partial}_{E^*}$ is naturally induced by $\overline{\partial}_E$.

Now, if a curve $\overline{\partial}_E(t) = \overline{\partial}_E + t\alpha$ in $\mathcal{C}$ for $|t| < \epsilon$ is obtained by a 1-parameter family of gauge transformations $g_t \in \mathcal{G}^c$, so that

$$\overline{\partial}_E(t) = g_t \overline{\partial}_E g_t^{-1}, \quad \text{with} \quad g_0 = I_{\mathbb{E}},$$

then $\alpha = -\overline{\partial}_E a$, where $a = \partial_t g_t|_{t=0}$. That is, a is an element of the tangent space at the identity of the infinite dimensional Lie group $\mathcal{G}^c$, which is isomorphic to $\mathcal{E}^0(X, \operatorname{End} \mathbb{E})$. Hence the "tangent space" at a point $[\overline{\partial}_E] \in \mathcal{C}/\mathcal{G}^c$ is isomorphic to

$$H^1(X, \operatorname{End} E) = \frac{\{\alpha \in \mathcal{E}^{0,1}(X, \operatorname{End} \mathbb{E})\}}{\{\overline{\partial}_E a\colon \ a \in \mathcal{E}^0(X, \operatorname{End} \mathbb{E})\}}.$$

Since stability is an open condition, this is indeed the tangent space at a point $[\overline{\partial}_E] \in M^s(n, d)$. To compute the dimension of $H^1(X, \operatorname{End} E)$ when E is stable, we first recall the *Riemann-Roch* theorem, (a special case of Theorem 5.8, Chap. IV, when X is a Riemann surface) which says that if E is a holomorphic vector bundle of degree d and rank n

$$\chi(E) = \dim H^0(X, E) - \dim H^1(X, E) = d - n(g-1),$$

where g is the genus of X. Applying this to the holomorphic vector bundle $\operatorname{End} E$ whose rank is n^2 and has 0 degree we obtain

$$\chi(\operatorname{End} E) = \dim H^0(X, \operatorname{End} E) - \dim H^1(X, \operatorname{End} E) = -n^2(g-1).$$

Here, we have used that the degree of $\operatorname{End} E$ vanishes, as follows from the first Chern class relation

$$\deg(E \otimes F) = \operatorname{rank} F \deg E + \operatorname{rank} E \deg F,$$

which can be deduced using Chern–Weil theory (Chap. III, Sec. 3).

A holomorphic vector bundle is said to be *simple* if $H^0(X, \operatorname{End} E) = \mathbf{C}$. It is not difficult to see that if E is stable, then it is simple (see e.g. [20]). Hence

$$\dim H^1(X, \operatorname{End} E) = 1 + n^2(g-1),$$

thus obtaining the dimension of $M^s(n, d)$ given by Theorem 2.2.

The manifold $M^s(n,d)$ is not compact in general. However it can be compactified by considering semistable vector bundles (it is indeed this compactification which is obtained by GIT). A vector bundle E is *semistable* if for every subbundle $F \subset E$

$$\mu(F) \leq \mu(E).$$

Another concept which is relevant here is that of polystability. The vector bundle E is *polystable* if $E = \oplus E_i$, where E_i is stable and $\mu(E_i) = \mu(E)$. A polystable bundle is in particular semistable.

Let E be be semistable vector bundle. If E is not stable, then there exists a stable subbundle $F \subset E$, such that $\mu(F) = \mu(E)$. Since E/F is semistable with $\mu(E/F) = \mu(E)$, this process can be iterated, to obtain a filtration of E (*Jordan–Hölder filtration*) by semistable vector bundles E_j,

$$0 = E_0 \subset E_1 \subset \ldots \subset E_m = E,$$

such that E_j/E_{j-1} is stable and $\mu(E_j/E_{j-1}) = \mu(E)$ for $1 \leq j \leq m$. The *graded vector bundle* associated to E is defined as

$$\operatorname{gr}(E) = \bigoplus_j E_j/E_{j-1}.$$

Although, this is not uniquely defined, its isomorphism class is. Two semistable vector bundles E and E' are said to be S-equivalent if $\operatorname{gr}(E) \cong \operatorname{gr}(E')$. The *moduli space of semistable vector bundles* $M(n,d)$ of rank n and degree d is defined as the set of S-equivalence classes of semistable bundles of rank n and degree d. Note that, since the graded vector bundle associated to a semistable bundle is polystable, $M(n,d)$ coincides with the set of isomorphism classes of polystable vector bundles of rank n and degree d.

One of the first applications of Mumford's Geometric Invariant Theory [25] was to the algebraic construction of $M(n,d)$.

Theorem 2.3: The moduli space $M(n,d)$ has the structure of an irreducible projective algebraic variety which contains $M^s(n,d)$ as an open smooth subvariety.

If $\mathrm{GCD}(n,d) = 1$, then there are no strictly semistable vector bundles and $M(n,d) = M^s(n,d)$, and hence we have the following.

Corollary 2.4: If $\mathrm{GCD}(n,d) = 1$, then $M(n,d)$ is a connected projective algebraic manifold of dimension $1 + n^2(g-1)$.

For the GIT construction of $M(n,d)$ and basic facts on the geometry of $M(n,d)$ the reader may look at [25, 28, 33, 29, 27, 26, 5].

2.2. Stable Bundles and Connections

We will see now that the stability of a holomorphic vector bundle emerges also from a differential-geometric point of view in relation to the existence of a certain class of Hermitian metrics on the bundle.

Let E be a holomorphic vector bundle over X and let h be a Hermitian metric on E. Let D be the canonical connection on E defined by h, and let $\Theta(D) = D^2$ be its curvature (see Chap. III, Sec. 2).

The connection D is said to be *flat* if $\Theta(D) = 0$. From Chern-Weil theory (Chap. III, Sec. 3), we know that the flatness of D implies that the first Chern class of E must be zero. If $c_1(E) \neq 0$, then the closest to flatness that we can have is that the curvature be central. To formulate this condition, it is convenient to fix a Hermitian metric on X. Since the dimension of X is one, such a metric is always Kähler (Chap. V, Sec. 4). Let Ω be its Kähler form. We will normalize this metric so that X has volume 2π, i.e., $\int_X \Omega = 2\pi$.

Now, we say that D has *constant central curvature* if

$$\Theta(D) = -i\mu I_E \Omega, \tag{2.1}$$

where I_E is the identity endomorphism of E and μ is a real constant. Taking the trace in (2.1), integrating and using that

$$c_1(E) = \frac{i}{2\pi}\int_X \mathrm{Tr}(\Theta(D))$$

(Chap. III, Sec. 3) we have that

$$\mu = \mu(E) = \deg E / \operatorname{rank} E.$$

Condition (2.1) can be better understood in terms of principal bundles. To do this, we briefly recall some basic facts about principal bundles. Let G be a Lie group. A principal G-bundle P over a smooth manifold X is a manifold with a smooth (right) G-action and orbit space $P/G = X$. We demand that the action admit local product structures, i.e., it is locally equivalent to the obvious action $U \times G$, where U is an open set in X. Then we have a fibration $\pi : P \to X$. We say that P has structure group G.

Now, a connection A for P is a G-invariant splitting of the natural exact sequence

$$0 \longrightarrow T_F P \longrightarrow TP \longrightarrow \pi^{-1}TX \longrightarrow 0 \tag{2.2}$$

of vector bundles over P. Here T_F denotes the tangent bundle along the fibres in P, and TX the tangent bundle of X. The group G acts on all terms of this sequence and so G-invariant splitting of (2.2) is a well-defined concept.

Given a principal G-bundle P over X and a representation $\rho : G \to \mathrm{GL}(\mathbf{V})$ of G on a complex vector space $\mathbf{V}$, one can associate to it a complex vector bundle defined by

$$\mathbb{V} = (P \times \mathbf{V})/G,$$

where G acts on P on the right, and on $\mathbf{V}$ on the left, via the representation ρ. Conversely, if $\mathbb{E}$ is a smooth complex vector bundle of rank n over X, we obtain a principal bundle P with structure group $\mathrm{GL}(n, \mathbf{C})$ by taking the set of all frames in $\mathbb{E}$. A point in the fibre of P over $x \in X$ is a set of basis vectors for E_x. A connection on the principal bundle P induces a connection on the vector bundle $\mathbb{E}$, and viceversa (see e.g. [11, 20]).

If the Lie group G is complex, all the constructions above on principal bundles can also be done in the holomorphic category.

Now, condition (2.1) is equivalent to saying that D is projectively flat, which means that D induces a flat connection on the $\mathrm{PGL}(n, \mathbf{C})$-principal bundle $\hat{P}$ associated to the $\mathrm{GL}(n, \mathbf{C})$-principal bundle P defined by E, where $\hat{P} = P/\mathbf{C}^* I_n$.

To analyse the relation with stability, we consider a holomorphic subbundle $F \subset E$ and the quotient vector bundle $Q = E/F$. These fit in an exact sequence

$$0 \longrightarrow F \longrightarrow E \longrightarrow Q \longrightarrow 0. \tag{2.3}$$

The Hermitian metric h on E defines a smooth splitting (not necessarily holomorphic) of this sequence

$$E \cong F \oplus Q,$$

with respect to which we can write

$$\bar{\partial}_E = \begin{pmatrix} \bar{\partial}_F & \beta \\ 0 & \bar{\partial}_Q \end{pmatrix}, \tag{2.4}$$

where $\bar{\partial}_F$ and $\bar{\partial}_Q$ are the corresponding $\bar{\partial}$ operators on F and Q, respectively, and $\beta \in \mathcal{E}^{0,1}(X, \mathrm{Hom}(Q, F))$. Since $\bar{\partial}_E^2 = 0$, we have that $\bar{\partial}_{\mathrm{Hom}(Q,F)}\beta = 0$ and hence β defines a class

$$[\beta] \in H^1(X, \mathrm{Hom}(Q, F)).$$

Conversely, given F and Q, the set of equivalence classes of extensions as (2.3) is given by $H^1(X, \mathrm{Hom}(Q, F))$, with $[\beta] = 0$ corresponding to the holomorphic trivial extension $E = F \oplus Q$.

The canonical connection D on E defined by h can be written as

$$D = \begin{pmatrix} D_F & \beta \\ -\beta^* & D_Q \end{pmatrix}, \tag{2.5}$$

where D_F and D_Q are the canonical connections defined by the Hermitian metrics on F and Q, respectively, induced by h, and $\beta^* \in \mathcal{E}^{1,0}(X, \mathrm{Hom}(F, Q))$ is obtained from β by combining the conjugation on the form part and the conjugate-linear bundle isomorphism between $\mathrm{Hom}(Q, F)$ and $\mathrm{Hom}(F, Q)$, defined by the Hermitian metric (Chap. V, Sec. 2).

Now, the curvature of D is given by

$$\Theta(D) = \begin{pmatrix} \Theta(D_F) - \beta \wedge \beta^* & D'\beta \\ -D''\beta^* & \Theta(D_Q) - \beta^* \wedge \beta \end{pmatrix}. \tag{2.6}$$

We can now prove the following.

Proposition 2.5: Let E be a holomorphic vector bundle over X. Let h be a Hermitian metric on E, such that the canonical connection satisfies (2.1). Then E is polystable.

Proof: Suppose that D satisfies (2.1). Let $F \subset E$ be a holomorphic subbundle and let $Q = E/F$. From (2.6) we have that

$$\Theta(D_F) - \beta \wedge \beta^* = -i\mu I_F \Omega.$$

Now, taking the trace of this equation, multiplying by $i/2\pi$ and integrating over X we have

$$\frac{i}{2\pi}\int_X \mathrm{Tr}(\Theta(D_F)) + \int_X |\beta|^2\Omega = \mu \operatorname{rank} F,$$

where we have used that $-i/2\pi\, \mathrm{Tr}(\beta \wedge \beta^*) = |\beta|^2\Omega$ (Chap. V, Sec. 2). Since $\deg(F) = i/2\pi \int_X \mathrm{Tr}(\Theta(D_F))$ we obtain that $\mu(F) \leq \mu = \mu(E)$. If E is indecomposable, then $\beta \neq 0$ for every $F \subset E$, and $\mu(F) < \mu(E)$. Hence E is stable. If $\beta = 0$ for some $F \subset E$, then $E = F \oplus Q$ and clearly $\mu(F) = \mu(Q) = \mu(E)$. Since D_F and D_Q have also central curvature, we can iterate the process (the rank of E is finite) until we obtain that $E = \oplus E_i$ with E_i stable and $\mu(E_i) = \mu(E)$ for every i, that is, E is polystable.

Q.E.D.

Note that if, in the proof above, we had considered the piece

$$\Theta(D_Q) - \beta^* \wedge \beta = -i\mu I_Q \Omega,$$

in (2.1) we would have obtained that $\mu(E) \leq \mu(Q)$, which is indeed an equivalent definition of semistability, and hence would have reached the same conclusion.

To explain the converse of Proposition 2.5, it is convenient to look at (2.1) from another point of view. Instead of fixing a holomorphic structure and looking for a Hermitian metric, we fix the Hermitian metric and look for the holomorphic structure or, what is equivalent, for the canonical connection.

To be more precise, let $\mathbb{E}$ be a smooth complex vector bundle over X. Let h be a Hermitian metric on $\mathbb{E}$. Let $\mathcal{A}$ be the set of all connections on $\mathbb{E}$, which are compatible with h. The gauge group $\mathcal{G}$ of $(\mathbb{E}, h)$ is the subgroup of $\mathcal{G}^c$ consisting of those automorphisms of $\mathbb{E}$ that preserve h. The group $\mathcal{G}$ acts on $\mathcal{A}$ as follows:

$$g \cdot D = gDg^{-1}, \quad \text{where} \quad g \in \mathcal{G} \quad \text{and} \quad D \in \mathcal{A}.$$

The connection $D \in \mathcal{A}$ has constant central curvature if

$$\Theta(D) = \lambda, \tag{2.7}$$

where

$$\lambda = -i\mu(\mathbb{E}) I_{\mathbb{E}} \Omega. \tag{2.8}$$

This equation is invariant under the action of $\mathcal{G}$ and then, if we set

$$\mathcal{A}_0 = \{D \in \mathcal{A} \ : \ \Theta(D) = \lambda\}, \tag{2.9}$$

the quotient $\mathcal{A}_0/\mathcal{G}$ is the moduli space of constant central curvature connections on $(\mathbb{E}, h)$.

Indeed, solving (2.1) for a Hermitian metric on a fixed holomorphic vector bundle E is equivalent to fixing a Hermitian metric on the smooth bundle $\mathbb{E}$, and then solving for a Hermitian connection satisfying (2.7).

To show this, we note that the space of Hermitian metrics on $\mathbb{E}$ is the space of global sections of the $\mathrm{GL}(n, \mathbf{C})/U(n)$-bundle associated to the principal $\mathrm{GL}(n, \mathbf{C})$-bundle of frames of $\mathbb{E}$, which in turn, by fixing h, can be identified with the symmetric space $\mathcal{G}^c/\mathcal{G}$, where $\mathcal{G}$ is the gauge group of $(\mathbb{E}, h)$. Now, if h' is a solution to (2.1) on the holomorphic bundle $E = (\mathbb{E}, \overline{\partial}_E)$, then if $h = g(h')$ for $g \in \mathcal{G}^c$, then the h-connection corresponding to $g(\overline{\partial}_E)$ solves (2.7) on the Hermitian bundle $(\mathbb{E}, h)$. Conversely, if on the $\mathcal{G}^c$-orbit of $\overline{\partial}_E$ we find an element $g(\overline{\partial}_E)$, whose corresponding h-connection satifies (2.7), then $h' = g(h)$ is the Hermitian metric solving (2.1) on E.

A connection $D \in \mathcal{A}$ is said to be *reducible* if $(\mathbb{E}, h) = (\mathbb{E}_1, h_1) \oplus (\mathbb{E}_2, h_2)$ and $D = D_1 \oplus D_2$, where D_1 and D_2 are connections on the Hermitian vector bundles $(\mathbb{E}_1, h_1)$ and $(\mathbb{E}_2, h_2)$, respectively. We say that D is *irreducible* if it is not reducible. We denote by $\mathcal{A}^*$ and $\mathcal{A}_0^*$ the subsets of $\mathcal{A}$ and $\mathcal{A}_0$, respectively, consisting of irreducible connections.

Using analytic methods similar to those used for the moduli space of stable vector bundles (see e.g. [20]), one has the following.

Theorem 2.6: The moduli space $\mathcal{A}_0^*/\mathcal{G}$ of irreducible constant central curvature connections on $(\mathbb{E}, h)$ has the structure of a smooth real manifold of dimension $2 + 2n^2(g-1)$, where n is the rank of $\mathbb{E}$.

Consider now the map

$$\begin{aligned} \mathcal{A} &\longrightarrow \mathcal{C} \\ D &\longmapsto D'' \end{aligned} \tag{2.10}$$

where D'' is the projection of D into the $(0,1)$-part (Chap. 3, Sec. 2). Theorem 2.1 in Chap. III. establishes that this map is an isomorphism.

Proposition 2.5 can be now reformulated as giving one direction of the following.

Theorem 2.7: The map (2.10) restricted to $\mathcal{A}_0$ has its image in the subspace $\mathcal{C}^{ps} \subset \mathcal{C}$ of polystable holomorphic structures, and descends to give a homeomorphism

$$\mathcal{A}_0/\mathcal{G} \cong \mathcal{C}^{ps}/\mathcal{G}^c. \tag{2.11}$$

Moreover, this restricts to give a homeomorphism

$$\mathcal{A}_0^*/\mathcal{G} \cong \mathcal{C}^s/\mathcal{G}^c. \tag{2.12}$$

This is the theorem of Narasimhan and Seshadri [28]. The formulation in [28] is in terms of representations of the fundamental group of the surface but, as we will see in Section 4, this is equivalent. Here we follow the approach of Atiyah–Bott [4] and Donaldson [10].

It is enough to prove surjectivity for (2.12). The polystable case can easily be reduced to the stable one. We just mention the main ideas of the proof given by Donaldson [10]. Let $D \in \mathcal{A}$ be the Hermitian connection corresponding to $\overline{\partial}_E$. The theorem says that, even though D may not satisfy (2.7), in the $\mathcal{G}^c$-orbit $\mathcal{O}(D)$ of D (where the action is via the identification of $\mathcal{A}$ with $\mathcal{C}$) we can find a connection which does satisfy (2.7), which is unique up to gauge transformations in $\mathcal{G}$. We suppose inductively that the result has been proved for bundles of lower rank (the case of line bundles being an easy consequence of Hodge theory). Then we choose a minimizing sequence in $\mathcal{O}(D)$ for a carefully constructed functional of the curvature, which is equivalent to the Yang-Mills functional, defined by $\|\Theta(D)\|_{L^2}$, and extract a weakly convergent subsequence. Either the limiting connection is in $\mathcal{O}(D)$ and we deduce the result by examining small variations within the orbit, or in another orbit and we deduce that E is not stable. The main ingredient in this approach is a result of Uhlenbeck on the weak compactness of the set of connections with L^2 bounded curvature. In the intermediate stages of the argument one has to allow generalized connections of class W^1, i.e., which differ from some fixed C^∞ connection D by an element of the Hilbert space with norm

$$\|\alpha\|_{W^1}^2 = \|\alpha\|^2 + \|D\alpha\|^2,$$

with curvature in L^2 and gauge transformations in W^2 (see Chap. IV, Sec. 1, for the definition of Sobolev norms). As explained in Atiyah–Bott [4] the group actions and properties of curvature that we use extend without essential change (in particular $W^2 \subset C^0$, so the topology of the bundle is preserved), and it is proved in [4] that each W^1 connection defines a holomorphic structure.

3. Higgs Bundles on Riemann Surfaces

3.1. Moduli Spaces of Higgs Bundles

There are many natural gauge-theoretic equations similar to (2.7) which involve also *Higgs fields*. These are simply sections of a certain vector bundle naturally associated to the original vector bundle. As for equation (2.7), the existence of solutions is related to some stability condition. Some of these equations arise from considering solutions to the Yang-Mills equations which are invariant under a certain symmetry group—a mechanism known in gauge theory as *dimensional reduction*. An important example is provided by the theory of Higgs bundles introduced by Hitchin [17].

Let X be a compact Riemann surface. A *Higgs bundle* over X is a pair consisting of a holomorphic vector bundle $E \to X$ together with a sheaf homomorphism — the *Higgs field* — $\Phi : E \to E \otimes K$, where K is the canonical line bundle of X, i.e.

$$\Phi \in H^0(X, \operatorname{End} E \otimes K).$$

The Higgs bundle (E, Φ) is said to be *stable* if

$$\mu(F) < \mu(E)$$

for every proper subbundle $F \subset E$ such that $\Phi(F) \subset F \otimes K$. Semistability, polystability, Jordan–Hölder filtrations and S-equivalence are defined in a similar way to vector bundles. Clearly, if E is stable, (E, Φ) is a stable Higgs bundle for every $\Phi \in H^0(X, \operatorname{End} E \otimes K)$.

The *moduli space of stable Higgs bundles* $\mathcal{M}^s(n, d)$ is defined as the set of isomorphism classes of stable Higgs bundles (E, Φ) with $\operatorname{rank} E = n$ and $\deg E = d$. In [17], Hitchin gave an analytic construction of this moduli space in the rank 2 case. Similarly, we define the *moduli space of semistable Higgs bundles* $\mathcal{M}(n, d)$ as the set of S-equivalence classes of semistable Higgs bundles of rank n and degree d. In contrast with the case of vector bundles, $\mathcal{M}(n, d)$ is not a projective algebraic variety, but only quasi-projective, i.e., it is an open subset of a projective variety. To be more precise, the GIT construction given by Nitsure [31] says the following.

Theorem 3.1: The moduli space of semistable Higgs bundles $\mathcal{M}(n, d)$ is a complex quasi-projective variety, which contains $\mathcal{M}^s(n, d)$ as an open smooth subvariety of dimension $2 + 2n^2(g - 1)$, where g is the genus of X.

Similarly to the vector bundle case, if n and d are coprime $\mathcal{M}(n,d) \cong \mathcal{M}^s(n,d)$. Notice that the dimension of $\mathcal{M}^s(n,d)$ is twice that of the moduli space of stable vector bundles $M^s(n,d)$. In fact $\mathcal{M}^s(n,d)$ contains as an open subset the cotangent bundle of $M^s(n,d)$. Indeed, let $[E] \in M^s(n,d)$. The cotangent space to $M^s(n,d)$ at $[E]$, as we saw in Section 2.1, is isomorphic to $H^1(X, \operatorname{End} E)^*$. But, by Serre duality

$$H^1(X, \operatorname{End} E)^* \cong H^0(X, \operatorname{End} E \otimes K),$$

and hence the Higgs field is an element of the cotangent space $T^*_{[E]}M^s(n,d)$. But, as mentioned above, if E is stable then (E, Φ) is stable for every $\Phi \in H^0(X, \operatorname{End} E \otimes K)$. In particular, $\mathcal{M}(1,d) = T^* \operatorname{Pic}^d(X)$.

The moduli space of stable Higgs bundles can also be described in differential-geometric terms, using $\overline{\partial}$ operators. To do that, let $\mathbb{E}$ be a smooth complex vector bundle over X of rank n and degree d. Consider the set of pairs

$$\mathcal{H} = \{(\overline{\partial}_E, \Phi) \in \mathcal{C} \times \mathcal{E}^{1,0}(X, \operatorname{End} \mathbb{E}) \ : \ \overline{\partial}_E \Phi = 0\}. \tag{3.1}$$

Using apropriate Sobolev metrics $\mathcal{C} \times \mathcal{E}^{1,0}(X, \operatorname{End} \mathbb{E})$ can be endowed with the structure of an infinite dimensional manifold and $\mathcal{H}$ is a closed subvariety, in general with singularities. The gauge group $\mathcal{G}^c$ acts on $\mathcal{E}^{1,0}(X, \operatorname{End} \mathbb{E})$ by

$$g \cdot \Phi = g\Phi g^{-1} \quad \text{where} \quad g \in \mathcal{G}^c \quad \text{and} \quad \Phi \in \mathcal{E}^{1,0}(X, \operatorname{End} \mathbb{E}),$$

and this, combined with the action on $\mathcal{C}$, gives the action on $\mathcal{H}$:

$$g \cdot (\overline{\partial}_E, \Phi) = (g\overline{\partial}_E g^{-1}, g\Phi g^{-1}) \quad \text{where} \quad g \in \mathcal{G}^c \quad \text{and} \quad (\overline{\partial}_E, \Phi) \in \mathcal{H}.$$

It is clear that if $(\overline{\partial}_E, \Phi) \in \mathcal{H}$, then the pair (E, Φ), where $E = (\mathbb{E}, \overline{\partial}_E)$, is a Higgs bundle, and $\mathcal{H}/\mathcal{G}^c$ can be identified with the set of isomorphism classes of Higgs bundles of rank n and degree d. If we define

$$\mathcal{H}^s = \{(\overline{\partial}_E, \Phi) \in \mathcal{H} \ : \ (E, \Phi) \ \text{ is stable}\},$$

Then

$$\mathcal{M}^s(n,d) \cong \mathcal{H}^s/\mathcal{G}^c.$$

To study the deformations of an element in $\mathcal{H}/\mathcal{G}^c$ we consider the maps

$$\mathcal{G}^c \xrightarrow{f_0} \mathcal{C} \times \mathcal{E}^{1,0}(X, \operatorname{End} \mathbb{E}) \xrightarrow{f_1} \mathcal{E}^{1,1}(X, \operatorname{End} \mathbb{E}), \tag{3.2}$$

where f_0 is defined for a fixed element $(\overline{\partial}_E, \Phi) \in \mathcal{C} \times \mathcal{E}^{1,0}(X, \operatorname{End} \mathbb{E})$ by

$$f_0(g) = g \cdot (\overline{\partial}_E, \Phi) \quad \text{where} \quad g \in \mathcal{G}^c;$$

and

$$f_1(\overline{\partial}_E, \Phi) = \overline{\partial}_E \Phi \quad \text{where} \quad (\overline{\partial}_E, \Phi) \in \mathcal{C} \times \mathcal{E}^{1,0}(X, \operatorname{End} \mathbb{E}).$$

We have that $\mathcal{H} = f_1^{-1}(0)$. Now, let $(\overline{\partial}_E, \Phi) \in \mathcal{H}$. Let f_0 be the map in (3.2) defined by this element. Let $D_0 = Df_0(I)$ and $D_1 = Df_1(\overline{\partial}_E, \Phi)$ be the differentials of f_0 and f_1 at $I \in \mathcal{G}^c$ and our fixed element $(\overline{\partial}_E, \Phi)$, respectively. We have the complex

$$(3.3)\qquad C^\bullet : 0 \longrightarrow \mathcal{E}^0(X, \operatorname{End}\mathbb{E}) \xrightarrow{D_0} \mathcal{E}^{0,1}(X, \operatorname{End}\mathbb{E}) \oplus \mathcal{E}^{1,0}(X, \operatorname{End}\mathbb{E}) \xrightarrow{D_1} \mathcal{E}^{1,1}(X, \operatorname{End}\mathbb{E}) \longrightarrow 0,$$

where

$$(3.4)\qquad D_0(\psi) = (\overline{\partial}_E\psi, [\Phi, \psi])$$

$$(3.5)\qquad D_1(\alpha, \varphi) = \overline{\partial}_E\varphi + [\alpha, \Phi].$$

This complex is elliptic since it is the sum of two elliptic complexes. Let $H^i(C^\bullet)$ be the cohomology groups of the complex (3.3). We have the following ([17, 31]).

Proposition 3.2:

(1) The space of endomorphisms of (E, Φ) is isomorphic to $H^0(C^\bullet)$.
(2) The space of infinitesimal deformations of (E, Φ) is isomorphic $H^1(C^\bullet)$.
(3) There is a long exact sequence

$$(3.6)\quad 0 \longrightarrow H^0(C^\bullet) \longrightarrow H^0(\operatorname{End} E) \longrightarrow H^0(\operatorname{End} E \otimes K) \longrightarrow H^1(C^\bullet) \longrightarrow H^1(\operatorname{End} E) \longrightarrow H^1(\operatorname{End} E \otimes K) \longrightarrow H^2(C^\bullet) \longrightarrow 0,$$

where the maps $H^i(\operatorname{End} E) \longrightarrow H^i(\operatorname{End} E \otimes K)$ are induced by $\operatorname{ad}(\Phi)$.

From (3.6) we deduce that

$$\chi(C^\bullet) - \chi(\operatorname{End} E) + \chi(\operatorname{End} E \otimes K) = 0,$$

hence by the Riemann-Roch theorem, we have that

$$h^1(C^\bullet) = h^0 + h^2 + 2n^2(g-1).$$

Now, if (E, Φ) is a stable Higgs bundle, then it is simple, i.e. $\mathbb{H}^0(C^\bullet) = \mathbf{C}$. On the other hand, from (3.6) we see that $H^i(C^\bullet) \cong \mathbb{H}^{2-i}(C^\bullet)^*$ for $i = 0, 1, 2$, and hence $\mathbb{H}^2(C^\bullet) = \mathbf{C}$. By deformation theory one has that $\mathcal{M}^s(n, d)$ is smooth [31] and has dimension $2 + 2n^2(g-1)$.

3.2. Hitchin Equations

The stability of a Higgs bundle is linked, as we have already mentioned, to the existence of solutions to a certain equation. Let (E, Φ) be a Higgs bundle. Let h be a Hermitian metric on E, with canonical connection D and curvature $\Theta(D)$. A natural condition to ask is that the metric satisfy the *Hitchin equation*:

$$\Theta(D) + [\Phi, \Phi^*] = -i\mu I_E \Omega, \tag{3.7}$$

where $[\Phi, \Phi^*] = \Phi\Phi^* + \Phi^*\Phi$ is the usual extension of the Lie bracket to Lie-algebra valued forms. Since $\mathrm{Tr}[\Phi, \Phi^*] = 0$, as in (2.7), $\mu = \mu(E)$. We have the following.

Theorem 3.3: Let (E, Φ) be a Higgs bundle such that there is a Hermitian metric on E satisfying (3.7). Then (E, Φ) is polystable. Conversely, if (E, Φ) is stable, then there exists a Hermitian metric on E satisfying (3.7).

This theorem is proved by Hitchin [17] for $n = 2$ and by Simpson [34, 35] for arbitrary rank (he also gives a generalization for a higher dimensional Kähler manifold, which we will mention in Section 8).

The proof of how polystability follows from (3.7) is very similar to the one given in Proposition 2.5. Suppose that h is a Hermitian metric satisfying (3.7). Let $F \subset E$ be a holomorphic subbundle such that $\Phi(F) \subset F \otimes K$, and let $Q = E/F$. In terms of the C^∞ splitting $E \cong F \oplus Q$ defined by h, the Higgs field can be written as

$$\Phi = \begin{pmatrix} \Phi_F & \theta \\ 0 & \Phi_Q \end{pmatrix}. \tag{3.8}$$

Using (2.6), from (3.7) we have

$$\Theta(D_F) - \beta \wedge \beta^* + [\Phi_F, \Phi_F^*] + \theta \wedge \theta^* = -i\mu I_E \Omega.$$

Now, taking the trace of this equation, multiplying by $i/2\pi$ and integrating over X we have

$$\frac{i}{2\pi}\int_X \mathrm{Tr}(\Theta(D_F)) + \int_X |\beta|^2\Omega + \int_X |\theta|^2\Omega = \mu \,\mathrm{rank}\, F,$$

and hence

$$\deg(F) + \|\beta\|^2 + \|\theta\|^2 = \mu \,\mathrm{rank}\, F.$$

We thus have $\mu(F) \leq \mu(E)$. The condition $\mu(F) = \mu(E)$ is equivalent to $\beta = 0$ and $\theta = 0$. This happens if and only if $(E, \Phi) = (F, \Phi_F) \oplus (Q, \Phi_Q)$, where both terms are Higgs bundles. Then we can iterate this process until we obtain that $(E, \Phi) = \oplus(E_i, \Phi_i)$ where (E_i, Φ_i) is stable and $\mu(E_i) = \mu(E)$.

To prove the existence result, it is convenient, as in the case of vector bundles, to look at Theorem 3.3 as a correspondence between moduli spaces. To do this, let $\mathbb{E}$ be a smooth complex vector bundle over X of rank n and degree d. Let h be a Hermitian metric on $\mathbb{E}$. Let $\mathcal{A}$ be the set of all connections on $\mathbb{E}$ which are compatible with h, and let $\mathcal{X} = \mathcal{A} \times \mathcal{E}^{1,0}(X, \operatorname{End}\mathbb{E})$. The Hitchin equation on a Higgs bundle (3.7) can be rephrased by requiring the pair $(D, \Phi) \in \mathcal{X}$ to satisfy

$$\begin{aligned} \Theta(D) + [\Phi, \Phi^*] &= \lambda \\ \overline{\partial}_E \Phi &= 0, \end{aligned} \tag{3.9}$$

where $\overline{\partial}_E = D''$ is the holomorphic structure defined by D, and λ is given by (2.8). The second equation simply says that Φ is holomorphic with respect to this holomorphic structure, and hence the pair (E, Φ) is a Higgs bundle, where $E = (\mathbb{E}, \overline{\partial}_E)$. Consider the set of all solutions to (3.9):

$$\mathcal{X}_0 = \{(D, \Phi) \in \mathcal{X} = \mathcal{A} \times \mathcal{E}^{1,0}(X, \operatorname{End}\mathbb{E}) \quad \text{satisfying} \quad (3.9)\}.$$

The set $\mathcal{X}_0$ is invariant under the action of $\mathcal{G}$ — the gauge group of $(\mathbb{E}, h)$, and the *moduli space of solutions to Hitchin equations* is defined as the quotient $\mathcal{X}_0/\mathcal{G}$.

A pair $(D, \Phi) \in \mathcal{X}$ is said to be *reducible* if $(\mathbb{E}, h) = (\mathbb{E}_1, h_1) \oplus (\mathbb{E}_2, h_2)$ and $D = D_1 \oplus D_2$, where D_1 and D_2 are connections on the Hermitian vector bundles $(\mathbb{E}_1, h_1)$ and $(\mathbb{E}_2, h_2)$, respectively, and $\Phi = \Phi_1 \oplus \Phi_2$, where Φ_1 and Φ_2 are Higgs fields on $\mathbb{E}_1$ and $\mathbb{E}_2$, respectively. We say that (D, Φ) is *irreducible* if it is not reducible. We denote by $\mathcal{X}_0^*$ the subset of $\mathcal{X}_0$ consisting of irreducible solutions.

Using analytic methods one can show the following (see [17]).

Theorem 3.4: The moduli space $\mathcal{X}_0^*/\mathcal{G}$ of irreducible solutions to (3.9) has the structure of a smooth real manifold of dimension $4 + 4n^2(g-1)$, where n is the rank of $\mathbb{E}$.

Let $\mathcal{H}$ be the set defined in (3.1), and let

$$\mathcal{H}^{ps} = \{(\overline{\partial}_E, \Phi) \in \mathcal{H} \;:\; (E, \Phi) \quad \text{is polystable}\}.$$

Theorem 3.3 can be restated as follows.

Theorem 3.5: There is a homeomorphism

$$\mathcal{X}_0/\mathcal{G} \cong \mathcal{H}^{ps}/\mathcal{G}^c,$$

which restricts to a homeomorphism

$$\mathcal{X}_0^*/\mathcal{G} \longleftrightarrow \mathcal{H}^s/\mathcal{G}^c.$$

We have proved one direction of this correspondence. To prove the converse, one has to show that in the $\mathcal{G}^c$-orbit of any element $(\overline{\partial}_{E_0}, \Phi_0) \in \mathcal{H}_s$ we can find an element $(\overline{\partial}_E, \Phi)$, unique up to gauge transformations in $\mathcal{G}$, whose corresponding (irreducible) pair (D, Φ) (under the correspondence between $\mathcal{A}$ and $\mathcal{C}$) satisfies (3.9). As in the case of Theorem 2.7, we need to work with Sobolev spaces and consider W^1 connections and W^1 Higgs fields, as well as W^2 gauge transformations. To find a solution one considers a minimizing sequence (D_n, Φ_n) for the functional $\|\Theta(D)+[\Phi, \Phi^*]\|_{L^2}^2$ defined on the $\mathcal{G}^c$-orbit of (D_0, Φ_0). In particular one has $\|\Theta(D)+[\Phi, \Phi^*]\|_{L^2}^2 < C$. From this, and from L^2 bounds that one can obtain for Φ_n, one obtains L^2 bounds for $\Theta(D_n)$, from which it follows, by a theorem of Uhlenbeck [37], that there are gauge transformations $g_n \in \mathcal{G}$ for which $g_n \cdot D_n$ has a weakly convergent subsequence. It is not difficult to find W^1 bounds for Φ_n, to conclude that (D_n, Φ_n) (after renaming) converges to a solution (D, Φ). To complete the proof one shows that (D, Φ) is in the same orbit as (D_0, Φ_0) (see [17, 34] for details).

4. Representations of the Fundamental Group

4.1. Connections and Representations

Let $\mathbb{E}$ be a C^∞ complex vector bundle of rank n over X, and let D be a connection on $\mathbb{E}$. A section $\xi \in \mathcal{E}(X, E)$ is said to be *parallel* if $D\xi = 0$. If $\gamma = \gamma(t)$, $0 \le t \le T$ is a curve in X, a section ξ defined along γ is said to be *parallel along* γ if

$$D\xi(\gamma'(t)) = 0 \quad \text{for} \quad 0 \le t \le T, \tag{4.1}$$

where $\gamma'(t)$ is the tangent vector of γ at $\gamma(t)$. If ξ_0 is an element of the initial fibre $\mathbb{E}_{\gamma(0)}$, by solving the system of ordinary differential equations (4.1) with initial condition ξ_0 we can extend ξ_0 uniquely to a parallel section ξ along γ, called the *parallel displacement of* ξ_0 *along* γ. If the initial and the end points of γ coincide so that $x_0 = \gamma(0) = \gamma(T)$ then the parallel displacement along γ defines a linear transformation of the fibre $\mathbb{E}_{x_0}$. We thus have a map

$$\{\text{closed paths based at } x_0\} \longrightarrow \mathrm{GL}(\mathbb{E}_{x_0}) \tag{4.2}$$

whose image is a subgroup of $\mathrm{GL}(\mathbb{E}_{x_0})$ called the holonomy group of D at x_0.

If D is flat the parallel displacement depends only on the homotopy class of the closed path and (4.2) defines a representation

$$\rho : \pi_1(X, x) \longrightarrow \mathrm{GL}(\mathbb{E}_x).$$

Conversely, given a representation $\rho : \pi_1(X, x_0) \to \mathrm{GL}(n, \mathbf{C})$, one can construct a vector bundle $\mathbb{E}$ of rank n with a flat connection by setting

$$\mathbb{E} = \tilde{X} \times_\rho \mathbf{C}^n,$$

where $\tilde{X}$ is the universal cover of X and $\tilde{X} \times_\rho \mathbf{C}^n$ is the quotient of $\tilde{X} \times \mathbf{C}^n$ by the action of $\pi_1(X, x)$ given by $(y, v) \mapsto (\gamma(y), \rho(\gamma)v)$ for $\gamma \in \pi_1(X, x)$ (regarded as the covering transformation group acting on $\tilde{X}$). The trivial connection on $\tilde{X} \times \mathbf{C}^n$ descends to give a flat connection on $\mathbb{E}$, whose holonomy is the image of ρ.

As we know, the existence of flat connections on $\mathbb{E}$ implies that the first Chern class of $\mathbb{E}$ must vanish. If $c_1(\mathbb{E}) \neq 0$, we can consider projectively flat connections. As we have seen in Section 2.2, these are connections on $\mathbb{E}$ which induce flat connections on the principal $\mathrm{PGL}(n, \mathbf{C})$-bundle associated to $\mathbb{E}$. The holonomy map of a projectively flat connection defines a homomorphism

$$\tilde{\rho} : \pi_1(X, x_0) \longrightarrow \mathrm{PGL}(\mathbb{E}_{x_0}),$$

i.e., a projective representation of $\pi_1(X, x_0)$.

Since we are assuming that X is connected we can drop the reference point $x_0 \in X$ from the notation. The fundamental group $\pi_1(X)$ is a finitely generated group generated by $2g$ generators, say $A_1, B_1, \ldots, A_g, B_g$, subject to the single relation

$$\prod_{i=1}^{g} [A_i, B_i] = 1.$$

It has a universal central extension

$$0 \longrightarrow \mathbf{Z} \longrightarrow \Gamma \longrightarrow \pi_1(X) \longrightarrow 1 \tag{4.3}$$

generated by the same generators as $\pi_1(X)$, together with a central element J subject to the relation $\prod_{i=1}^{g} [A_i, B_i] = J$. By the universal property of Γ (see [4]), we can lift every $\tilde{\rho} : \pi_1(X) \longrightarrow \mathrm{PGL}(n, \mathbf{C})$ to a representation $\rho : \Gamma \longrightarrow \mathrm{GL}(n, \mathbf{C})$ such that

$$\begin{array}{ccccccccc}
0 & \longrightarrow & \mathbf{Z} & \longrightarrow & \Gamma & \longrightarrow & \pi_1(X) & \longrightarrow & 1 \\
 & & \downarrow & & \downarrow{\scriptstyle\rho} & & \downarrow{\scriptstyle\tilde{\rho}} & & \\
1 & \longrightarrow & \mathbf{C}^* & \longrightarrow & \mathrm{GL}(n, \mathbf{C}) & \longrightarrow & \mathrm{PGL}(n, \mathbf{C}) & \longrightarrow & 1.
\end{array}$$

Since a projective connection is equivalent to a connection with constant central curvature, we conclude that there is a correspondence between constant central curvature connections and representations of Γ in $\mathrm{GL}(n, \mathbf{C})$. To be more precise, let $\mathrm{Hom}(\Gamma, \mathrm{GL}(n, \mathbf{C}))$ be the set of all representations of Γ in $\mathrm{GL}(n, \mathbf{C})$. Given an element $\rho \in \mathrm{Hom}(\Gamma, \mathrm{GL}(n, \mathbf{C}))$ we can associate to it a topological invariant given by the first Chern class of the vector bundle E_ρ with central curvature associated to ρ. Fixing this invariant, we define

$$\text{(4.4)} \qquad \mathrm{Hom}_d(\Gamma, \mathrm{GL}(n, \mathbf{C})) = \{\rho \in \mathrm{Hom}(\Gamma, \mathrm{GL}(n, \mathbf{C})) \ : \ c_1(E_\rho) = d\}.$$

The group $\mathrm{GL}(n, \mathbf{C})$ acts on $\mathrm{Hom}(\Gamma, \mathrm{GL}(n, \mathbf{C}))$ by conjugation, that is,

$$(g \cdot \rho)(\gamma) = g\rho(\gamma)g^{-1} \quad \text{for} \quad g \in \mathrm{GL}(n, \mathbf{C}), \rho \in \mathrm{Hom}(\Gamma, \mathrm{GL}(n, \mathbf{C})), \gamma \in \Gamma,$$

preserving $\mathrm{Hom}_d(\Gamma, \mathrm{GL}(n, \mathbf{C}))$.

Now, let $\mathbb{E}$ be a C^∞ vector bundle of rank n and degree d. Let $\mathcal{D}$ be the set of all connections on $\mathbb{E}$, and let $\mathcal{D}_0$ be the set of connections with constant central curvature, that is,

$$\text{(4.5)} \qquad \mathcal{D}_0 = \{D \in \mathcal{D} \ : \ \Theta(D) = \lambda\},$$

where λ is given by (2.8).

Proposition 4.1: There is a homeomorphism

$$\mathcal{D}_0/\mathcal{G}^c \cong \mathrm{Hom}_d(\Gamma, \mathrm{GL}(n, \mathbf{C}))/\, \mathrm{GL}(n, \mathbf{C}).$$

Here, the gauge group $\mathcal{G}^c$ acts on $\mathcal{D}$ as usual, by $g \cdot D = gDg^{-1}$ for $g \in \mathcal{G}^c$ and $D \in \mathcal{D}$.

4.2. Theorem of Narasimhan–Seshadri Revisited

Let $\mathbb{E}$ be a C^∞ vector bundle of rank n and degree d. Let h be a Hermitian metric on $\mathbb{E}$. It is clear that the holonomy group of a connection compatible with h is a subgroup of the unitary group $\mathrm{U}(n)$. Let $\mathrm{Hom}_d(\Gamma, \mathrm{U}(n))$ be defined in a similar way to (4.4). Hence, if $\mathcal{A}$ is the set of all connections on $\mathbb{E}$ which are compatible with h, and $\mathcal{A}_0$ is the set of constant central curvature connections in $\mathcal{A}$ defined in (2.9), then similarly to Proposition 4.1, we have the following.

Proposition 4.2: There is a homeomorphism

$$\mathcal{A}_0/\mathcal{G} \cong \mathrm{Hom}_d(\Gamma, \mathrm{U}(n))/\, \mathrm{U}(n),$$

which restricts to a homeomorphism

$$\mathcal{A}_0^*/\mathcal{G} \longleftrightarrow \mathrm{Hom}_d^*(\Gamma, \mathrm{U}(n))/\, \mathrm{U}(n),$$

where $\mathcal{A}_0^*$ and $\mathrm{Hom}_d^*(\Gamma, \mathrm{U}(n))$ are irreducible connections and representations respectively.

We define the *moduli space of representations of degree d of* Γ *in* $\mathrm{U}(n)$ as

$$R(n,d) = \mathrm{Hom}_d(\Gamma, \mathrm{U}(n))/\,\mathrm{U}(n),$$

and the *moduli space of irreducible representations* of degree d of Γ in $\mathrm{U}(n)$ as

$$R^*(n,d) = \mathrm{Hom}_d^*(\Gamma, \mathrm{U}(n))/\,\mathrm{U}(n).$$

Combining Proposition 4.2 with Theorem 2.7 we obtain the original formulation of the theorem of Narasimhan and Seshadri [28]:

Theorem 4.3: There is a homeomorphism

$$R(n,d) \cong M(n,d),$$

which restricts to a homeomorphism

$$R^*(n,d) \cong M^s(n,d).$$

5. Non-abelian Hodge Theory

5.1. Harmonic Metrics

Our next goal is to show that there is a similar correspondence to that given by Theorem 4.3 between Higgs bundles and representations of Γ in $\mathrm{GL}(n,\mathbf{C})$. This requires, though, another existence theorem due to Donaldson [14] for $n = 2$ and to Corlette [8] for arbitrary n (and in fact in much more generality).

The set $\mathrm{Hom}(\Gamma, \mathrm{GL}(n,\mathbf{C}))$ can be embedded in $\mathrm{GL}(n,\mathbf{C})^{2g+1}$ via the map

$$\mathrm{Hom}(\Gamma, \mathrm{GL}(n,\mathbf{C})) \longrightarrow \mathrm{GL}(n,\mathbf{C})^{2g+1} \tag{5.1}$$

$$\rho \mapsto (\rho(A_1), \ldots \rho(B_g), \rho(J)). \tag{5.2}$$

We can then give $\mathrm{Hom}(\Gamma, \mathrm{GL}(n,\mathbf{C}))$ the subspace topology and consider on

$$\mathrm{Hom}(\Gamma, \mathrm{GL}(n,\mathbf{C}))/\,\mathrm{GL}(n,\mathbf{C})$$

the quotient topology. In contrast to what happen in the unitary case, this topology is non-Hausdorff. This phenomenon is very similar to what we have already seen in the study of holomorphic structures. In order to obtain a Hausdorff space we have to restric our attention to reductive representations.

A representation $\rho \in \mathrm{Hom}(\Gamma, \mathrm{GL}(n,\mathbf{C}))$ is said to be *reductive* if it is the direct sum of irreducible representations. Under the correspondence given by Proposition 4.1, a reductive representation is obtained from the holonomy representation of a reductive connection with constant central curvature, where a connection D is called *reductive* if every D-invariant subbundle admits a D-invariant complement. Due to the compactness of $\mathrm{U}(n)$, every unitary representation is reductive.

Restricting to the sets of reductive and irreducible representations, denoted respectively by $\mathrm{Hom}_d^+(\Gamma, \mathrm{GL}(n, \mathbf{C}))$ and $\mathrm{Hom}_d^*(\Gamma, \mathrm{GL}(n, \mathbf{C}))$, we obtain the *moduli space of reductive representations* of degree d of Γ in $\mathrm{GL}(n, \mathbf{C})$,

$$\mathcal{R}(n, d) = \mathrm{Hom}_d^+(\Gamma, \mathrm{GL}(n, \mathbf{C}))/\,\mathrm{GL}(n, \mathbf{C}), \tag{5.3}$$

and the *moduli space of irreducible representations* of degree d of Γ in $\mathrm{GL}(n, \mathbf{C})$,

$$\mathcal{R}^*(n, d) = \mathrm{Hom}_d^*(\Gamma, \mathrm{GL}(n, \mathbf{C}))/\,\mathrm{GL}(n, \mathbf{C}), \tag{5.4}$$

respectively. In particular, $\mathrm{Hom}_0(\Gamma, \mathrm{GL}(n, \mathbf{C})) = \mathrm{Hom}(\pi_1(X), \mathrm{GL}(n, \mathbf{C}))$ and hence

$$\mathcal{R}(n, 0) = \mathrm{Hom}^+(\pi_1(X), \mathrm{GL}(n, \mathbf{C}))/\,\mathrm{GL}(n, \mathbf{C}) \tag{5.5}$$

is the moduli space of reductive representations of the fundamental group of X in $\mathrm{GL}(n, \mathbf{C})$.

Let $\mathbb{E}$ be a C^∞ vector bundle of rank n and degree d over X. We denote the sets of reductive and irreducible connections in $\mathcal{D}_0$ by $\mathcal{D}_0^+$ and $\mathcal{D}_0^*$, respectively, and define the corresponding moduli spaces as $\mathcal{D}_0^+/\mathcal{G}^c$ and $\mathcal{D}_0^*/\mathcal{G}^c$. It is clear that the correspondence given by Propostion 4.1 restricts to give the following.

Proposition 5.1: There are homeomorphisms

$$\mathcal{R}(n, d) \cong \mathcal{D}_0^+/\mathcal{G}^c,$$

and

$$\mathcal{R}^*(n, d) \cong \mathcal{D}_0^*/\mathcal{G}^c.$$

In analogy to what happens with polystable bundles and Higgs bundles, a C^∞ complex vector bundle $\mathbb{E}$ equipped with a reductive connection with constant central curvature admits a special type of Hermitian metric. To explain this, let h be a Hermitian metric on $\mathbb{E}$. Using h we can decompose D uniquely as

$$D = \nabla + \Psi, \tag{5.6}$$

where ∇ is a connection on $\mathbb{E}$ compatible with h and Ψ is a 1-form with values in the bundle of self-adjoint endomorphisms of $\mathbb{E}$. That is, under the decomposition

$$\mathrm{End}\,\mathbb{E} = \mathrm{ad}\,\mathbb{E} \oplus i\,\mathrm{ad}\,\mathbb{E},$$

where $\operatorname{ad}\mathbb{E} = \operatorname{End}(\mathbb{E}, h)$ is the bundle of skew-Hermitian endomorphisms of $(\mathbb{E}, h)$, the connection ∇ takes values in $\operatorname{ad}\mathbb{E}$, while Ψ takes values in $i\operatorname{ad}\mathbb{E}$.

The metric h is said to be *harmonic* if

$$\nabla^*\Psi = 0, \tag{5.7}$$

where we use the metric on X (in fact the conformal structure is enough here), and the metric on $\mathbb{E}$ to define ∇^* (see Chap. V, Sec. 2). To explain why the word "harmonic" is used here, recall that a Hermitian metric h on $\mathbb{E}$ is simply a section of the $\mathrm{GL}(n, \mathbf{C})/\mathrm{U}(n)$-bundle over X naturally associated to $\mathbb{E}$. This can be viewed as a $\pi_1(X)$-equivariant function

$$\tilde{h} : \tilde{X} \to \mathrm{GL}(n, \mathbf{C})/\mathrm{U}(n),$$

where $\tilde{X}$ is the universal cover of X. It turns out that $\nabla^*\Psi = 0$ is equivalent to the condition that the map $\tilde{h}$ be harmonic, in the sense that it minimizes the energy $\mathcal{E}(\tilde{h}) = \int_{\tilde{X}} |d\tilde{h}|^2$. In fact the one-form Ψ can be identified with the differential of $\tilde{h}$, and ∇ with the pull-back of the Levi–Civita connection on $\mathrm{GL}(n, \mathbf{C})/\mathrm{U}(n)$ ([14, 8]).

The theorem proved by Donaldson [14] in rank 2 and by Corlette [8] in general is the following.

Theorem 5.2: Let D be a connection on $\mathbb{E}$ with constant central curvature. Then $(\mathbb{E}, D)$ admits a harmonic metric if and only if D is reductive.

Remark: Corlette's version of this theorem includes the case in which the base manifold is a compact Riemannian manifold of arbitrary dimension and $\mathrm{GL}(n, \mathbf{C})$ – the structure group of the bundle — is replaced by any reductive non-compact Lie group.

As in the previous existence theorems, we can formulate Theorem 5.2 as a correspondence between moduli spaces. To do that, we fix a Hermitian metric h on $\mathbb{E}$. We have a bijection

$$\begin{array}{ccc} \mathcal{D} & \longrightarrow & \mathcal{A} \times \mathcal{E}^1(X, \operatorname{ad}\mathbb{E}) \\ D & \mapsto & (\nabla, \Psi), \end{array}$$

defined by (5.6). Now, the condition for a connection D to have central curvature, i.e. $\Theta(D) = \lambda$, where λ is given by (2.8), and the harmonicity equation (5.7) are equivalent to the following set of equations for (∇, Ψ):

$$\begin{array}{l} \Theta(\nabla) + \frac{1}{2}[\Psi, \Psi] = \lambda \\ \nabla\Psi = 0 \\ \nabla^*\Psi = 0. \end{array} \tag{5.8}$$

Let $\mathcal{Y} = \mathcal{A} \times \mathcal{E}^1(X, \operatorname{ad}\mathbb{E})$ and let

$$\mathcal{Y}_0 = \{(\nabla, \Psi) \in \mathcal{Y} = \mathcal{A} \times \mathcal{E}^1(X, \operatorname{ad}\mathbb{E}) \quad \text{satisfying} \quad (5.8)\}.$$

The gauge group $\mathcal{G}$ acts on $\mathcal{Y}_0$, and $\mathcal{Y}_0/\mathcal{G}$ is the moduli space of solutions to (5.8). Theorem 5.2 can now be restated as follows.

Theorem 5.3: There is a homeomorphism between the moduli spaces of solutions to (5.8) and the moduli space of reductive connections with constant central curvature on $\mathbb{E}$, i.e.,

$$\mathcal{Y}_0/\mathcal{G} \cong \mathcal{D}_0^+/\mathcal{G}^c.$$

This restricts to a homeomorphism between the corresponding moduli spaces of irreducible objects.

5.2. Representations and Higgs Bundles

Consider now the correspondence

$$\begin{aligned} \mathcal{E}^{1,0}(X, \operatorname{End}\mathbb{E}) &\longrightarrow \mathcal{E}^1(X, \operatorname{ad}\mathbb{E}) \\ \Phi &\mapsto \Psi = \Phi + \Phi^*. \end{aligned}$$

We have the following.

Proposition 5.4: The pair (∇, Ψ) satisfies (5.8) if and only if (∇, Φ) satisfies Hitchin equations (3.9), where $\Psi = \Phi + \Phi^*$. Moreover, there is a homeomorphism between the moduli spaces of solutions

$$\mathcal{Y}_0/\mathcal{G} \cong \mathcal{X}_0/\mathcal{G},$$

which restricts to a homeomorphism between the moduli spaces of irreducible solutions.

Combining previous results in this section with results in Section 2, we obtain the following.

Theorem 5.5: There is a homeomorphism

$$\mathcal{R}(n, d) \cong \mathcal{M}(n, d)$$

which restricts to

$$\mathcal{R}^*(n, d) \cong \mathcal{M}^s(n, d).$$

This correspondence can be viewed as a Hodge theorem for non-abelian cohomology. To see this, consider first the abelian cohomology: $H^1(X, \mathbf{C})$ can be regarded as the space of homomorphisms from $\pi_1(X)$ into $\mathbf{C}$, or equivalently the space of closed one-forms modulo exact one-forms. Since X is Kähler, the Hodge theorem (Chap. V) gives a decomposition

$$H^1(X, \mathbf{C}) = H^1(X, \mathcal{O}_X) \oplus H^0(X, K).$$

In other words, a cohomology class can be thought of as a pair (e, Φ) with $e \in H^1(X, \mathcal{O}_X)$ and Φ a holomorphic one-form. The correspondence between Higgs bundles and representations of $\pi_1(X)$ in $\mathrm{GL}(n, \mathbf{C})$ is analogous. If $\pi_1(X)$ acts trivially on $\mathrm{GL}(n, \mathbf{C})$, then the non-abelian cohomology set $H^1(\pi_1(X), \mathrm{GL}(n, \mathbf{C}))$ is the set of representations $\pi_1(X) \to \mathrm{GL}(n, \mathbf{C})$ modulo conjugation. Equivalently, it is the set of isomorphism classes of C^∞ vector bundles of rank n with flat connections. What we have described is a correspondence between the set of reductive representations and the set of polystable pairs (E, Φ) where E is a holomorphic vector bundle, i.e. an element in the non-abelian cohomology set $H^1(X, \mathrm{GL}(n, \mathcal{O}_X))$, and Φ is an endomorphism valued one-form.

6. Representations in $\mathrm{U}(p, q)$ and Higgs Bundles

6.1. Representations in $\mathrm{U}(p, q)$ and Harmonic Metrics

Let X be a compact Riemann surface. We have seen how stable vector bundles correspond to irreducible representations of $\pi_1(X)$ in $\mathrm{U}(n)$, and stable Higgs bundles correspond to irreducible representations of $\pi_1(X)$ in $\mathrm{GL}(n, \mathbf{C})$. The group $\mathrm{U}(n)$ is the compact real form of $\mathrm{GL}(n, \mathbf{C})$. We can also consider a non-compact real form G of $\mathrm{GL}(n, \mathbf{C})$ and ask whether we can use complex geometry to study representations of $\pi_1(X)$ in G. We illustrate in this section how to do this for the group $\mathrm{U}(p, q)$ (see [6, 7]).

The group $\mathrm{U}(p, q)$, with $p+q = n$, is defined as the group of linear transformations of $\mathbf{C}^n$ which preserve the Hermitian inner product of signature (p, q) defined by

$$\langle z, w \rangle = z_1 \overline{w}_1 + \cdots + z_p \overline{w}_p - \cdots - z_{p+1} \overline{w}_{p+1} - z_{p+q} \overline{w}_{p+q},$$

for $z = (z_1, \cdots, z_n) \in \mathbf{C}^n$ and $w = (w_1, \cdots, w_n) \in \mathbf{C}^n$. That is,

$$\mathrm{U}(p, q) = \{A \in \mathrm{GL}(n, \mathbf{C}) \ : \ \langle Az, Aw \rangle = \langle z, w \rangle, \ \text{for every } z, w \in \mathbf{C}^n\}.$$

If

$$I_{p,q} = \begin{pmatrix} I_p & 0 \\ 0 & I_q \end{pmatrix},$$

we have that

$$\mathrm{U}(p, q) = \{A \in \mathrm{GL}(n, \mathbf{C}) \ : \ A I_{p,q} \overline{A}^t = I_{p,q}\}.$$

As is done for an ordinary Hermitian metric on a C^∞ complex vector bundle $\mathbb{E}$ we can consider a Hermitian metric H of signature (p, q) and study connections which are compatible with H. This is equivalent to having a reduction of the structure group of the principal $\mathrm{GL}(n, \mathbf{C})$-bundle P of frames associated to $\mathbb{E}$ to a $\mathrm{U}(p,q)$-bundle and considering connections on P. A vector bundle $\mathbb{E}$ equipped with a $\mathrm{U}(p,q)$-structure H has a finer topological invariant than its degree d. To show this, we first observe that $\mathrm{U}(p)\times\mathrm{U}(q)\subset\mathrm{U}(p,q)$ is the maximal compact subgroup of $\mathrm{U}(p,q)$. Now, since the symmetric space $\mathrm{U}(p,q)/\,\mathrm{U}(p)\times\mathrm{U}(q)$ is simply connected, there is no obstruction to further reduce the structure group of P to the group $\mathrm{U}(p)\times\mathrm{U}(q)$. This is equivalent to saying that $\mathbb{E}\cong\mathbb{V}\oplus\mathbb{W}$, where $\mathbb{V}$ and $\mathbb{W}$ are vector bundles with $\operatorname{rank}\mathbb{V}=p$ and $\operatorname{rank}\mathbb{W}=q$, naturally equipped with Hermitian metrics $h_{\mathbb{V}}$ and $h_{\mathbb{W}}$, respectively. The topological invariant naturally associated to $(\mathbb{E},H)$ is the pair of integers (a,b), where $a=\deg\mathbb{V}$ and $b=\deg\mathbb{W}$, which does not depend on the reduction to $\mathrm{U}(p)\times\mathrm{U}(q)$. Notice that $d=a+b$.

Let $\rho\in\operatorname{Hom}(\Gamma,\mathrm{U}(p,q))$ be a representation of Γ in $\mathrm{U}(p,q)$. As in the case of $\mathrm{U}(n)$ and $\mathrm{GL}(n,\mathbf{C})$, to ρ we can associate a smooth vector bundle E_ρ equipped with a $\mathrm{U}(p,q)$ structure and a $\mathrm{U}(p,q)$-connection with constant central curvature. We can in this way attach a topological invariant $c(\rho)=(a,b)$ to ρ, corresponding to the invariant of the $\mathrm{U}(p,q)$-bundle E_ρ. Consider

$$\operatorname{Hom}_{a,b}(\Gamma,\mathrm{U}(p,q))=\{\rho\in\operatorname{Hom}(\Gamma,\mathrm{U}(p,q))\ :\ c(\rho)=(a,b)\},$$

and define the moduli spaces of reductive and irreducible representations of Γ in $\mathrm{U}(p,q)$ with invariant $(a,b)\in\mathbb{Z}\times\mathbb{Z}$ as

$$\mathcal{R}(p,q,a,b)=\operatorname{Hom}^{+}_{a,b}(\Gamma,\mathrm{U}(p,q))/\,\mathrm{U}(p,q),$$

and

$$\mathcal{R}^{*}(p,q,a,b)=\operatorname{Hom}^{*}_{a,b}(\Gamma,\mathrm{U}(p,q))/\,\mathrm{U}(p,q),$$

respectively.

The representations for which $a+b=0$ correspond to representations of the fundamental group of X.

Let $(\mathbb{E},H)$ be a C^∞ vector bundle on X equipped with a $\mathrm{U}(p,q)$-structure and topological invariant $(a,b)\in\mathbb{Z}\times\mathbb{Z}$. Let $\mathcal{B}$ be the set of all connections on $\mathbb{E}$ compatible with H. Let

$$\mathcal{B}_0=\{D\in\mathcal{B}\ :\ \Theta(D)=\lambda\},\tag{6.1}$$

where λ is given by (2.8). Let $\mathcal{G}_H$ be the gauge group of $(\mathbb{E},H)$. We denote the sets of reductive and irreducible connections in $\mathcal{B}_0$ by $\mathcal{B}_0^{+}$ and $\mathcal{B}_0^{*}$, respectively, and define the corresponding moduli spaces as $\mathcal{B}_0^{+}/\mathcal{G}_H$ and $\mathcal{B}_0^{*}/\mathcal{G}_H$.

Proposition 6.1: There are homeomorphisms

$$\operatorname{Hom}_{a,b}(\Gamma,\mathrm{U}(p,q))/\,\mathrm{U}(p,q)\cong\mathcal{B}_0/\mathcal{G}_H,$$

$$\mathcal{R}(p,q,a,b)\cong\mathcal{B}_0^{+}/\mathcal{G}_H,$$

and

$$\mathcal{R}^*(p,q,a,b) \cong \mathcal{B}_0^*/\mathcal{G}_H.$$

Now Theorem 5.2 can be generalized to $U(p,q)$ to show that if $D \in \mathcal{B}_0^+$, then there is a harmonic reduction of the $U(p,q)$ structure to $U(p) \times U(q)$. Any reduction defines a decomposition

$$D = \nabla + \Psi,$$

where ∇ is a $U(p) \times U(q)$-connection on the reduced $U(p) \times U(q)$-bundle and Ψ is a one-form with values in the associated bundle with fibre $\mathfrak{m}$, where

$$\mathfrak{u}(p,q) = \mathfrak{u}(p) \oplus \mathfrak{u}(q) \oplus \mathfrak{m}$$

is the Cartan decomposition of the Lie algebra of $U(p,q)$, and $\mathfrak{m}$ is the set of matrices

$$\begin{pmatrix} 0 & A \\ -\overline{A}^t & 0 \end{pmatrix},$$

with A a complex matrix with p columns and q rows. The harmonicity condition is, as for $GL(n, \mathbf{C})$,

$$\nabla^* \Psi = 0,$$

where we use now the reduction to $U(p) \times U(q)$ to define ∇^*. Similar to the $GL(n, \mathbf{C})$ case, a reduction to $U(p) \times U(q)$ can be viewed as a $\pi_1(X)$-equivariant function

$$\tilde{X} \to U(p,q)/U(p) \times U(q),$$

where $\tilde{X}$ is the universal cover of X, and $\nabla^*\Psi = 0$ is equivalent to the condition that this map be harmonic.

We thus have the following [8].

Theorem 6.2: Let D be a connection on a $U(p,q)$ bundle $(\mathbb{E}, h)$ with constant central curvature. Then $(\mathbb{E}, D)$ admits a harmonic reduction to $U(p) \times U(q)$ if and only if D is reductive.

6.2. $\mathbf{U}(p,q)$-Higgs Bundles and Hitchin Equations

There is a special class of $GL(n, \mathbf{C})$-Higgs bundles, related to representations in $U(p,q)$ given by the requirements that

$$\left(E = V \oplus W, \Phi = \begin{pmatrix} 0 & \beta \\ \gamma & 0 \end{pmatrix}\right), \tag{6.2}$$

where V and W are holomorphic vector bundles of rank p and q respectively and the non-zero components in the Higgs field are $\beta \in H^0(\mathrm{Hom}(W, V) \otimes$

K), and $\gamma \in H^0(\mathrm{Hom}(V, W) \otimes K)$. We say (E, Φ) is a *stable* $\mathrm{U}(p,q)$-Higgs bundle if the slope stability condition $\mu(E') < \mu(E)$, is satisfied for all Φ-invariant subbundles $E' = V' \oplus W'$, i.e. for all subbundles $V' \subset V$ and $W' \subset W$ such that

$$\beta : W' \longrightarrow V' \otimes K$$
$$\gamma : V' \longrightarrow W' \otimes K.$$

Semistability and polystability are defined as for $\mathrm{GL}(n, \mathbf{C})$-Higgs bundles. Let $(a, b) \in \mathbb{Z} \times \mathbb{Z}$. We define the moduli space of stable $\mathrm{U}(p,q)$-Higgs bundles $\mathcal{M}^s(p, q, a, b)$ as the set of isomorphism classes of stable $\mathrm{U}(p,q)$-Higgs bundles with $\deg(V) = a$ and $\deg W = b$. Similarly, we define the moduli space of polystable $\mathrm{U}(p,q)$-Higgs bundles $\mathcal{M}(p, q, a, b)$. The basic relation with representations of Γ in $\mathrm{U}(p,q)$ is given by the following [6].

Theorem 6.3: There is a homeomorphism

$$\mathcal{M}(p, q, a, b) \cong \mathcal{R}(p, q, a, b),$$

which restricts to

$$\mathcal{M}^s(p, q, a, b) \cong \mathcal{R}^*(p, q, a, b).$$

The scheme of the proof is very similar to that of Theorem 5.5 for $\mathrm{GL}(n, \mathbf{C})$. A key ingredient is an existence theorem for solutions to the $\mathrm{U}(p,q)$-Hitchin equations that we explain now.

Let (E, Φ) be a $\mathrm{U}(p,q)$-Higgs bundle as in (6.2). Hitchin equations are now equations for Hermitian metrics h_V and h_W on V and W, respectively. If $\Theta(D_V)$ and $\Theta(D_W)$ are the curvatures of the corresponding canonical connections, the equations are

$$\begin{aligned} \Theta(D_V) + \beta \wedge \beta^* + \gamma^* \wedge \gamma &= \lambda \\ \Theta(D_W) + \gamma \wedge \gamma^* + \beta^* \wedge \beta &= \lambda, \end{aligned} \tag{6.3}$$

where λ is given by (2.8). We have the following [6].

Theorem 6.4: Let (E, Φ) be a $\mathrm{U}(p,q)$-Higgs bundle as in (6.2), such that there are Hermitian metrics on V and W satisfying (6.3). Then (E, Φ) is polystable. Conversely, if (E, Φ) is stable, then there exist Hermitian metrics on V and W satisfying (3.7).

Following a similar scheme to that of the $\mathrm{GL}(n, \mathbf{C})$ case, Theorem 6.4 combined with Theorem 6.2 give the proof of Theorem 6.3.

In contrast with the case of stable vector bundles and stable $\mathrm{GL}(n, \mathbb{C})$-Higgs bundles, which exist for any value of the degree of the bundle, the topological invariant (a, b) of a polystable $\mathrm{U}(p,q)$-Higgs bundle has to satisfy a certain constraint. This is expressed in terms of the Toledo invariant.

Given a representation of Γ in $\mathrm{U}(p,q)$ with topological invariant $c(\rho) = (a,b)$, the *Toledo invariant* of ρ is defined by

$$\tau(\rho) = \tau(p,q,a,b) = 2\frac{qa - pb}{p+q}.$$

This invariant satisfies the inequality

$$|\tau(p,q,a,b)| \leq \min\{p,q\}(2g-2).$$

proved by Domic and Toledo [9] (this is a generalization of the Milnor inequality for the Euler class of a $\mathrm{PSL}(2,\mathbb{R})$-flat connection [24]). This inequality can also be proved for a reductive representation $\rho \in \mathcal{R}(p,q,a,b)$ using the polystability condition of the corresponding Higgs bundle $(E,\Phi) \in \mathcal{M}(p,q,a,b)$ [6].

One of the main results in [6] is the following.

Theorem 6.5: The moduli space $\mathcal{M}^s(p,q,a,b)$ (and hence $\mathcal{R}^s(p,q,a,b)$) is a connected smooth Kähler manifold of complex dimension $1+(p+q)^2(g-1)$, which is non-empty if and only if $|\tau(p,q,a,b)| \leq \min\{p,q\}(2g-2)$.

7. Moment Maps and Geometry of Moduli Spaces

7.1. Symplectic and Kähler Quotients

In this section we review some standard facts about the moment map for the symplectic action of a Lie group G on a symplectic manifold, and the special situation in which the manifold has a Kähler structure which is preserved by the action of the group.

A symplectic manifold is by definition a differentiable manifold X together with a non-degenerate closed 2-form Ω. A Kähler manifold with its Kähler form is an example of a symplectic manifold. A transformation f of X is called *symplectic* if it leaves invariant the 2-form, i.e., $f^*\Omega = \Omega$.

Suppose now that a Lie group G acts symplectically on (X,Ω). If v is a vector field generated by the action, then the Lie derivative $L_v\Omega$ vanishes. Now for Ω, as for any differential form,

$$L_v\Omega = i(v)d\Omega + d(i(v)\Omega);$$

hence $d(i(v)\Omega) = 0$, and so, if $H^1(X,\mathbf{R}) = 0$, there exists a function $\mu_v : X \to \mathbf{R}$ such that

$$d\mu_v = i(v)\Omega.$$

The function μ_v is said to be a *Hamiltonian function* for the vector field v. As v ranges over the set of vector fields generated by the elements of the

Lie algebra $\mathfrak{g}$ of G, these functions can be chosen to fit together to give a map to the dual of the Lie algebra,

$$\mu : X \longrightarrow \mathfrak{g}^*,$$

defined by

$$\langle \mu(x), a \rangle = \mu_{\tilde{a}}(x),$$

where $\tilde{a}$ is the vector field generated by $a \in \mathfrak{g}$, $x \in X$ and $\langle \cdot, \cdot \rangle$ is the natural pairing between $\mathfrak{g}$ and its dual. There is a natural action of G on both sides and a constant ambiguity in the choice of μ_v. If this can be adjusted so that μ is G-equivariant, i.e.

$$\mu(g(x)) = (\mathrm{Ad}\, g)^*(\mu(x)) \quad \text{for } g \in G \ x \in X,$$

then μ is called a *moment map* for the action of G on X. The remaining ambiguity in the choice of μ is the addition of a constant abelian character in $\mathfrak{g}^*$. If μ is a moment map then

$$d\mu_{\tilde{a}}(x)(v) = \Omega(\tilde{a}(x), v) \quad \text{for } \ a \in \mathfrak{g} \ , \ v \in TX_x \ , \ x \in X.$$

An important feature of the moment map is that it gives a way of constructing new symplectic manifolds. More precisely, suppose that G acts freely and discontinuously on $\mu^{-1}(0)$ (recall that $\mu^{-1}(0)$ is G-invariant), then

$$\mu^{-1}(0)/G$$

is a symplectic manifold of dimension $\dim M - 2\dim G$. This is the Marsden–Weinstein *symplectic quotient* of a symplectic manifold acted on by a group [23]. There is a more general construction by taking μ^{-1} of a coadjoint orbit. In particular if λ is a central element in $\mathfrak{g}^*$ we can consider the symplectic quotient

$$\mu^{-1}(\lambda)/G.$$

Suppose now that X has a Kähler structure. It is convenient to describe a Kähler structure on the manifold X as a triple (g, J, Ω) consisting of a Riemannian metric g, an integrable almost complex structure (a complex structure) J and a symplectic form Ω on X which satisfies

$$\Omega(u, v) = g(Ju, v), \quad \text{for } x \in X \text{ and } u, v \in T_xX.$$

Any two of these structures determines the third one. This is equivalent to the definition given in Chap. V.

Let G now be a Lie group acting on (X, g, J, Ω) preserving the Kähler structure. Then if $\mu : X \longrightarrow \mathfrak{g}^*$ is a moment map, and G acts freely and discontinuously on $\mu^{-1}(\lambda)$, for a central element $\lambda \in \mathfrak{g}^*$, the quotient $\mu^{-1}(\lambda)/G$ is also a Kähler manifold. This process is called *Kähler reduction* [25].

A very basic example is the following. Let $X = \mathbf{C}^n$ be equipped with its natural Kähler structure and let U(1) act on X by multiplication. The action of U(1) preserves the symplectic structure and has a moment map $\mu : \mathbf{C}^n \to \mathbf{R}$ given by $z \mapsto \sum |z_i|^2$. We can consider $\mu^{-1}(1) = S^{2n-1}$. We then have the symplectic quotient

$$\mu^{-1}(1)/\,\mathrm{U}(1) = S^{2n-1}/\,\mathrm{U}(1) \cong \mathbf{P}_{n-1}(\mathbf{C}).$$

Since the action of U(1) preserves also the complex structure, this construction exhibits $\mathbf{P}_{n-1}(\mathbf{C})$ as a Kähler quotient whose induced Kähler structure is in fact the standard one. Note that $\mu^{-1}(1)/\,\mathrm{U}(1)$ is hence isomorphic to the "good" quotient $\mathbf{C}^n - \{0\}/\mathbf{C}^*$. This relation turns out to be true in a more general context as we will see below.

When X is a projective algebraic manifold there is a very important relation between the symplectic quotient and the algebraic quotient defined by Mumford's Geometric Invariant Theory (GIT) [25]. Suppose that $i : X \subset \mathbf{P}_{n-1}(\mathbf{C})$ is a projective algebraic manifold acted on by a reductive algebraic group which we can assume to be the complexification G^c of a compact subgroup $G \subset \mathrm{U}(n)$. Then, following [25], we say that $x \in X$ is *semistable* if there is a non-constant invariant polynomial f with $f(x) \neq 0$. This is equivalent to saying that if $\tilde{x} \in \mathbf{C}^n$ is any representative of x, then the closure of the G^c-orbit of $\tilde{x}$ does not contain the origin. Let $X^{ss} \subset X$ the set of all semistable points. There is a subset $X^s \subset X^{ss}$ of *stable* points which satisfy the stronger condition that the G^c-orbit of $\tilde{x}$ is closed in $\mathbf{C}^n$. The *algebraic quotient* is by definition the orbit space X^{ss}/G^c. That this is the right quotient in this setup is confirmed by the fact that if $A(X)$ is the graded coordinate ring of X, then the invariant subring $A(X)^{G^c}$ is finitely generated and has X^{ss}/G^c as its corresponding projective variety. The quotient X^s/G^c gives a dense open set of X^{ss}/G^c.

To relate to symplectic quotients, consider the action of $\mathrm{U}(n)$ on $\mathbf{P}_{n-1}(\mathbf{C})$ induced by the standard action on $\mathbf{C}^n$. This action is symplectic and has a moment map $\mu : \mathbf{P}_{n-1}(\mathbf{C}) \to \mathfrak{u}(n)^*$ given by

$$\mu(x) = \frac{1}{2\pi}\frac{xx^*}{\|x\|^2},$$

where we are using the Killing form of $\mathfrak{u}(n)$ to identify $\mathfrak{u}(n)$ with $\mathfrak{u}(n)^*$. Let X and G be as above. Then $p \circ \mu \circ i$, where $p : \mathfrak{u}(n)^* \to \mathfrak{g}^*$ is the projection induced by the inclusion $\mathfrak{g} \subset \mathfrak{u}(n)$, is a moment map for the action of G on X. We can then consider the symplectic quotient $\mu^{-1}(0)/G$. The relation between this quotient and the algebraic quotient is given by the following result due to Mumford, Kempf–Ness, Guillemin and Sternberg and others (see [25]).

Theorem 7.1: There is an isomorphism

$$\mu^{-1}(0)/G \cong X^{ss}/G^c.$$

7.2. Moduli Spaces as Kähler Quotients

The symplectic and Kähler quotient constructions explained above can also be extended to the context of infinite dimensional manifolds (see [23]). We show now how this can be used to endow our moduli spaces with symplectic and Kähler structures.

Coming back to the setup of Section 2.2, let $(\mathbb{E}, h)$ be a smooth complex Hermitian vector bundle over a compact Riemann surface X. The set $\mathcal{A}$ of connections on $(\mathbb{E}, h)$ is an affine space modelled on $\mathcal{E}^1(X, \operatorname{ad}\mathbb{E})$, which is equipped with a symplectic structure defined by

$$\Omega_{\mathcal{A}}(\psi, \eta) = \int_X \operatorname{Tr}(\psi \wedge \eta), \quad \text{for } D \in \mathcal{A} \text{ and } \psi, \eta \in T_D\mathcal{A} = \mathcal{E}^1(\operatorname{ad}\mathbb{E}).$$

This is obviously closed since it is independent of $D \in \mathcal{A}$.

Now, the set $\mathcal{C}$ of holomorphic structures on $\mathbb{E}$ is an affine space modelled on $\mathcal{E}^{0,1}(X, \operatorname{End}\mathbb{E})$, and it has a complex structure $J_{\mathcal{C}}$, induced by the complex structure of the Riemann surface, which is defined by

$$J_{\mathcal{C}}(\alpha) = i\alpha, \quad \text{for } \bar{\partial}_E \in \mathcal{C} \text{ and } \alpha \in T_{\bar{\partial}_E}\mathcal{C} = \mathcal{E}^{0,1}(X, \operatorname{End}\mathbb{E}).$$

The complex structure $J_{\mathcal{C}}$ defines a complex structure $J_{\mathcal{A}}$ on $\mathcal{A}$ via the identification $\mathcal{A} \cong \mathcal{C}$ given by (2.10). The symplectic structure $\Omega_{\mathcal{A}}$ and the complex structure $J_{\mathcal{A}}$ define a Kähler structure on $\mathcal{A}$, which is preserved by the action of the gauge group $\mathcal{G}$. We have that $\operatorname{Lie}\mathcal{G} = \mathcal{E}^0(X, \operatorname{ad}\mathbb{E})$ and hence its dual $(\operatorname{Lie}\mathcal{G})^*$ can be identified with $\mathcal{E}^2(X, \operatorname{ad}\mathbb{E})$. One has the following [4].

Proposition 7.2: There is a moment map for the action of $\mathcal{G}$ on $\mathcal{A}$ given by

$$\begin{aligned} \mathcal{A} &\longrightarrow \mathcal{E}^2(X, \operatorname{ad}\mathbb{E}) \\ D &\longmapsto \Theta(D). \end{aligned}$$

To prove this, let $a \in \operatorname{Lie}\mathcal{G} = \mathcal{E}^0(X, \operatorname{ad}\mathbb{E})$, and let $\tilde{a}$ be the vector field generated by a. We have to show that the function $\mu_{\tilde{a}} : \mathcal{A} \to \mathbf{R}$ given by

$$\mu_{\tilde{a}}(D) = \int_X \operatorname{Tr}(a \wedge \Theta(D))$$

is Hamiltonian. But this follows simply from the following:

$$\begin{aligned} d\mu_{\tilde{a}}(D)(v) &= \int_X \operatorname{Tr}(a \wedge Dv) \\ &= -\int_X \operatorname{Tr}(Da \wedge v) \\ &= \Omega_{\mathcal{A}}(v, \tilde{a}), \end{aligned}$$

where we have used that $\tilde{a} = Da$. In order to have a non-empty symplectic reduction, we take the central element $\lambda \in \mathcal{E}^2(X, \operatorname{ad}\mathbb{E})$ given by (2.8) and consider $\mu^{-1}(\lambda)$. This coincides with the set defined in (2.9) and hence the Kähler quotient $\mu^{-1}(\lambda)/\mathcal{G}$ is precisely the moduli space of constant central curvature connections on $(\mathbb{E}, h)$, which in this way is shown to have a Kähler structure. In view of this, the correspondence given by Theorem 2.7 is formally an infinite dimensional version of the isomorphism between the symplectic and the algebraic quotients in finite dimensions given by Theorem 7.1. Note that even though $\mathcal{A}$ is infinite dimensional, the quotient obtained has finite dimension, as shown in Section 2.2. It should be mentioned that in order to perform the quotient construction in the infinite dimensional set up, all the spaces and the gauge group have to be naturally completed to have Banach manifold structures (see e.g. [4, 20]).

Similarly, the moduli spaces of Higgs bundles can be endowed with a Kähler structure. To explain this, let us denote $\mathcal{E} = \mathcal{E}^{1,0}(X, \operatorname{End}\mathbb{E})$. The linear space $\mathcal{E}$ has a natural complex structure $J_{\mathcal{E}}$ defined by multiplication by i, and a symplectic structure given by

$$\Omega_{\mathcal{E}}(\psi, \eta) = i \int_X \operatorname{Tr}(\psi \wedge \eta^*), \quad \text{for } \Phi \in \mathcal{E} \text{ and } \psi, \eta \in T_\Phi \mathcal{E} = \mathcal{E}.$$

We can now consider $\mathcal{X} = \mathcal{A} \times \mathcal{E}$ with the symplectic structure $\Omega_{\mathcal{X}} = \Omega_{\mathcal{A}} + \Omega_{\mathcal{E}}$ and complex structure $J_{\mathcal{X}} = J_{\mathcal{A}} + J_{\mathcal{E}}$. The action of $\mathcal{G}$ on $\mathcal{X}$ preserves $\Omega_{\mathcal{X}}$ and $J_{\mathcal{X}}$ and there is a moment map given by

$$\begin{array}{rccc} (7.1) \qquad & \mu_{\mathcal{X}} : \mathcal{X} & \longrightarrow & \mathcal{E}^2(X, \operatorname{ad}\mathbb{E}) \\ & (D, \Phi) & \mapsto & \Theta(D) + [\Phi, \Phi^*]. \end{array}$$

This follows from Proposition 7.2 and the fact that $\Phi \mapsto [\Phi, \Phi^*]$ is a moment map for the action of $\mathcal{G}$ on $\mathcal{E}$, as can be easily proved [17].

We now consider the subvariety

$$(7.2) \qquad \mathcal{N} = \{(D, \Phi) \in \mathcal{X} \ : \ D''\Phi = 0\},$$

which corresponds to the space $\mathcal{H}$ defined in (3.1) under the identification between $\mathcal{A}$ and $\mathcal{C}$. Avoiding difficulties with possible singularities, $\mathcal{N}$ inherits a Kähler structure from $\mathcal{X}$ and, since it is $\mathcal{G}$-invariant, the moment map is the restriction $\mu = \mu_{\mathcal{X}}|_{\mathcal{N}} : \mathcal{N} \to \mathcal{E}^2(X, \operatorname{ad}\mathbb{E})$. Now, the moduli space of solutions to Hitchin equations (3.9) is the Kähler quotient $\mu^{-1}(\lambda)/\mathcal{G}$.

To show this construction in the case of $\mathrm{U}(p, q)$-Higgs bundles, we go back to the setup in Section 6, and consider the set

$$\mathcal{Y} = \mathcal{A}_{\mathbb{V}} \times \mathcal{A}_{\mathbb{W}} \times \mathcal{E}^+ \times \mathcal{E}^-,$$

where $\mathcal{A}_{\mathbb{V}}$ and $\mathcal{A}_{\mathbb{W}}$ are the sets of connections on the Hermitian bundles $(\mathbb{V}, h_{\mathbb{V}})$ and $(\mathbb{W}, h_{\mathbb{W}})$ respectively, and $\mathcal{E}^+ = \mathcal{E}^{1,0}(X, \operatorname{Hom}(\mathbb{W}, \mathbb{V}))$ and $\mathcal{E}^- = \mathcal{E}^{1,0}(X, \operatorname{Hom}(\mathbb{V}, \mathbb{W}))$. Let $(\mathbb{E}, h) = (\mathbb{V} \oplus \mathbb{W}, h_{\mathbb{V}} \oplus h_{\mathbb{W}})$ and $\mathcal{A}$, $\mathcal{E}$ and $\mathcal{G}$ be the corresponding set of connections, Higgs fields and gauge group. The space $\mathcal{Y}$ is a Kähler submanifold of $\mathcal{X} = \mathcal{A} \times \mathcal{E}$ which is invariant by the subgroup

$\mathcal{G}_{\mathbb{V}} \times \mathcal{G}_{\mathbb{W}} \subset \mathcal{G}$, and hence the moment map is given by projecting onto $(\mathrm{Lie}\,\mathcal{G}_{\mathbb{V}})^* \oplus (\mathrm{Lie}\,\mathcal{G}_{\mathbb{W}})^*$ in (7.2). The moment map for the action of $\mathcal{G}_{\mathbb{V}} \times \mathcal{G}_{\mathbb{W}}$ on $\mathcal{Y}$ is thus

$$\mu_{\mathcal{Y}} : \mathcal{Y} \longrightarrow \mathcal{E}^2(X, \mathrm{ad}\,\mathbb{V}) \oplus \mathcal{E}^2(X, \mathrm{ad}\,\mathbb{W})$$

$$(D_{\mathbb{V}}, D_{\mathbb{W}}, \beta, \gamma) \longmapsto (\Theta(D_{\mathbb{V}}) + \beta \wedge \beta^* + \gamma^* \wedge \gamma, \Theta(D_{\mathbb{W}}) + \gamma \wedge \gamma^* + \beta^* \wedge \beta).$$

We can then restrict this to obtain a moment map μ on the $(\mathcal{G}_{\mathbb{V}} \times \mathcal{G}_{\mathbb{W}})$-invariant Kähler submanifold $\mathcal{N}_{\mathcal{Y}} = \mathcal{N} \cap \mathcal{Y}$, where $\mathcal{N}$ is given by (7.2). The quotient $\mu^{-1}(\lambda)/\mathcal{G}_{\mathbb{V}} \times \mathcal{G}_{\mathbb{W}}$ is the moduli space of solutions to the $\mathrm{U}(p,q)$-Hitchin equations, which is isomorphic to the moduli space of $\mathrm{U}(p,q)$-Higgs bundles $\mathcal{M}(p,q,a,b)$ where a and b are the Chern classes of $\mathbb{V}$ and $\mathbb{W}$ respectively.

7.3. Hyperkähler Quotients and Moduli Spaces

A hyperkähler manifold is a differentiable manifold X equipped with a Riemannian metric g and complex structures J_i, $i = 1, 2, 3$ satisfying the quaternion relations $J_i^2 = -I$, $J_3 = J_1 J_2$, etc., such that if we define $\Omega_i(\cdot,\cdot) = g(J_i\cdot,\cdot)$, then (g, J_i, Ω_i) is a Kähler structure on X. As for Kähler manifolds, there is a natural quotient construction for hyperkähler manifolds [18].

Let G be a Lie group acting on X preserving the Kähler structure (g, J_i, Ω_i) and having moment maps $\mu_i : X \to \mathfrak{g}^*$ for $i = 1, 2, 3$. We can combine these moment maps in a map

$$\boldsymbol{\mu} : X \longrightarrow \mathfrak{g}^* \otimes \mathbf{R}^3$$

defined by $\boldsymbol{\mu} = (\mu_1, \mu_2, \mu_3)$. Let $\lambda_i \in \mathfrak{g}^*$ for $i = 1, 2, 3$ be central elements and consider the G-invariant submanifold $\boldsymbol{\mu}^{-1}(\boldsymbol{\lambda})$ where $\boldsymbol{\lambda} = (\lambda_1, \lambda_2, \lambda_3)$. Then G acts on $\boldsymbol{\mu}^{-1}(\boldsymbol{\lambda})$ freely and discontinuously and the quotient

$$\boldsymbol{\mu}^{-1}(\boldsymbol{\lambda})/G$$

is a hyperkähler manifold.

One way to understand the non-abelian Hodge theory correspondence in Section 5 is through the analysis of the hyperkähler structures of the moduli spaces involved. We explain how these can be obtained as hyperkähler quotients. For this, let us go back to the setup of Section 7.2, and let $(\mathbb{E}, h)$ be a smooth complex Hermitian vector bundle over a compact Riemann surface X. As we have seen in Section 7.2, the space $\mathcal{X} = \mathcal{A} \times \mathcal{E}$ has a Kähler structure defined by $J_{\mathcal{X}}$ and $\Omega_{\mathcal{X}}$. Let us rename $J_1 = J_{\mathcal{X}}$. Via the identification $\mathcal{A} \cong \mathcal{C}$, we have for $\alpha \in \mathcal{E}^{0,1}(X, \mathrm{End}\,\mathbb{E})$ and $\psi \in \mathcal{E}^{1,0}(X, \mathrm{End}\,\mathbb{E})$ the following three complex structures on $\mathcal{X}$:

$$\begin{aligned} J_1(\alpha, \psi) &= (i\alpha, i\psi) \\ J_2(\alpha, \psi) &= (i\psi^*, -i\alpha^*) \\ J_3(\alpha, \psi) &= (-\psi^*, \alpha^*), \end{aligned}$$

where α^* and ψ^* is defined using the Hermitian metric h on $\mathbb{E}$. Clearly, J_i, $i = 1, 2, 3$ satisfy the quaternion relations, and define a hyperkähler structure on $\mathcal{X}$, with symplectic structures Ω_i, $i = 1, 2, 3$, where $\Omega_1 = \Omega_{\mathcal{X}}$. The action of the gauge group $\mathcal{G}$ on $\mathcal{X}$ preserves the hyperkähler structure and there are moment maps given by

$$\mu_1(D, \Phi) = \Theta(D) + [\Phi, \Phi^*], \quad \mu_2(D, \Phi) = \mathrm{Re}(\overline{\partial}_E \Phi), \quad \mu_3(D, \Phi) = \mathrm{Im}(\overline{\partial}_E \Phi).$$

Taking $\boldsymbol{\lambda} = (\lambda, 0, 0)$, where λ is given by (2.8) we have that $\boldsymbol{\mu}^{-1}(\boldsymbol{\lambda})/\mathcal{G}$ is the moduli space of solutions to Hitchin equations (3.9). In particular, if we consider the irreducible solutions $\boldsymbol{\mu}_*^{-1}(\boldsymbol{\lambda})$ we have that

$$\boldsymbol{\mu}_*^{-1}(\boldsymbol{\lambda})/\mathcal{G}$$

is a hyperkähler manifold which, by Theorem 3.5, is isomorphic to the moduli space $\mathcal{M}^s(n, d)$ of stable Higgs bundles of rank n and d.

Let us now see how the moduli of harmonic flat connections on $(\mathbb{E}, h)$ can be realized as a hyperKähler quotient. As in Section 4.1, let $\mathcal{D}$ be the set of all complex connections on $\mathbb{E}$. This is an affine space modelled on $\mathcal{E}^1(X, \mathrm{End}\,\mathbb{E}) = \mathcal{E}^0(X, T^*X \otimes_{\mathbf{R}} \mathrm{End}\,\mathbb{E})$. The space $\mathcal{D}$ has a complex structure $I_1 = 1 \otimes i$, which comes from the complex structure of the bundle. Using the complex structure of X we have also the complex structure $I_2 = i \otimes \tau$, where $\tau(\psi) = \psi^*$ is the involution defined by the Hermitian metric h. We can finally consider the complex structure $I_3 = I_1 I_2$.

The Hermitian metric on $\mathbb{E}$ together with a Riemannian metric in the conformal class of X defines a flat Riemannian metric $g_{\mathcal{D}}$ on $\mathcal{D}$ which is Kähler for the above three complex structures. Hence $(\mathcal{D}, g_{\mathcal{D}}, I_1, I_2, I_3)$ is also a hyperkähler manifold. As in the previous case, the action of the gauge group $\mathcal{G}$ on $\mathcal{D}$ preserves the hyperkähler structure and there are moment maps

$$\mu_1(D) = \nabla^*\Psi, \quad \mu_2(D) = \mathrm{Im}(\Theta(D)), \quad \mu_3(D) = \mathrm{Re}(\Theta(D)),$$

where $D = \nabla + \Psi$ is the decomposition of D defined by (5.6). Hence the moduli space of solutions to the harmonicity equations (5.8) is the hyperkähler quotient defined by

$$\boldsymbol{\mu}^{-1}(0, \lambda, 0)/\mathcal{G},$$

where $\boldsymbol{\mu} = (\mu_1, \mu_2, \mu_3)$ and λ given by (2.8).

The homeomorphism between the moduli spaces of solutions to the Hitchin and the harmonicity equations is induced from the hypercomplex affine map

$$\begin{aligned} \mathcal{A} \times \mathcal{E} &\longrightarrow \mathcal{D} \\ (D, \Phi) &\longmapsto D + \Phi + \Phi^*. \end{aligned}$$

One can see easily, for example, that this map sends $\mathcal{A} \times \mathcal{E}$ with complex structure J_2 to $\mathcal{D}$ with complex structure I_1 (see [17]).

Now, Theorems 3.5 and 5.2 can be regarded as existence theorems, establishing the non-emptiness of the hyperkähler quotient, obtained by focusing on different complex structures. For Theorem 3.5 one gives a special status to the complex structure J_1. Combining the symplectic forms determined by J_2 and J_3 one has the J_1-holomorphic symplectic form $\Omega_c = \Omega_2 + i\Omega_3$ on $\mathcal{A} \times \mathcal{E}$. The complex gauge group $\mathcal{G}^c$ acts on $\mathcal{A} \times \mathcal{E}$ preserving Ω_c. The symplectic quotient construction can also be extended to the holomorphic situation (see e.g. [20]) to obtain the holomorphic symplectic quotient $\{(\overline{\partial}_E, \Phi) \ : \ \overline{\partial}_E \Phi = 0\}/\mathcal{G}^c$. What Theorem 3.5 says is that for a class $[(\overline{\partial}_E, \Phi)]$ in this quotient to have a representative (unique up to unitary gauge) satisfying $\mu_1 = \lambda$ it is necessary and sufficient that the pair $(\overline{\partial}_E, \Phi)$ be polystable. This identifies the hyperkähler quotient to the set of equivalence classes of polystable pairs on $\mathbb{E}$. If one now takes J_2 on $\mathcal{A} \times \mathcal{E}$ or equivalently $\mathcal{D}$ with I_1 and argues in a similar way, one gets Theorem 5.2 identifying the hyperkähler quotient to the set of equivalence classes of reductive central curvature connections on $\mathbb{E}$.

Note that for $n = 1$, the moduli space with complex structure defined by J_1 is $T^* \operatorname{Pic}^d(X)$ and with complex structure defined by J_2 is $(\mathbf{C}^*)^g$.

8. Higher Dimensional Generalizations

8.1. Hermitian–Einstein Connections and Stable Bundles

Mumford's stability condition for a holomorphic vector bundle over a compact Riemann surface was generalized by Takemoto to higher dimensional Kähler manifolds (see [36, 20]). Let X be a compact Kähler manifold of dimension n, with Kähler form Ω. Let E be a holomorphic vector bundle over X. Associated to E one has the sheaf of its holomorphic sections. This is a locally free sheaf, which is thus a coherent sheaf (see Chap. II, Sec. 1), that will also be denoted by E.

The degree of a coherent sheaf F over X is defined as

$$\deg F = \frac{1}{(n-1)!} \int_X c_1(F) \wedge \Omega^{n-1},$$

where $c_1(F) = c_1(\det F)$, and $\det F$ is a line bundle associated to F, which coincides with the determinant line bundle when F is locally free (see [20], for instance). As in the Riemann surface case, the *slope* of F is defined as

$$\mu(F) = \deg F / \operatorname{rank} F,$$

where $\operatorname{rank} F$ is the rank of the vector bundle that the coherent sheaf F determines outside of a subset of X, called the singularity set of F, that has codimension at least one (see [20]). We say that E is *stable* with respect to Ω if for every coherent subsheaf $F \subset E$ with $0 < \operatorname{rank} F < \operatorname{rank} E$,

$$\mu(F) < \mu(E).$$

Semistability and polystability are defined as in the one-dimensional case. Note that, in contrast to the Riemann surface case, in higher dimensions the stability condition depends on the Kähler metric of X.

We will see now how stability on a Kähler manifold is also related to the existence of a special Hermitian metric on E. Let h be a Hermitian metric on E with canonical connection and curvature D and $\Theta(D)$, respectively. The metric h is called *Hermitian-Einstein* if

$$\Theta(D) \wedge \Omega^{n-1} = -i\mu I_E \Omega^n, \tag{8.1}$$

where, after the normalization $\operatorname{Vol} X = 2\pi$, $\mu = \mu(E)$. When $n = 1$, (8.1) reduces to the constant central curvature condition given by (2.1). The following generalizes to higher dimensions the theorem of Narasimhan and Seshadri.

Theorem 8.1: Let E be a holomorphic vector bundle over a compact Kähler manifold (X, Ω). Then E has a Hermitian–Einstein metric if and only if it is polystable.

The polystability of a vector bundle admiting a Hermitian–Einstein metric was proved by Lübke [22] (see also [20]). The existence of a Hermitian–Einstein metric on a stable vector bundle was first proved by Donaldson for algebraic surfaces [12] and then by Uhlenbeck and Yau [38] for arbitrary Kähler manifolds. In [13] Donaldson gave another proof for algebraic manifolds of any dimension.

Theorem 8.1, can also be viewed as a correspondence between moduli spaces. To see this, let $(\mathbb{E}, h)$ be a smooth Hermitian vector bundle over X. A connection D on $(\mathbb{E}, h)$ is called Hermitian–Einstein if it satisfies

$$\begin{aligned} &D''^2 = 0 \\ &\Theta(D) \wedge \Omega^{n-1} = -i\mu(\mathbb{E}) I_{\mathbb{E}} \Omega^n. \end{aligned} \tag{8.2}$$

The first equation simply means that the $(0, 1)$-part of the connection D, defines a holomorphic structure on $\mathbb{E}$. The moduli space of Hermitian–Eintein connections is defined as the set of all connections satisfying (8.2) modulo the action of the gauge group of $(\mathbb{E}, h)$.

Theorem 8.1 is equivalent to the *Hitchin–Kobayashi correspondence*:

Theorem 8.2: There is a bijection between the moduli space of Hermitian–Einstein connections on $(\mathbb{E}, h)$ and the moduli space of polystable holomorphic vector bundles whose underlying smooth vector bundle is isomorphic to $\mathbb{E}$.

In the case of Kähler surfaces, the Hermitian–Einstein equation is equivalent to the *anti-self-dual instanton* equation for a connection on a Hermitian vector bundle over a real 4-dimensional Riemannian manifold. Using the moduli space of instantons, Donaldson defined topological invariants for four-manifolds. One of the main applications of the correspondence given by Theorem 8.2 has been in the computation of the Donaldson invariants for Kähler surfaces (see [11]).

8.2. Higgs Bundles and Representations of Kähler Groups

As above, let (X, Ω) be a compact Kähler manifold of dimension n, whose volume is normalized such that $\operatorname{Vol} X = 2\pi$. A *Higgs bundle* over X is a pair (E, Φ) consisting of a holomorphic vector bundle E over X and a Higgs field $\Phi \in H^0(X, \operatorname{End} E \otimes \Omega^1)$, satisfying $\Phi \wedge \Phi = 0$, where Ω^1 is the bundle of holomorphic one-forms on X. Note that on a Riemann surface Ω^1 coincides with the canonical line bundle, and the condition $\Phi \wedge \Phi = 0$ is trivialy satisfied for dimensional reasons.

A Higgs bundle (E, Φ) is said to be *stable* if and only if $\mu(E') < \mu(E)$ for every coherent subsheaf $E' \subset E$ invariant under Φ, i.e. $\Phi(E') \subset E' \otimes \Omega^1$. Semistability and polystability are defined as usual. As in the one-dimensional case, the notion of stability is related to the existence of a special Hermitian metric on E. More precisely, one has the following theorem proved by Simpson [34, 35].

Theorem 8.3: Let (E, Φ) be a Higgs bundle over X. The existence of a Hermitian metric h on E satisfying

$$(\Theta(\nabla) + [\Phi, \Phi^*]) \wedge \Omega^{n-1} = -i\mu(E) I_E \Omega^n, \tag{8.3}$$

is equivalent to the polystability of (E, Φ). Here ∇ is the canonical connection determined by h and the holomorphic structure of E.

To relate Higgs bundles to representations of the fundamental group of the Kähler manifold X (what is called a Kähler group) one has to impose topological conditions on E. Namely, one needs

$$\int_X c_1(E) \wedge \Omega^{n-1} = 0, \quad \text{and} \quad \int_X (c_1(E)^2 - 2c_2(E)) \wedge \Omega^{n-2} = 0. \tag{8.4}$$

Under conditions (8.4), if (E, Φ) is polystable, then the Hermitian metric h on E in Theorem 8.3 satisfies the stronger equation

$$\Theta(\nabla) + [\Phi, \Phi^*] = 0. \tag{8.5}$$

Note that the first condition in (8.4) is simply that $\deg E = 0$ and hence $\mu(E) = 0$. We can then consider the pair $(\mathbb{E}, D)$, taking $\mathbb{E}$ to be the underlying C^∞ bundle to E and $D = \nabla + \Phi + \Phi^*$. From $\bar{\partial}_E \Phi = 0$, $\Phi \wedge \Phi = 0$ and (8.5) one easily sees that $\Theta(D) = 0$, i.e. D is a flat connection on $\mathbb{E}$. In fact setting $\Psi = \Phi + \Phi^*$, we can see that $\nabla^* \Psi = 0$, i.e. h is a harmonic metric in the sense of Sec. 5. But Theorem 5.2 holds also in higher dimensions (see [8]) and hence we have the following.

Theorem 8.4: Let D be a flat connection on $\mathbb{E}$. Then $\mathbb{E}$ admits a harmonic metric if and only if D is reductive.

The construction of a polystable Higgs bundle from a flat reductive connection follows along the same lines as the one dimensional case explained in Sec. 5 (see also [35]), establishing the following.

Theorem 8.5: There is a one-to-one correspondence between the moduli space of polystable Higgs bundles (E, Φ) with rank $E = r$ and E satisfying (8.4), and the moduli space of reductive representations of $\pi_1(X)$ in $\mathrm{GL}(r, \mathbf{C})$.

Higgs bundles can also be regarded from the point of view of Tannakian categories. This is a point of view taken by Simpson in [35], which we follow (see also [15] and the references given there for details on Tannakian categories). A *tensor category* is a category $\mathcal{C}$ with a functorial binary operation $\otimes : \mathcal{C} \times \mathcal{C} \to \mathcal{C}$. An *associative* and *commutative* tensor category is a tensor category provided with additional natural isomorphisms expressing associativity and commutativity of the tensor product that have to satisfy certain *canonical axioms*. A *unit* 1 is an object 1 provided with natural isomorphisms $1 \otimes V \cong V$ satisfying canonical axioms. A functor $\mathcal{F}$ between associative and commutative categories with unit is a functor provided with natural isomorphisms $\mathcal{F}(U \otimes V) \cong \mathcal{F}(U) \otimes \mathcal{F}(V)$. A *neutral Tannakian category* $\mathcal{C}$ is an associative and commutative tensor category with unit, which is *abelian, rigid* (duals exist), $\mathrm{End}(1) = \mathbf{C}$, and which is provided with an *exact, faithful fibre functor* $\mathcal{F} : \mathcal{C} \to \mathrm{Vect}$, where Vect is the tensor category of complex, finite dimensional vector spaces.

If G is an affine group scheme over $\mathbf{C}$ the category $\mathrm{Rep}(G)$ of complex representations of G is a neutral Tannakian category. The fibre functor $\mathcal{F}_G$ is given by sending a representation of G to the underlying vector space. The group G is recovered as the group $G = \mathrm{Aut}^{\otimes}(\mathcal{F}_G)$ of tensor automorphisms of the fibre functor. The converse is given by the fundamental duality theorem of Tannaka–Grothendieck–Saavedra.

Theorem 8.6: Let $(\mathcal{C}, \mathcal{F})$ be a neutral Tannakian category and let $G = \mathrm{Aut}^{\otimes}(\mathcal{F})$ be the group of tensor automorphisms of the fibre funtor. Then $(\mathcal{C}, \mathcal{F}) \cong (\mathrm{Rep}(G), \mathcal{F}_G)$.

To briefly describe the group $\mathrm{Aut}^{\otimes}(\mathcal{F})$ — referred sometimes as the *Tannaka group* of the Tannakian category $(\mathcal{C}, \mathcal{F})$ —, let $\mathrm{End}(\mathcal{F})$ be the algebra of endomorphisms of the the fibre functor. Its elements are collections $\{f_V\}$ with $f_V \in \mathrm{End}(\mathcal{F}(V))$ such that for any morphism $\psi : V \to W$, one has

$\mathcal{F}(\psi)f_V = f_W\mathcal{F}(\psi)$. Let $\mathrm{Aut}^{\otimes}(\mathcal{F})$ be the set of elements $\{f_V\}$ of $\mathrm{End}(\mathcal{F})$ satisfying

$$f_1 = 1 \quad f_{V\otimes W} = f_V \otimes f_W.$$

The existence of duals in $\mathcal{C}$ implies that any element in $\mathrm{Aut}^{\otimes}(\mathcal{F})$ consists entirely of automorphisms, and hence there is no need to include a condition for invertibility. The algebra $\mathrm{End}(\mathcal{F})$ is a projective limit of finite dimensional algebras and it is endowed with a projective limit topology. The subset $\mathrm{Aut}^{\otimes}(\mathcal{F})$ has a structure of projective limit of algebraic varieties.

We come now to the Tannakian nature of Higgs bundles. Let (E, Φ) and (F, Ψ) be two Higgs bundles. Its tensor product is given by the Higgs bundle $(E \otimes F, \Phi \otimes I_F + I_E \otimes \Psi)$. One has the following (see [35, 15]).

Proposition 8.7: The tensor category of polystable Higgs bundles (E, Φ) over X, satisfying (8.4) with fibre functor defined by sending a Higgs bundle to the fibre of the bundle at a fixed point of X, is a neutral Tannakian category.

The category of reductive complex representations of the fundamental group of X (with the obvious fibre functor) is also a neutral Tannakian category. It follows from the Tannaka duality theorem that the Tannaka group of this category is the pro-reductive completion of $\pi_1(X)$ (see [35]). This is a group G, that comes equipped with a homomorphism $\pi_1(X) \to G$ such that for every reductive group H and a homomorphism $\pi_1(X) \to H$ there exists a unique extension $G \to H$ such that the following diagram

$$\begin{array}{ccc} \pi_1(X) & \longrightarrow & H \\ \downarrow & \nearrow & \\ G & & \end{array}$$

commutes.

Now, the correspondence given in Theorem 8.5 gives actually an equivalence of Tannakian categories between the category of complex representations of $\pi_1(X)$ and the category of polystable Higgs bundles over (X, Ω) satisfying (8.4), and hence the Tannaka group of the latter category is isomorphic to the G — the pro-reductive completion of $\pi_1(X)$.

An central ingredient in non-abelian Hodge theory is the action of the group $\mathbf{C}^*$ on the category of polystable Higgs bundles given by

$$(E, \Phi) \mapsto (E, \lambda\Phi) \quad \text{for every } \lambda \in \mathbf{C}^*.$$

This action induces an action of $\mathbf{C}^*$ on the category of reductive representations of the fundamental group. It should be pointed out that, while this action is very clear and explicit from the point of view of Higgs bundles, its explicit effect on a representation of the fundamental group is not easy to describe.

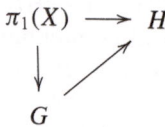

One can formalise the action of $\mathbf{C}^*$ on a Tannakian category $(\mathcal{C}, \mathcal{F})$ in terms of certain tensor functors satisfying canonical axioms. If the action preserves the fibre functor $\mathcal{F}$ one has an action of $\mathbf{C}^*$ on $\mathrm{End}(\mathcal{F})$ by sending the element $\{f_V\}$ of $\mathrm{End}(\mathcal{F})$ to $\{f_V^\lambda\}$ with $f_V^\lambda = f_{\lambda V}$ for every $\lambda \in \mathbf{C}^*$, and hence one has an action on the Tannaka group $\mathrm{Aut}^{\otimes}(\mathcal{F})$. The action of $\mathbf{C}^*$ on the category of polystable Higgs bundles preserves clearly the fibre functor since the bundle is unchanged, and one can then transfer this action to G — the pro-reductive completion of the fundamental group. More precisely one has the following theorem proved by Simpson in [35].

Theorem 8.8: There exists a unique action of $\mathbf{C}^*$ on G, each $\lambda \in \mathbf{C}^*$ acting by a homomorphism of pro-reductive groups, such that if $\rho : G \to \mathrm{GL}(r, \mathbf{C})$ is the representation corresponding to (E, Φ), then $\rho \circ \lambda$ is the representation corresponding to $(E, \lambda\Phi)$.

In some sense the action of $\mathbf{C}^*$ on G is the essence of non-abelian Hodge theory, replacing the $\mathbf{C}^*$-action on cohomology explained in Chap. V for the abelian theory.

References

[1] Arbarello, Cornalba, Griffiths, Harris, Springer, *Geometry of Algebraic Curves* Vol. I, Springer-Verlag, New York 1985.

[2] M.F. Atiyah, Vector bundles over an elliptic curve, *Proc. London Math. Soc.* **7** (1957), 414–452.

[3] M.F. Atiyah, The moment map in symplectic geometry. *Durham Symposium on Global Riemannian Geometry*. Ellis Horwood Ltd. (1984), 43–51.

[4] M.F. Atiyah and R. Bott, The Yang-Mills equations over Riemann surfaces, *Phil. Trans. R. Soc. Lond. A* **308** (1982), 523–615.

[5] I. Biswas and U.N. Bhosle, *Notes on Vector Bundles on Curves*, 2006, to appear.

[6] S.B. Bradlow, O. García-Prada and P.B. Gothen, Surface group representations and U(p, q)-Higgs bundles, *J. Differential Geom.* **64** (2003), 111–170.

[7] S.B. Bradlow, O. García-Prada and P.B. Gothen, Moduli spaces of holomorphic triples over compact Riemann surface. *Math. Ann.* **328** (2004), 299–351.

[8] K. Corlette, Flat G–bundles with canonical metrics, *J. Diff. Geom.* **28** (1988), 361–382.

[9] A. Domic and D. Toledo, The Gromov norm of the Kaehler class of symmetric domains, *Math. Ann.* **276** (1987), 425–432.

[10] S.K. Donaldson, A new proof a theorem of Narasimhan and Seshadri, *J. Diff. Geom.* **18**, (1982), 269–278.

[11] S.K. Donaldson and P.B. Kronheimer, *The Geometry of Four Manifolds*, Oxford Mathematical Monographs, Oxford University Press, 1990.

[12] S.K. Donaldson, Anti-self-dual Yang–Mills connections on a complex algebraic surface and stable vector bundles, *Proc. Lond. Math. Soc.* **3** (1985), 1–26.

[13] S.K. Donaldson, Infinite determinants, stable bundles and curvature, Duke Math. J. **54** (1987), 231–247.

[14] S.K. Donaldson, Twisted harmonic maps and the self-duality equations, *Proc. London Math. Soc.* (3) **55** (1987), 127–131.

[15] O. García-Prada, and S. Ramanan, Twisted Higgs bundles and the fundamental group of compact Kähler manifolds, *Mathematical Research Letters* **7** (2000), 517–535.

[16] A. Grothendieck, Sur la classification des fibrés holomorphes sur la sphère de Riemann, *Amer. J. Math.* **79** (1957), 121–138.

[17] N.J. Hitchin, The self-duality equations on a Riemann surface, *Proc. Lond. Math. Soc.* **55** (1987), 59–126.

[18] N.J. Hitchin, A. Karlhede, U. Lindström, and M. Roçek, Hyperkähler metrics and supersymmetry, *Comm. Math. Phys.* **108** (1987), 535–589.

[19] J.-L. Koszul, B. Malgrange, Sur certaines structures fibrés complexes, *Arch. Math.* **9** (1958), 102–109.

[20] S. Kobayashi, *Differential Geometry of Complex Vector Bundles*, Princeton University Press, New Jersey, 1987.

[21] J. Le Potier, *Lectures on Vector Bundles*, Cambridge University Press, 1997.

[22] M. Lübke, Stability of Einstein–Hermitian vector bundles, *Manuscripta Mathematica* **42** (1983), 245–257.

[23] J. Marsden and A.D. Weinstein, Reduction of symplectic manifolds with symmetry, *Reports on Math. Physics* **5** (1974), 121–130.

[24] J. Milnor, On the existence of a connection with curvature zero, *Comment. Math. Helv.* **32** (1958), 215–223.

[25] D. Mumford, J. Fogarty and F. Kirwan, *Geometric Invariant Theory*, 3rd edition, Springer, 1994.

[26] S. Mukai, *An Introduction to Invariants and Moduli*, Cambridge studies in advanced mathematics, 81, CUP, 2003.

[27] M.S. Narasimhan, S. Ramanan, Moduli of vector bundles on a compact Riemann surface, *Ann. of Math.* **89** (1969), 19–51.

[28] M.S. Narasimhan and C.S. Seshadri, Stable and unitary bundles on a compact Riemann surface, *Ann. of Math.* **82** (1965), 540–564.

[29] P.E. Newstead, Stable bundles of rank 2 and odd degree over a curve of genus 2, *Topology* **7** (1968), 205–215.

[30] P.E. Newstead, *Introduction to moduli problems and orbit spaces*, Tata Institute Lecture Notes, Springer, 1978.

[31] N. Nitsure, Moduli spaces of semistable pairs on a curve, *Proc. London Math. Soc.* **62** (1991), 275–300.

[32] B. Riemann, Theorie der Abel'schen Funktionen, *J. Reine angew. Math.* **54** (1857), 115–155.

[33] C.S. Seshadri, Theory of moduli, *Proc. Symp. Pure Math.*, Algebraic Geometry, Amer. Math. Soc., 1975.

[34] C.T. Simpson, Constructing variations of Hodge structure using Yang–Mills theory and applications to uniformization, *J. Amer. Math. Soc.* **1** (1988), 867–918.

[35] C.T. Simpson, Higgs bundles and local systems, *Inst. Hautes Études Sci. Publ. Math.* **75**, 5–95 (1992).

[36] F. Takemoto, Stable vector bundles on algebraic surfaces, *Nagoya Math. J.* **47** (1973), 29–48; II, ibid. **52** (1973), 173–195.

[37] K.K. Uhlenbeck, Connections with L^p bounds on curvature, *Comm. Math. Phys.* **83** (1982), 31-42.

[38] K.K. Uhlenbeck and S.T. Yau, On the existence of Hermitian-Yang-Mills connections in stable vector bundles *Comm. Pure. Appl. Math.* **39** (1986), 5257–5293

REFERENCES

Michael F. Atiyah

1. *K-Theory*, W. A. Benjamin, Inc., Reading, Mass., 1967.

M. F. Atiyah and R. Bott

1. "A Lefschetz fixed point formula for elliptic complexes, I, II," *Ann. of Math.*, **86** (1967), 347–407; **88** (1968), 451–491.

M. F. Atiyah and F. Hirzebruch

1. "Vector bundles and homogeneous spaces," in *Proceedings of Symposia in Pure Mathematics*, Vol. III, American Mathematical Society, Providence, 1961.
2. "The Riemann-Roch theorem for analytic embeddings," *Topology*, **1** (1962), 151–166.

M. F. Atiyah and I. Singer

1. "The index of elliptic operators on compact manifolds," *Bull. Amer. Math. Soc.*, **69** (1963), 422–433.
2. "The index of elliptic operators: I," *Ann. of Math.*, **87** (1968), 484–530.

Stefan Bergman

1. *The Kernel Function and Conformal Mapping*, American Mathematical Society, Providence, 1968 (Mathematical Surveys, No. 5, 2nd ed.).

Bruno Bigolin

1. "Gruppi di Aeppeli," *Ann. Scuola Norm. Sup. Pisa*, **23** (3) (1969), 259–287.

R. L. Bishop and R. J. Crittenden

1. *Geometry of Manifolds*, Academic Press, Inc., New York, 1964.

A. Borel and F. Hirzebruch

1. "Characteristic classes and homogeneous spaces, I, II," *Am. J. Math.*, **80** (1958), 458–538; **81** (1959), 315–382.

A. Borel and J. P. Serre

1. "Groupes de Lie et puissances réduites de Steenrod," *Am. J. Math.*, **75** (1953), 409–448.
2. "Le théorème de Riemann-Roch (d'après Grothendieck)," *Bull Soc. Math. France*, **86** (1958), 97–136.

Raoul Bott

1. "Homogenous vector bundles," *Ann. of Math.*, **68** (1957), 203–248.

Raoul Bott and S. S. Chern

1. "Hermitian vector bundles and the equidistribution of the zeroes of their holomorphic sections," *Acta Math.*, **114** (1965), 71–112.

Glen E. Bredon

1. *Sheaf Theory*, McGraw-Hill Book Company, New York, 1967.

H. Cartan

1. "Quotients of analytic spaces," in *Contributions to Function Theory*, Tata Institute of Fundamental Research, Bombay, 1960.

S. S. Chern

1. "Characteristic classes of Hermitian manifolds," *Ann. of Math.*, **47** (1946), 85–121.
2. *Complex Manifolds Without Potential Theory*, Van Nostrand Reinhold Company, New York, 1967.
3. "On a generalization of Kähler geometry," *Algebraic Geometry and Topology—A Symposium in Honor of S. Lefschetz*, Princeton Univ. Press, Princeton, N. J. (1957), 103–121.

C. Chevalley

1. *Theory of Lie Groups*, I. Princeton University Press, Princeton, 1946.

Fabio Conforto

1. *Abelsche Funktionen and Algebraische Geometrie*, Springer-Verlag, Berlin, 1956.

Georges de Rham

1. *Variétés Differentiables*, Hermann & Cie, Paris, 1955.

P. Dolbeault

1. "Sur la cohomologie des variétés analytiques complexes," *C. R. Acad. Sci. Paris*, **236** (1953), 175–177.

L. P. Eisenhart

1. *Differential Geometry*, Princeton University Press, Princeton, N.J., 1947.

A. Fröhlicher

1. "Relations between the cohomology groups of Dolbeault and topological invariants," *Proc. Nat. Acad. Sci. U.S.A.*, **41** (1955), 641–644.

Roger Godement

1. *Topologie Algébrique et Théorie des Faisceaux*, Hermann & Cie, Paris, 1964.

S. I. Goldberg

1. *Curvature and Homology*, Academic Press, New York, (1962).

Hans Grauert

1. "On Levi's problem and the imbedding of real-analytic manifolds," *Ann. of Math.*, **68** (1958), 460–472.
2. "Über Modifikationen und exceptionelle analytische Mengen," *Math. Ann.*, **146** (1962), 331–368.

Hans Grauert and Oswald Riemenschneider

1. "Kählersche Mannigfaltigkeiten mit hyper-q-konvexem Rand," in *Problems in Analysis: A Symposium in Honor of Salomon Bochner*, Princeton University Press, Princeton, N.J., 1970.

2. "Verschwindungssätze für analytische Kohomologiegruppen auf Komplexen Räumen," *Invent. Math.*, **11** (1970), 263–292.

Marvin Greenberg
1. *Lectures on Algebraic Topology*, W. A. Benjamin, Inc., Reading, Mass., 1967.

Phillip A. Griffiths
1. "Periods of integrals on algebraic manifolds, I, II," *Amer. J. Math.*, **90** (1968), 568–626; 805–865.
2. "Hermitian differential geometry, Chern classes, and positive vector bundles," in *Global Analysis*, Princeton University Press, Princeton, N.J., 1969.
3. "Periods of integrals on algebraic manifolds: summary of main results and discussion of open problems," *Bull. Amer. Math. Soc.*, **76** (1970), 228–296.
4. "Some results on algebraic cycles on algebraic manifolds," *Bombay Colloquium, 1968*, Tata Institute of Fundamental Research, Bombay, 1969, 93–191.
5. "On the periods of certain rational integrals, I, II," *Ann. of Math.*, **90** (1969), 460–495; 498–541.
6. "Periods of integrals of algebraic manifolds, III (some global differential-geometric properties of the period mapping)," *Publ. Math. I.H.E.S.*, **38** (1970), 125–180.

Phillip A. Griffiths and Wilfried Schmid
1. "Locally homogeneous complex manifolds," *Acta. Math.*, **123** (1970), 253–302.

A. Grothendieck
1. "Standard conjectures on algebraic cycles," *Algebraic Geometry*, Bombay Colloquium, 1968, Oxford University Press, 1969.

R. C. Gunning
1. *Lectures on Riemann Surfaces*, Princeton University Press, Princeton, N.J., 1966.

R. C. Gunning and Hugo Rossi
1. *Analytic Functions of Several Complex Variables*, Prentice-Hall, Inc., Englewood Cliffs, N.J., 1965.

Robin Hartshorne
1. "Ample vector bundles," *Publ. I.H.E.S.*, **29** (1966), 319–350.

H. Hecht
1. "On Kähler identities," preprint, Univ. of Utah, 1978 (to be published).

Sigurdur Helgason
1. *Differential Geometry and Symmetric Spaces*, Academic Press, Inc., New York, 1962.

F. Hirzebruch
1. *Topological Methods in Algebraic Geometry*, Springer Verlag New York, Inc., New York, 1966.

W. V. D. Hodge

1. *The Theory and Application of Harmonic Integrals*, Cambridge University Press, New York, 1941 (2nd ed., 1952).
2. "The topological invariants of algebraic varieties," in *Proceedings International Congress of Mathematians*, **1950**, Vol. I, American Mathematical Society, Providence, 1952, 182–191.

Lars Hörmander

1. *Linear Partial Differential Operators*, Springer-Verlag New York Inc., New York, and Academic Press, Inc., New York, 1963.
2. *An Introduction to Complex Analysis in Several Variables*, Van Nostrand Reinhold Company, New York, 1966.
3. "Pseudo-differential operators," *Comm. Pure Appl. Math.*, **18** (1965), 501–517.
4. "Pseudodifferential operators and hypoelliptic equations," in *Singular Integrals*, Proceedings of Symposia in Pure Mathematics, Vol. X, American Mathematical Society, Providence, 1967, 138–183.

Shoshichi Kobayashi and Katsumi Nomizu

1. *Foundations of Differential Geometry*, John Wiley & Sons, Inc. (Interscience Division), New York, Vol. I, 1963; Vol. II, 1969.

K. Kodaira

1. "On a differential-geometric method in the theory of analytic stacks," *Proc. Nat. Acad. Sci. U.S.A.*, **39** (1953), 1268–1273.
2. "On Kähler varieties of restricted type (an intrinsic characterization of algebraic varieties)," *Ann. of Math.*, **60** (1954), 28–48.
3. "On compact complex analytic surfaces, I," *Ann. of Math.*, **71** (1960), 111–152.
4. "Complex structures on $S^1 \times S^3$," *Proc. Nat. Acad. Sci. U.S.A.*, **55** (1966), 240–243.
5. "On the structure of compact complex analytic surfaces," I, *Am. J. Math.*, **86** (1964), 751–798.

K. Kodaira and D. C. Spencer

1. "On deformations of complex analytic structures, I, II," *Ann. of Math.*, **67** (1958), 328–466.
2. "On deformations of complex analytic structures, III. Stability theorems for complex structures," *Ann. of Math.*, **71** (1960), 43–76.

J. J. Kohn

1. "Harmonic integrals on strongly pseudo-convex manifolds I," *Ann. Math.* **78** (1963), 112–148.

J. J. Kohn and L. Nirenberg

1. "On the algebra of pseudo-differential operators," *Comm. Pure Appl. Math.*, **18** (1965), 269–305.

Serge Lang

1. *Analysis II*, Addison-Wesley Publishing Company, Inc., Reading, Mass., 1969.

S. Lefschetz

1. *L'Analysis Situs et la Géométrie Algébrique*, Gauthier-Villars, Paris, 1924.

S. MacLane
1. *Homology Theory*, Springer-Verlag New York, Inc., New York, 1967 (2nd ed.).

John Milnor
1. *Morse Theory*, Annals of Mathematics Studies, No. 51, Princeton University Press, Princeton, N. J., 1963.
2. *Lectures on Differential Topology*, Princeton University, Princeton, N.J., 1958.

C. B. Morrey, Jr.
1. "The analytic embedding of abstract real-analytic manifolds," *Ann. of Math.*, **68** (1958), 159–201.

J. Morrow and K. Kodaira
1. *Complex Manifolds*, Holt, Rinehart and Winston, Inc. New York, 1971.

S. Nakano
1. "On complex analytic vector bundles," *J. Math. Soc. Japan*, **7** (1955), 1–12.

R. Narasimhan
1. *Analysis on Real and Complex Manifolds*, North-Holland Publishing Company, Amsterdam, 1968.
2. *Introduction to the Theory of Analytic Spaces*, Lecture Notes in Mathematics, Vol. 25, Springer-Verlag New York, Inc., New York, 1966.

A. Newlander and L. Nirenberg
1. "Complex analytic coordinates in almost complex manifolds," *Ann. of Math.*, **65** (1957), 391–404.

L. Nirenberg
1. "Pseudo-differential operators," in *Global Analysis*, Proceedings of Symposia in Pure Mathematics, Vol. 16, American Mathematical Society, Providence, pp. 149–167.

K. Nomizu
1. *Lie Groups and Differential Geometry*, Mathematical Society of Japan, Tokyo, 1956.

R. Palais
1. *Seminar on the Atiyah-Singer Index Theorem*, Annals of Mathematics Studies, No. 57, Princeton University Press, Princeton, N.J., 1965.

J. Peetre
1. 'Rectification à l'article "Une caractérisation abstraite des operateurs,"' *Math. Scand.*, **8** (1960), 116–120.

F. Riesz and B. Sz. Nagy
1. *Functionat Analysis*, Frederick Ungar Publishing Co., Inc., New York, 1955.

B. Riemann
1. *Gesammelte Mathematische Werke und Wissentschaftliche Nachlass*, Dover, New York (1953).

Oswald Riemenschneider
1. "Characterizing Moišezon spaces by almost positive coherent analytic sheaves," *Math. Zeitschrift* **123** (1971), 263–284.

W. Rudin

1. *Functional Analysis*, McGraw-Hill, New York, 1973.

I. R. Šafarevič

1. *Algebraic Surfaces*, American Mathematical Society, Providence, 1967 (English translation of Russian ed.: Proceedings of Steklov Inst. of Mathematics, No. 75, Moscow, 1965).

Wilfried Schmid

1. *Homogeneous Complex Manifolds and Representations of Semisimple Lie Groups*, Ph.D. Thesis, University of California, Berkeley, 1967.

L. Schwartz

1. *Theorie des Distributions* (2nd edition), Hermann, Paris, (1966).

R. T. Seeley

1. "Integro-differential operators on vector bundles," *Trans. Amer. Math. Soc.*, **117** (1965), 167–204.

J. P. Serre

1. "Un théorème de dualité," *Comment. Math. Helv.*, **29** (1955), 9–26.
2. "Géométrie algébraique et géométrie analytique," *Ann. Inst. Fourier*, **6** (1956), 1–42.
3. *Algèbres de Lie semi-simples complexes*, W. A. Benjamin, Inc., Reading, Mass., 1966.

C. L. Siegel

1. *Analytic Functions of Several Complex Variables*, Institute for Advanced Study, Princeton, N. J., 1948 (reprinted with corrections, 1962).

Michael Spivak

1. *Calculus on Manifolds*, W. A. Benjamin, Inc., Reading, Mass., 1965.

N. Steenrod

1. *The Topology of Fibre Bundles*, Princeton University Press, Princeton, N. J., 1951.

Shlomo Sternberg

1. *Lectures on Differential Geometry*, Prentice-Hall, Inc., Englewood Cliffs, N. J., 1965.

V. S. Varadarajan

1. *Lie Groups, Lie Algebras, and Their Representations*, Prentice Hall, Englewood Cliffs, New Jersey, 1974.

André Weil

1. *Introduction à l'Étude des Variétés Kählériennes*, Hermann & Cie., Paris, 1958.

R. O. Wells, Jr.

1. "Parameterizing the compact submanifolds of a period matrix domain by a Stein manifold," *Proceedings of Conference on Several Complex Variables*, Park City, Utah (1970), Lecture Notes in Mathematics, Vol. 184, Springer-Verlag New York, Inc., New York, 1971, 121–150.
2. "Automorphic cohomology of homogeneous complex manifolds" Proceedings of the Conference on Complex Analysis, Rice University, March 1972, *Rice Univ. Studies*, **59** (1973), 147–155.

R. O. Wells, Jr. and Joseph A. Wolf

1. "Poincaré series and automorphic cohomology on flag domains," *Ann. of Math.* **105** (1977), 397–448.

Hassler Whitney

1. "Differentiable manifolds," *Ann. of Math.*, **37** (1936), 645–680.

A. Zygmund

1. *Trigonometric Series*, Cambridge University Press, New York, 1968 (rev. ed.).

AUTHOR INDEX

M

N

O

P

R

S

V

W

Z

SUBJECT INDEX

[1] The page numbers in the index which are *italicized* correspond to references in the text which have been italicized either for purposes of definition or of emphasis.

D

I

J

K

Graduate Texts in Mathematics

(continued from page ii)

64 Edwards. Fourier Series. Vol. I. 2nd ed.
65 Wells. Differential Analysis on Complex Manifolds. 3rd ed.
66 Waterhouse. Introduction to Affine Group Schemes.
67 Serre. Local Fields.
68 Weidmann. Linear Operators in Hilbert Spaces.
69 Lang. Cyclotomic Fields II.
70 Massey. Singular Homology Theory.
71 Farkas/Kra. Riemann Surfaces. 2nd ed.
72 Stillwell. Classical Topology and Combinatorial Group Theory. 2nd ed.
73 Hungerford. Algebra.
74 Davenport. Multiplicative Number Theory. 3rd ed.
75 Hochschild. Basic Theory of Algebraic Groups and Lie Algebras.
76 Iitaka. Algebraic Geometry.
77 Hecke. Lectures on the Theory of Algebraic Numbers.
78 Burris/Sankappanavar. A Course in Universal Algebra.
79 Walters. An Introduction to Ergodic Theory.
80 Robinson. A Course in the Theory of Groups. 2nd ed.
81 Forster. Lectures on Riemann Surfaces.
82 Bott/Tu. Differential Forms in Algebraic Topology.
83 Washington. Introduction to Cyclotomic Fields. 2nd ed.
84 Ireland/Rosen. A Classical Introduction to Modern Number Theory. 2nd ed.
85 Edwards. Fourier Series. Vol. II. 2nd ed.
86 van Lint. Introduction to Coding Theory. 2nd ed.
87 Brown. Cohomology of Groups.
88 Pierce. Associative Algebras.
89 Lang. Introduction to Algebraic and Abelian Functions. 2nd ed.
90 Brøndsted. An Introduction to Convex Polytopes.
91 Beardon. On the Geometry of Discrete Groups.
92 Diestel. Sequences and Series in Banach Spaces.
93 Dubrovin/Fomenko/Novikov. Modern Geometry—Methods and Applications. Part I. 2nd ed.
94 Warner. Foundations of Differentiable Manifolds and Lie Groups.
95 Shiryaev. Probability. 2nd ed.
96 Conway. A Course in Functional Analysis. 2nd ed.
97 Koblitz. Introduction to Elliptic Curves and Modular Forms. 2nd ed.
98 Bröcker/tom Dieck. Representations of Compact Lie Groups.
99 Grove/Benson. Finite Reflection Groups. 2nd ed.
100 Berg/Christensen/Ressel. Harmonic Analysis on Semigroups: Theory of Positive Definite and Related Functions.
101 Edwards. Galois Theory.
102 Varadarajan. Lie Groups, Lie Algebras and Their Representations.
103 Lang. Complex Analysis. 3rd ed.
104 Dubrovin/Fomenko/Novikov. Modern Geometry—Methods and Applications. Part II.
105 Lang. $SL_2(\mathbf{R})$.
106 Silverman. The Arithmetic of Elliptic Curves.
107 Olver. Applications of Lie Groups to Differential Equations. 2nd ed.
108 Range. Holomorphic Functions and Integral Representations in Several Complex Variables.
109 Lehto. Univalent Functions and Teichmüller Spaces.
110 Lang. Algebraic Number Theory.
111 Husemöller. Elliptic Curves. 2nd ed.
112 Lang. Elliptic Functions.
113 Karatzas/Shreve. Brownian Motion and Stochastic Calculus. 2nd ed.
114 Koblitz. A Course in Number Theory and Cryptography. 2nd ed.
115 Berger/Gostiaux. Differential Geometry: Manifolds, Curves, and Surfaces.
116 Kelley/Srinivasan. Measure and Integral. Vol. I.
117 J.-P. Serre. Algebraic Groups and Class Fields.
118 Pedersen. Analysis Now.
119 Rotman. An Introduction to Algebraic Topology.
120 Ziemer. Weakly Differentiable Functions: Sobolev Spaces and Functions of Bounded Variation.

121 Lang. Cyclotomic Fields I and II. Combined 2nd ed.
122 Remmert. Theory of Complex Functions. *Readings in Mathematics*
123 Ebbinghaus/Hermes et al. Numbers. *Readings in Mathematics*
124 Dubrovin/Fomenko/Novikov. Modern Geometry—Methods and Applications Part III.
125 Berenstein/Gay. Complex Variables: An Introduction.
126 Borel. Linear Algebraic Groups. 2nd ed.
127 Massey. A Basic Course in Algebraic Topology.
128 Rauch. Partial Differential Equations.
129 Fulton/Harris. Representation Theory: A First Course. *Readings in Mathematics*
130 Dodson/Poston. Tensor Geometry.
131 Lam. A First Course in Noncommutative Rings. 2nd ed.
132 Beardon. Iteration of Rational Functions.
133 Harris. Algebraic Geometry: A First Course.
134 Roman. Coding and Information Theory.
135 Roman. Advanced Linear Algebra.2nd ed.
136 Adkins/Weintraub. Algebra: An Approach via Module Theory.
137 Axler/Bourdon/Ramey. Harmonic Function Theory. 2nd ed.
138 Cohen. A Course in Computational Algebraic Number Theory.
139 Bredon. Topology and Geometry.
140 Aubin. Optima and Equilibria. An Introduction to Nonlinear Analysis.
141 Becker/Weispfenning/Kredel. Gröbner Bases. A Computational Approach to Commutative Algebra.
142 Lang. Real and Functional Analysis. 3rd ed.
143 Doob. Measure Theory.
144 Dennis/Farb. Noncommutative Algebra.
145 Vick. Homology Theory. An Introduction to Algebraic Topology. 2nd ed.
146 Bridges. Computability: A Mathematical Sketchbook.
147 Rosenberg. Algebraic *K*-Theory and Its Applications.
148 Rotman. An Introduction to the Theory of Groups. 4th ed.
149 Ratcliffe. Foundations of Hyperbolic Manifolds. 2nd ed.
150 Eisenbud. Commutative Algebra with a View Toward Algebraic Geometry.
151 Silverman. Advanced Topics in the Arithmetic of Elliptic Curves.
152 Ziegler. Lectures on Polytopes.
153 Fulton. Algebraic Topology: A First Course.
154 Brown/Pearcy. An Introduction to Analysis.
155 Kassel. Quantum Groups.
156 Kechris. Classical Descriptive Set Theory.
157 Malliavin. Integration and Probability.
158 Roman. Field Theory.
159 Conway. Functions of One Complex Variable II.
160 Lang. Differential and Riemannian Manifolds.
161 Borwein/Erdélyi. Polynomials and Polynomial Inequalities.
162 Alperin/Bell. Groups and Representations.
163 Dixon/Mortimer. Permutation Groups.
164 Nathanson. Additive Number Theory: The Classical Bases.
165 Nathanson. Additive Number Theory: Inverse Problems and the Geometry of Sumsets.
166 Sharpe. Differential Geometry: Cartan's Generalization of Klein's Erlangen Program.
167 Morandi. Field and Galois Theory.
168 Ewald. Combinatorial Convexity and Algebraic Geometry.
169 Bhatia. Matrix Analysis.
170 Bredon. Sheaf Theory. 2nd ed.
171 Petersen. Riemannian Geometry. 2nd ed.
172 Remmert. Classical Topics in Complex Function Theory.
173 Diestel. Graph Theory. 2nd ed.
174 Bridges. Foundations of Real and Abstract Analysis.
175 Lickorish. An Introduction to Knot Theory.
176 Lee. Riemannian Manifolds.
177 Newman. Analytic Number Theory.
178 Clarke/Ledyaev/ Stern/Wolenski. Nonsmooth Analysis and Control Theory.
179 Douglas. Banach Algebra Techniques in Operator Theory. 2nd ed.
180 Srivastava. A Course on Borel Sets.
181 Kress. Numerical Analysis.